T0231803

# *Solar Air Systems – Built Examples*

First published by James & James (Science Publishers) Ltd. in 1999

This edition published 2013 by Earthscan

For a full list of publications please contact:

Earthscan
2 Park Square, Milton Park, Abingdon, Oxon OX14 4RN
Simultaneously published in the USA and Canada by Earthscan
711 Third Avenue, New York, NY 10017

First issued in hardback 2017

*Earthscan is an imprint of the Taylor & Francis Group, an informa business*



A catalogue record for this book is available from the British Library

ISBN 13: 978-1-138-40911-8 (hbk)
ISBN 13: 978-1-873936-85-6 (pbk)

---

## ACKNOWLEDGEMENTS

*Editor*
S. Robert Hastings
Forschungsstelle Solararchitektur
ETH - Hönggerberg
CH-8093 Zürich
Switzerland

*Foreword and Introduction*
S. Robert Hastings
Forschungsstelle Solararchitektur

*Production*
Annuscha Gassler-Schmidt
Yvonne Kaiser
Katalin Ilosvay
Forschungsstelle Solararchitektur

*Foreword and Introduction*
S. Robert Hastings

*Air Collectors*
Sture Larsen
Lindauerstrasse 33
A-6912 Hörbranz
Austria

*Storage Elements*
Johann Reiss
Fraunhofer-Institut für Bauphysik
Nobelstrasse 12
D-70569 Stuttgart
Germany

*Distribution*
Helmut Meyer
Transsolar
Ingenieurgeschaft GmbH
Nobelstrasse 15
D-70569 Stuttgart
Germany

*Controls*
Joachim Morhenne
Ingenieurbüro Morhenne GbR
Schülkestrasse 10
D-42277Wuppertal
Germany

*Air to Water Heat Exchangers*
Søren Østergaard Jensen
Solar Energy Centre Denmark
Danish Technological Institute
Postbox 141
DK-2630 Taastrup
Denmark

*Fire Protection*
Christer Nordström
Ch. Nordström Arkitektkontor AB
Asstigen 14
S-43645 Askim
Sweden

# Contents

# Foreword

Solar air systems have unique advantages for space heating and tempering ventilation air. Air, unlike water, needs no protection against freezing. Nor are leaks damaging to the building structure or its contents. In contrast to passive systems, active air systems provide better heat distribution and hence improved comfort and fuller use of solar gains. Furthermore solar air systems fit naturally into mechanically ventilated buildings and mechanical ventilation is increasingly common, not only in commercial and institutional buildings, but also in very-low-energy residences.

The economics of air systems improves when they serve additional uses, such as admitting daylight, inducing cooling, providing sunshading, generating electricity or preheating domestic hot water. They may also have non-energy functions, such as providing a usable zone, a load-bearing element, weather protection or a barrier against street noise.

Unfortunately, designers lack experience in planning, analysing and constructing such systems. Furthermore, documentation on built prototypes that could be used to convince building clients is scarce. The purpose of this book is to share the experiences from a wide range of building projects. In addition, the multitude and diversity of projects should help convince building clients of the maturity and reliability of solar air systems.

This documentation is the work of 17 experts from the diverse climates of nine countries in Europe and North America. Applications include single-family houses, apartment buildings, schools, gymnasiums, large industrial buildings and commercial buildings. Six different types of solar air systems are reported.

*S. Robert Hastings*

# Terminology and Units

## TERMINOLOGY

**Close loop system:** A system in which solar collector air circulates in a closed channel between the collector and heat storage or radiating surface. Room air does not enter into the system.

**COP:** The coefficient of performance for a system is the energy output divided by the energy input, i.e. for a heat pump this would be kWh heat output per kWh electricity input (3 to 4 is a realistic range). For a solar air system the COP is kWh hot air output per kWh electricity consumed by the fan/s (15 to 30 is possible according to monitored projects).

**Design Temperature:** The coldest ambient temperature which per the local building code, a building must cope with in guaranteeing the minimum room temperature is achieved. The design temperature under night conditions (no solar) is a basis to size the capacity of the heating plant for a given building design.

**Heating degree days:** A unit which characterizes how cold a climate is. There is unfortunately no consensus how this should be calculated.

When one base number is cited, i.e. 18.3°C in the USA, then heating degree days are the sum of (18.3°C – **Tambient)** for all days where **Tambient <** 18.3°C. When two numbers are cited, i.e. 20°C/12 in middle Europe, then the heating degrees are the sum of (20°C – **Tambient)** for all days where the average **Tambient** ≤ 12°C.

**Hybrid System:** Solar air systems are often referred to as hybrid systems vs. "passive systems" because they use a fan to assist in moving the air from the collector to the point of use. Hence the system is hybrid in that it makes use of both solar energy and electricity to power the fan.

**Hypocaust:** An old Roman construction in which smoke from a wood fire was circulated through a space beneath a massive floor to provide radiant heating. Here, hypocausts refer to massive floors with channels through which solar heated air passes thereby warming the floor mass and providing a delayed release of radiant heat to the room.

**Murocaust:** Similar to a hypocaust, except that a massive wall with air channels is used and the air flow is vertical instead of horizontal.

**Open loop system:** A system in which the room is part of the path through which air passes going to or coming from the collector and or storage.

**Thermosiphon:** Flow induced by a warm fluid or gas being less dense then a cool medium and hence rising. In a thermosiphon solar air system, solar warmed air in the collector rises and so can be self-transported to either a room or storage.

## UNITS

1 Joule (J) = 1 Watt second (W s)
3.6MJ= 1 kWh
1 K = 1 °C temperature difference as used here
1 $m^2$= 10.76 $ft^2$
1 litre = 1.06 quarts
$ = US dollars unless otherwise noted

# I Introduction

# I Introduction

## WHAT IS A SOLAR AIR SYSTEM?

A solar air system is a system that collects solar energy, converts it to heat and then transports the heat by means of air either to storage or for direct use.

The collector may be similar to a conventional water collector except that instead of circulating water through the absorber, air is circulated behind, through or above the absorber. The collector may be:

- a facade or roof-integrated panel (glazed or unglazed), which also provides the weather skin of the building;
- a double window with an internal adjustable blind (an absorber that also offers glare control and privacy);
- a transparent second facade over the primary facade;
- a glazed space, such as an atrium, sun space or attic, which has a spatial function as well as preheating ventilation air.

The storage may be the building itself, for example concrete floor slabs with channels through which the sun-warmed air is circulated (hypocausts), hollow-core brick walls (murocausts), a rock bed located in the basement or core of the building or a phase-change material. Alternatively, the sun-warmed air can be supplied directly into the space to provide heat, fresh air or dehumidification.

Circulation of the air can be by the natural flow of warm air rising, but is most often fan-forced. In a well designed system, for each kWh of electricity consumed by the fan, between 15 and 25 kWh of heat can be provided (compared to a factor of 1:3 or 1:4 for most heat pumps). Optionally, the power can be provided by PV panels and this has the advantage that the solar power is proportional to solar heat generation, so the system is self-regulating.

The distribution system can make use of the conventional system in mechanically ventilated buildings, the above-mentioned hollow-core construction or the rooms themselves.

## USES OF SOLAR AIR SYSTEMS

Systems that can serve more than one function are inherently more economical. Typical energy functions of solar air systems include:

- space heating;
- heating of ventilation air;
- water preheating;
- inducing cooling (i.e. solar chimneys);
- electricity generation (hybrid PV systems);
- sunshading (i.e. window collectors).

### Advantages and limitations

Compared to a purely passive solar concept, a solar air system affords:

- better collection of solar heat, without the comfort restrictions of direct solar gains in living spaces;
- better timing of solar heat, for example heat release from storage in the evening, when there are no longer direct solar gains through windows;
- better distribution of solar heat, for example when it is channelled to north-facing rooms.

Compared to solar water systems, air systems also offer important advantages:

- *Security.* A leak results in lost air, not water damage to the building structure or its contents.
- *Environment.* No antifreeze chemicals are needed.

Further plus points include:

- *Secondary uses.* Solar air systems can also provide a habitable zone, a load-bearing element, weather protection or a barrier against street noise.
- *Integration.* System components can be part of the conventional building envelope, of its structure or of a HVAC system. Solar air systems fit naturally into mechanically ventilated buildings and mechanical ventilation is increasingly common, not only in commercial and institutional buildings, but also in very-low-energy residences.

Limitations include:

- Air has a very small heat capacity compared to water (0.0003 kWh/m$^3$K versus 1.16 kWh/m$^3$K).
- A large volume of air must be moved to transport a small amount of heat.

## Variations

*System Type 1. (ambient/collector/room – Figure I.1)*
Outside air is circulated through an unglazed or glazed collector directly into the space to be ventilated and heated. This system can achieve very high efficiencies because cool air is supplied to the collector. In the summer the collector can be vented to the outside.

Appropriate applications range from vacation cottages, to keep them from becoming damp and musty when unoccupied, to large industrial spaces needing much ventilation.

*System Type 2. (collector/room/collector – Figure I.1)*
This system, called the Bara Costantini System after its Italian inventor, circulates room air into the collector, where it is heated. It then rises and returns via a thermal storage ceiling back into the room, all by natural convection. The storage ceiling provides radiant heat after sunset. In summer, the collector can be vented at the top to the outside, thus extracting room air, which can then be replaced with cooler air from an earth storage device or from open north-facing windows.

This system has mostly been used for apartment buildings.

*System Type 3. (collector/shell/collector – Figure I.1)*
Collector-warmed air is circulated through a cavity between an outer, insulated wall and an inner wall of the building envelope. Thereby, a sun-buffered climate is created, which reduces room heat losses through the envelope. Because even a slight rise in the temperature of this cavity reduces losses, the collector can be inexpensive and can also operate with very weak solar radiation.

Appropriate applications include apartment buildings. This system is particularly suitable for retrofitting in poorly insulated existing buildings.

*System Type 4. (collector/storage/collector – Figure I.1)*
In this, the classical type of solar heating system, collector-warmed air is circulated through channels in a massive floor or wall, which then radiates the heat into the room after a time delay of four to six hours. This system has the advantage of large radiating surfaces and hence provides comfort. Fan-forced circulation provides the best overall system efficiency and output.

Applications include all building types that have large surfaces available to act as radiant sources.

*System Type 5. (collector/storage/collector plus room/storage/room – Figure I.1)*
Here, system type 4 is actively discharged by circulating room air through separate channels in the storage. Thereby, the heat can be stored longer and released when it is needed. Furthermore, the storage can be located remotely from the rooms to be heated.

Relatively few buildings with this system exist because of the expense involved in having two independent systems of air channels and fans.

*System Type 6. (collector/heat exchanger/collector plus heat-exchanger/load/heat exchanger – Figure I.1)*
The advantages of a solar air collector are combined here with the compactness of using water for heat distribution to the point of need. Thus, conventional radiators or radiant floors or walls can be used. Furthermore, domestic hot water can be heated.

Applications where the heat must be transported over a distance are particularly suited to this system. This system can also be used for retrofit of existing buildings with radiators or radiant floors.

## THE EXAMPLES IN THIS BOOK

The buildings presented in the remainder of this book are taken from throughout Europe and North America, as indicated in Figure I.7.

The climates with which the solar air systems of these buildings must cope are diverse. For simplification, the climates can be characterized in three groupings as follows:

- *Middle European.* Winter temperatures average around 7°C with the heating season beginning in October and ending in April. A design temperature of –12°C is typical. Wind is not an important factor. Much of midwinter is overcast. Summer temperatures range from 20 to 35°C.
- *Scandinavian coastal.* Winter temperatures average around 5°C with the heating season running from September to May. A design temperature of –15°C is typical. Wind is an important factor. In midwinter the sun is low in the sky and weak during the short days. But because the winter is longer, it extends to the months when there are even more hours of sunlight than in more southern locations. Temperatures in the summer rarely reach 30°C.
- *North American continental.* Cold winters and hot summers are the case here. The heating season runs from October to May and a design temperature of -17°C is typical. Wind is an important factor. Solar radiation greatly exceeds that of Scandinavia or middle Europe. Summer temperatures can reach 38°C.

The solar air heated buildings presented here include a wide range of building types including single-family and multi-family housing, schools, sports halls, large industrial buildings and commercial buildings. Table I.1 gives an overview of building and system types.

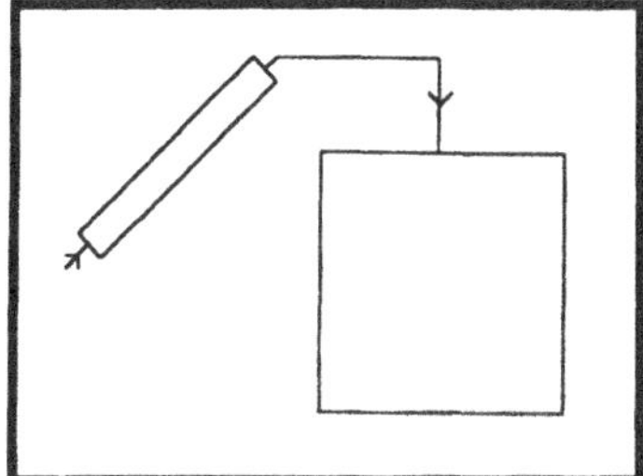

*System 1. Ambient/collector/room*

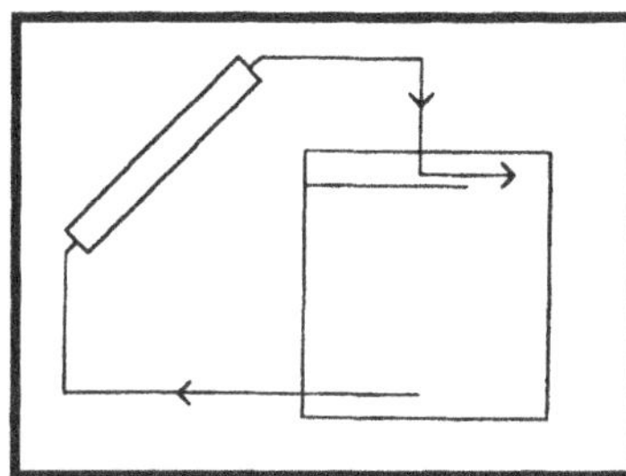

*System 2. Collector/room/collector*

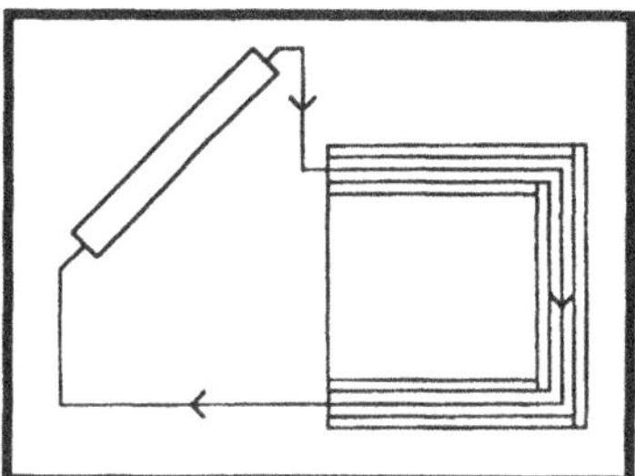

*System 3. Collector/shell/collector*

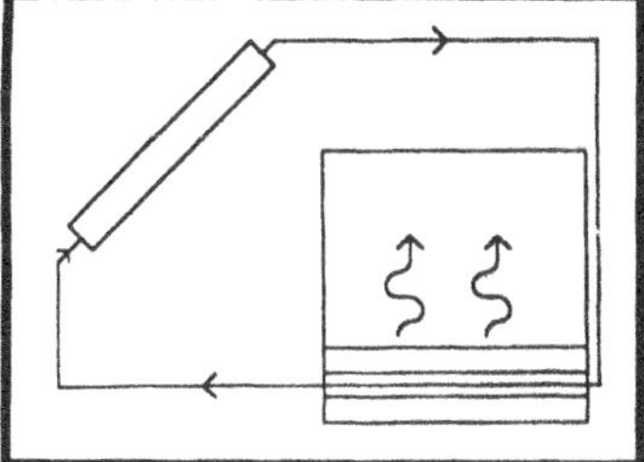

*System 4. Collector/storage/collector*

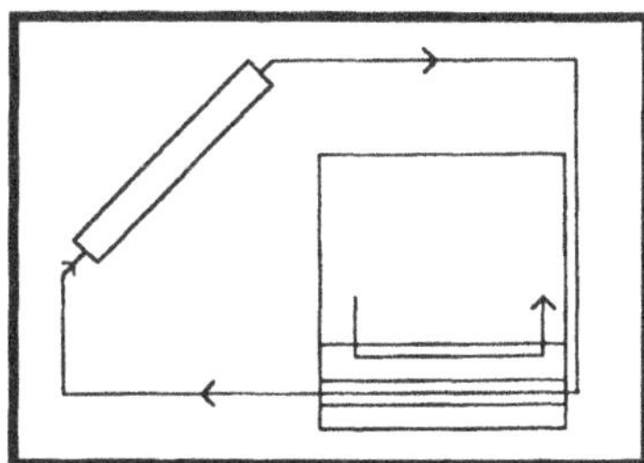

*System 5. Collector/storage/collector plus room/storage/room*

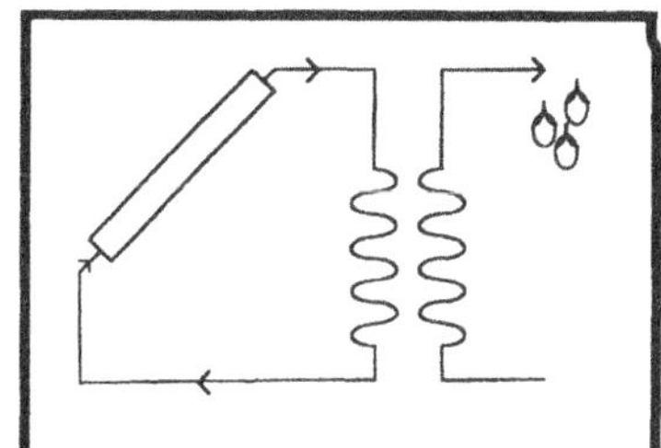

*System 6. Collector/heat exchanger/collector plus heat-exchanger/load/heat exchanger*

*Figure I.1. System types*

*Table I.1. Buildings and the systems used*

| Chapter | Building | Country | Systems 1 | 2 | 3 | 4 | 5 | 6 |
|---|---|---|---|---|---|---|---|---|
| Single-family houses | | | | | | | | |
| II.2 | The Frei house | A | | | | × | | |
| II.3 | The Gwadt house | CH | | | | × | | |
| II.4 | The Lenherr house | CH | | | | × | | |
| II.5 | The Schüpfen house | CH | | | | | | × |
| II.6 | The Morhenne house | D | | | | × | | |
| II.7 | The Vigander house | N | | | | × | | |
| II.8 | The summer-house package | DK | × | | | | | |
| II.9 | The Humlum house | D | × | | | | | |
| II.10 | The OKA *Haus der Zukunft* | A | | | | × | | |
| II.11 | The CSU Solar House II | USA | × | | | | | |
| Apartment buildings (multi-family residential) | | | | | | | | |
| III.2 | The Marostica passive solar dwelling | I | | × | | | | |
| III.3 | The Luino Apartments | I | | × | | | | |
| III.4 | The multi-family building in Munich | D | × | | | | × | |
| III.5 | The Lützowstrasse building | D | | | | × | | |
| III.6 | Toftegård | DK | | | | | | × |
| III.7 | Havrevangen | DK | | | | | | × |
| III.8 | The Rødovre Apartments | DK | | | × | | | |
| III.9 | The Solar Air Apartment Block, Gothenburg | S | | | × | | | |
| III.10 | The Senior Citizens Building, Windsor | CDN | × | | | | | |
| III.11 | The Weinmeisterhornweg row houses | D | | | | | × | |
| Schools | | | | | | | | |
| IV.2 | Secondary Modern School, Koblach | A | × | | | | | |
| IV.3 | Kindergarten, Lochau | A | | | | × | | |
| IV.4 | Kindergarten Schopfloch | D | | | | × | | |
| IV.5 | Green Park School | UK | × | | | | | |
| Sports halls | | | | | | | | |
| V.2 | Karl High School gym | D | | | | | | × |
| V.3 | Odenwald School gym | D | × | | | | | |
| V.4 | Stavanger Squash Center | N | | | | | | × |
| Large industrial buildings | | | | | | | | |
| VI.2 | Mätzler garage | A | | | | | × | |
| VI.3 | Steel warehouse Kägi | CH | | × | | | | |
| VI.4 | Wewer's brickyard | DK | × | | | | | |
| VI.5 | Bombardier Inc. factory | CDN | × | | | | | |
| VI.6 | JRC Research building, ISPRA | I | × | | | | | |
| VI.7 | US Army hangar | USA | × | | | | | |
| Office buildings | | | | | | | | |
| VII.2 | WAT office building | D | × | | | | | |

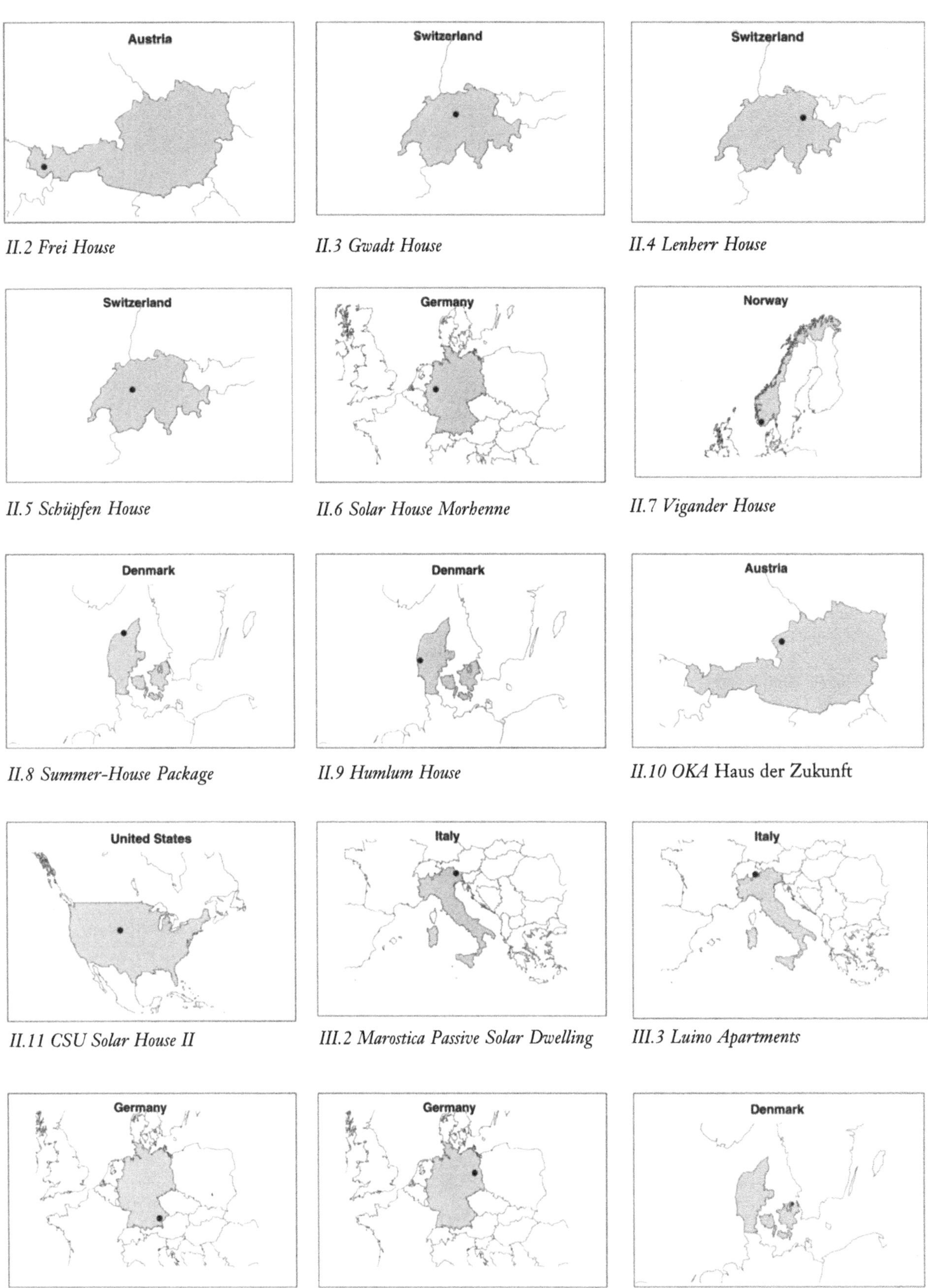


*II.2 Frei House* — *II.3 Gwadt House* — *II.4 Lenherr House*

*II.5 Schüpfen House* — *II.6 Solar House Morhenne* — *II.7 Vigander House*

*II.8 Summer-House Package* — *II.9 Humlum House* — *II.10 OKA* Haus der Zukunft

*II.11 CSU Solar House II* — *III.2 Marostica Passive Solar Dwelling* — *III.3 Luino Apartments*

*III.4 Multi-Family Residential Building* — *III.5 Lützowstrasse Residential Building* — *III.6 Toftegård Multi-Family Houses*

*Figure I.7. The maps show the location of some of the buildings, presented in this book, throughout Europe and North America*

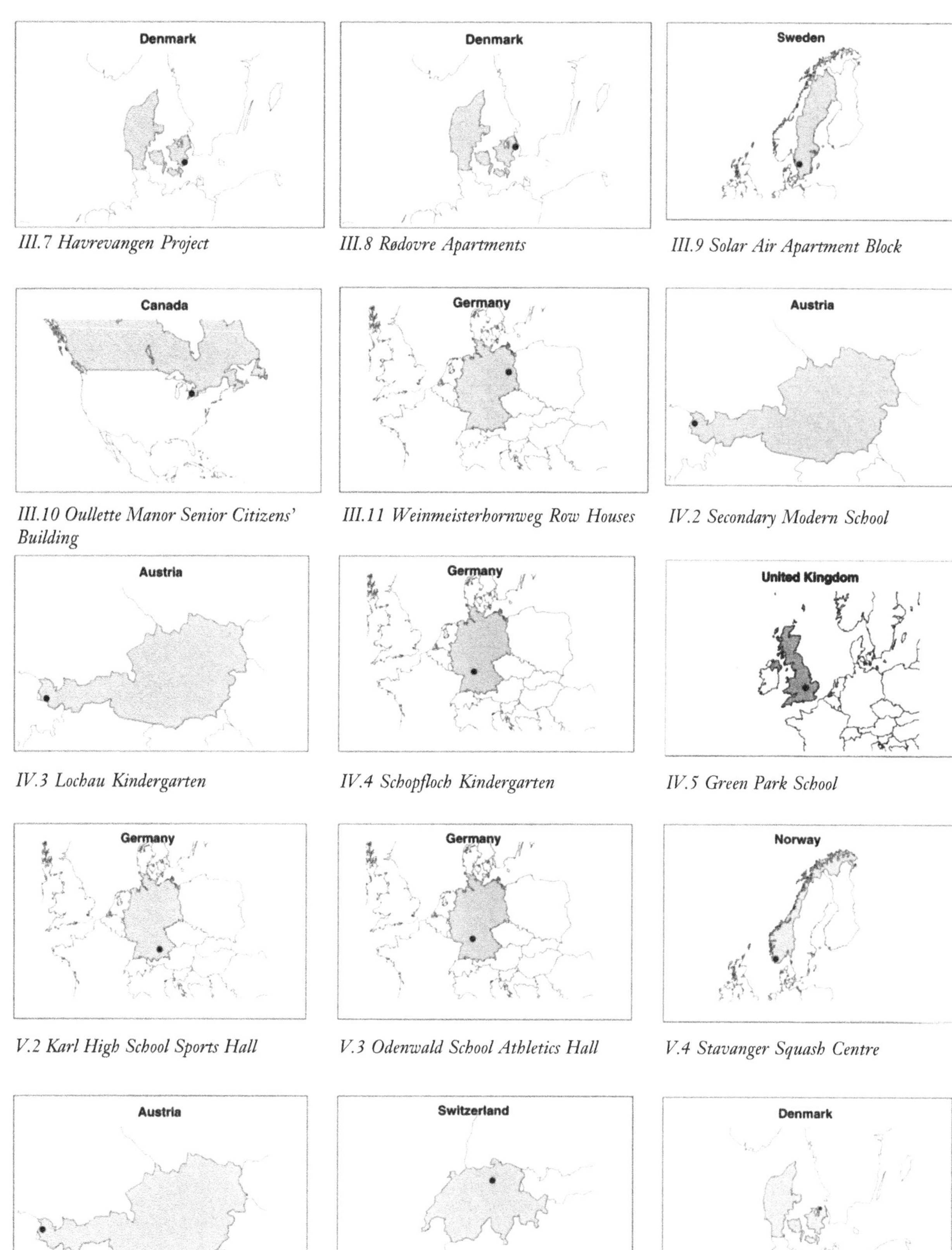


*III.7 Havrevangen Project*

*III.8 Rødovre Apartments*

*III.9 Solar Air Apartment Block*

*III.10 Oullette Manor Senior Citizens' Building*

*III.11 Weinmeisterhornweg Row Houses*

*IV.2 Secondary Modern School*

*IV.3 Lochau Kindergarten*

*IV.4 Schopfloch Kindergarten*

*IV.5 Green Park School*

*V.2 Karl High School Sports Hall*

*V.3 Odenwald School Athletics Hall*

*V.4 Stavanger Squash Centre*

*VI.2 Mätzler Garage*

*VI.3 Kägi Steel Warehouse*

*VI.4 Wewer's Brickyard*

# II Single-family houses

# II.1 Introduction

## BUILDING CHARACTERISTICS

The single-family house is the building type that has profited most from experiments in using solar energy. It also has the greatest envelope energy losses because small buildings have large ratios of envelope surface to building volume. The resulting relatively large heating demand makes it a good candidate for solar air systems. In the case of older houses, heat losses by transmission are considerably larger than by air infiltration, whereas in newer and well insulated houses the inverse is true. In either case, there are appropriate types of solar air system.

The rules for designing residential buildings are completely different from the rules for schools or offices, especially concerning internal gains, ventilation and scheduling. Solar single-family houses generally share the following commonalities:

- They must be well insulated, given their small size.
- Internal gains, when considered over 24 hours, are small and unevenly spatially distributed.
- Thermal storage can even out temperature swings, improving comfort and energy use, given the night-time occupancy.

Table II.1.1 shows some typical aspects of the solar air systems used in the houses described in this book.

*Table II.1.1. Typical aspects of the solar air systems used in the houses described in this book*

| System type | 1 | 2 | 3 | 4 | 5 | 6 |
|---|---|---|---|---|---|---|
| Single-family houses | × | × | | × | | |
| Multi-storey residential | | | × | | × | |
| Collector air enters into room | × | | | | × | × |
| No contact with room air | | × | × | × | | |
| Radiant heat distribution only | | | × | × | | |
| Large-scale space heating | × | | | | | |
| Mainly small-scale for residential | | | | | | × |
| DHW subsystem | | × | | | | |
| Strongly affects building design | × | | | × | × | |

## COMBINING DIFFERENT SYSTEMS

### Direct gain

Since direct gains can easily lead to overheating, it is important that the solar air system does not exacerbate this situation. Therefore, solar air systems that either include storage or deliver heat to non-south-facing rooms are beneficial.

### Sun spaces

These are unlikely to conflict with solar air systems and may even serve as the collector. It is critical to keep the costs of fans, ducting and controls as low as possible.

### Active solar water systems

An air collector can readily also serve for heating domestic hot water (DHW) if an air-to-water heat exchanger is incorporated in the ducting.

### Auxiliary heating

This can be readily integrated into systems 1 and 4.

## ACKNOWLEDGEMENTS

Author: Sture Larsen (Austria)

# II.2 Frei House

## Nüziders, Austria

*Nüziders*

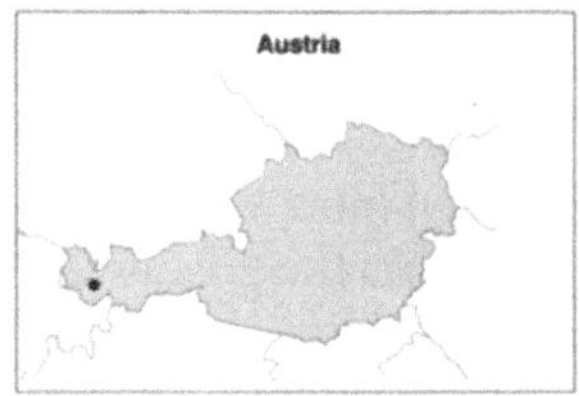


*System type 4*

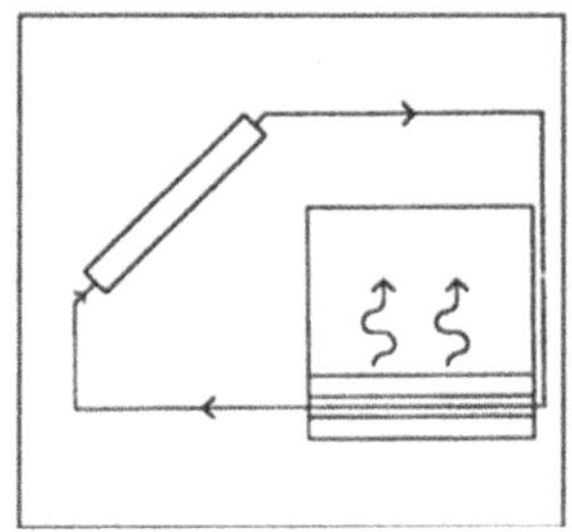

## PROJECT SUMMARY

The Frei house incorporates an air collector, a rock bed and hollow-core constructions for storage and distribution, leaving a requirement for only 17 kWh/m²a (kWh/m² per annum) for auxiliary heat in an average season. The system has minimal technical equipment, simple controls and only one fan. The back-draught dampers are self-operating. Another attribute is the high level of thermal comfort provided by the radiant heat from the hypocausts and murocausts. A wood stove in the cellar can supply hot air to the hypo/murocausts if required. The larger part of the domestic hot water is solar heated by a heat exchanger in the duct from the air collector.

### Summary statistics

| | |
|---|---|
| System type: | type 4 |
| Auxiliary heating demand: | 17.6 kWh/m²a |
| Heated floor area: | 239 m² |
| Collector type: | opaque, roof integrated |
| Collector area: | 58.20 m² |
| Storage type: | Rock bed and hypocausts |
| Rockbed storage capacity: | 12.75 kWh/K |
| Solar system output: | 28 kWh/m²a |
| Year solar system built: | 1992 |

## SITE DESCRIPTION

The house is situated in the foothills on the sunny side of a major alpine valley, in Nüziders:

| | |
|---|---|
| latitude | 47°N |
| longitude | 10°E |
| altitude | 570 m above sea level. |

The climate is typical for the main alpine valleys in Austria with average monthly temperatures ranging from −1.9°C to 18.5°C. The seasonal heat degree day average is 3865 relative to the Austrian base of 20°C/12°C. In terms of the same standards the number of heating days is 231, but the Frei house registered only 55 days auxiliary heating during 1995/96. Autumn fog is not as frequent as in the neighbouring Rhine valley.

## BUILDING PRESENTATION

The outer envelope of the house is mostly of lightweight wooden construction (24 cm rockwool), while a large part of the inner structure is masonry and concrete, providing thermal mass. Table II.2.1 gives the building statistics and Figure II.2.1 shows the floor plan and section, while Table II.2.2 gives the *U*-values of the building elements.

*Table II.2.1. Building statistics*

| | |
|---|---|
| Gross heated floor area | 239 m² |
| Gross heated volume | 831 m³ |
| Envelope specifications | |
| Building heating coefficients: | |
| conduction load | 210W/K |
| air-change load | 83 W/K |
| auxiliary space heating | 6.7 W/K |
| auxiliary water heating | 533 kWh/a |

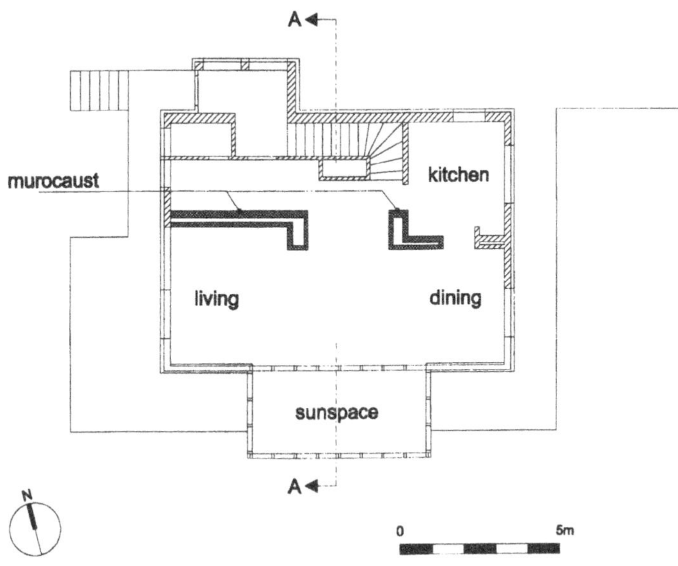


*(a) Ground floor plan*

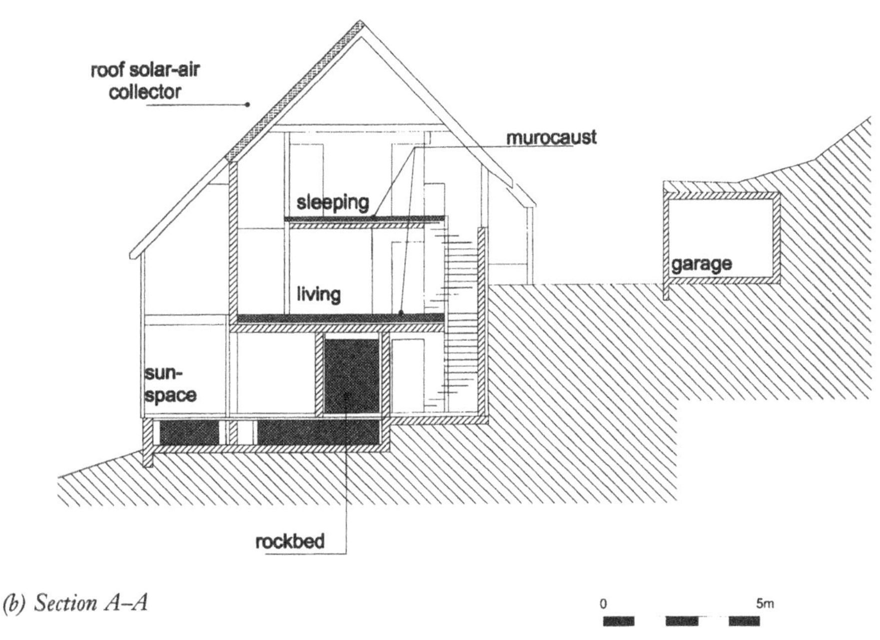


*(b) Section A–A*

*Figure II.2.1. (a) Ground floor plan and (b) section A–A*

*Table II.2.2.* U*-values of building elements*

| Element | *U*-values (W/m²K) |
|---|---|
| Walls: | |
| light-weight, wooden | 0.18 |
| masonry/wood | 0.29 |
| concrete/polystyrene | 0.34 |
| Roof: | |
| above heated space | 0.19 |
| above attic | 0.24 |
| Glazing: | |
| windows | 1.50 |
| interior sun-space glazing | 1.50 |
| exterior sun-space glazing | 2.90 |

## SOLAR SYSTEM

The solar air heating system of this house is a variation of System 4 and is shown in Figure II.2.2. Its main characteristics are:

- *Active charge.* The fan-forced air collector delivers the solar-heated air to the rock bed. During the main heating season, a damper to the hollow core circuit is left open, allowing a part of hot air to enter the hypocausts directly.

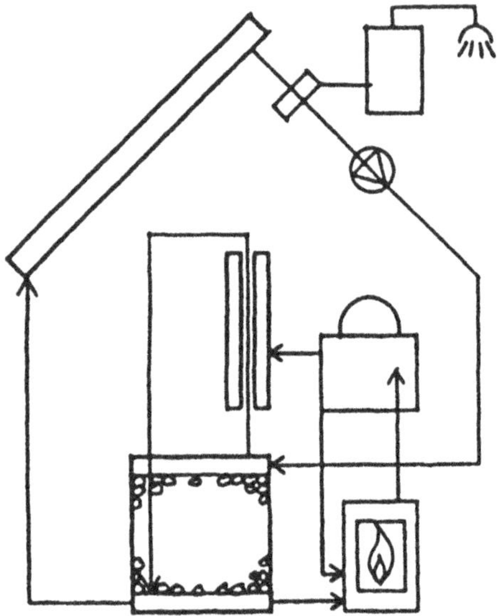

*Figure II.2.2. System diagram: main system with subsystems*

- *Passive discharge*. During the sunless hours, warm air thermosiphons from the rock bed to the hypocausts. The hypocausts deliver heat passively by radiation.
- *Auxiliary heating*. A tile oven in the living room is supplied with hot air from a wood stove in the basement, via a thermosiphon closed loop. On demand, a damper can be opened to let the hot air thermosiphon into the general hypocaust system.
- *DHW*. Domestic hot water is heated by heat exchanger in the charge loop. The heat exchanger is also in use during the summer venting of the air collector. Backup heat is provided electrically.

## Components

*Air collector*

The air collector is single glazed with underflow, a corrugated aluminium absorber and a selective surface. Its properties are given in Table II.2.3.

*Table II.2.3. Properties of the collector*

| | |
|---|---|
| Net collector area: | 58.2 $m^2$ |
| Tilt: | 45° |
| Orientation: | 190° (10° west of south) |
| Solar contribution: | $kWh/m^2a$ |
| Air flow rate through the collector: | $50m^3/m^2_{collector\ area}h$ |
| Fan power: | 900 W |

The site-assembled air collector is roof-integrated and covers the total area of the south-facing roof. The collector covers the whole roof for practical, economic and aesthetic reasons. In this way construction of the surrounding building components is simpler and cheaper. Site assembly is efficient because a standardized mounting procedure and partial prefabrication have been used. Figure II.2.3 shows a section of the air collector.

*Storage*

The storage (Table II.2.4) consists of the rock bed, the concrete-floor hypocausts and the murocausts of the masonry walls. Thus much of the internal construction contributes to the total storage capacity. A section of the hypocaust and murocaust is shown in Figure II.2.4.

*Distribution*

During collector operation there is partial fan-forced charging of the muro/hypocausts. During the remaining time there is thermosiphoning from the rock bed to the hypocausts and passive radiation from the hypocausts.

*Table II.2.4. Storage*

| | |
|---|---|
| Rock bed: | |
| total volume | 34.5 $m^3$ |
| total capacity | 12.75 kWh/K |
| temperature range | < 60°C |
| Hypo/murocausts: | |
| temperature range | Slightly above room temperature |

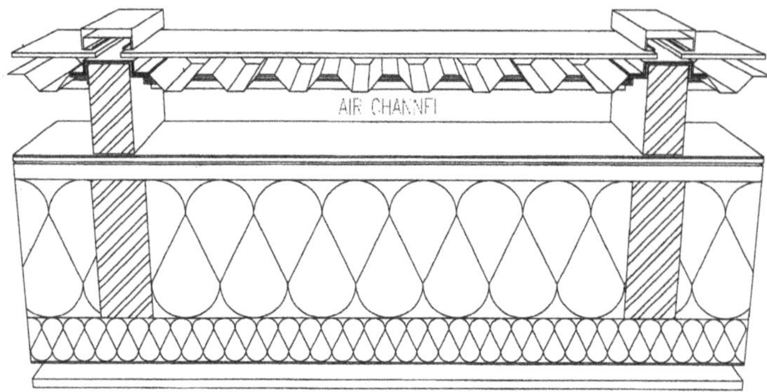


*Figure II.2.3. Detail section of the air collector*

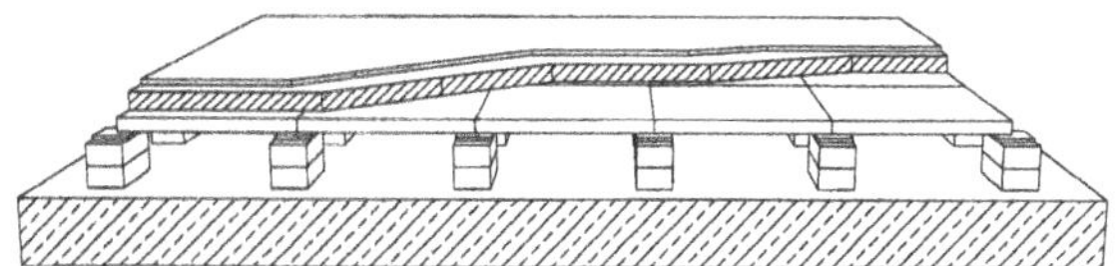

*Figure II.2.4. Detail section of the hypocaust and murocaust*

*Controls*

The fan for the charging loop is automatically controlled with reference to the difference in temperature between the collector and the rock bed. The back-draught dampers are self-operating and need no controls; if there is a heat demand, a damper is opened to let solar-heated air into the hollow core circuit. This damper is left open for most of the heating season and needs little attention. Additionally dampers can be opened to release air to the upper floor. There seems to be no reason for automatically controlling the discharge loop. The thermal mass of the hypocausts contributes to an even room-temperature level.

### Sun space

The two-storey sun space covers the central part of the south façade, which provides a major part of the daylighting of the living area. The sun space is not connected to the solar air system of the heated area of the house, but does have an open loop delivering surplus heat to its own rock bed. This can either be used for prolonging comfort in the sun space or to keep it free of frost. The sun space's rock bed has a limited energy-saving effect.

## PERFORMANCE

The house was monitored over the heating season 1995/96

### Comfort

The measurements confirm that the house is very comfortable with an average temperature in the living room during the heating season of 22°C. In spite of this relatively high average temperature (with peaks of 25° and 27°C), the auxiliary heating demand was only 18 kWh/m²a. Because the family has become accustomed to the efficient solar heating during autumn and spring, they also desire higher temperatures during the heating season. A further contribution to comfort is the high surface temperature of the hypocausts and the well-insulated exterior envelope. There have been no problems with summer overheating, since unwanted solar gain can be avoided by shutting off the system whenever desired.

### System performance

The system has been in operation since autumn 1992. The fan operating the air collector had to be replaced because the fan installed initially shut off occasionally in early autumn and late spring as a result of overheating. Installing a second fan solved the problem. The controls of the fan have functioned without any problems.

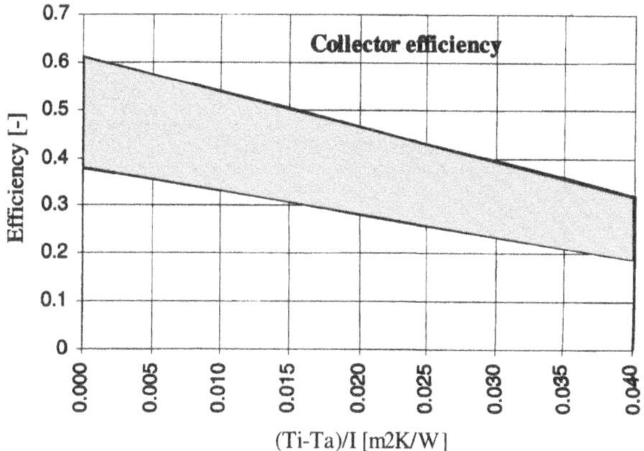


*Figure II.2.5. Collector efficiency*

### The collector

The air flow volume was planned at approximately 50 m³/h per m² of collector. Although some difficulty was encountered while measuring the air flow, it was found to be 34 m³/h per m² of collector. The range of the collector efficiency is shown in Figure II.2.5.

### Energy consumption

During the monitoring period, the auxiliary heating was limited to 55 days within a period of 12 weeks. The auxiliary heating energy demand during the 1995/96 season was 4200 kWh (1250 kg of wood) or 17.6 kW/m²a. Water heating consumed 533 kWh of auxiliary electricity and 2016 kWh of solar energy over a 12 month period. The electricity consumption is shown in Table II.2.5, while the energy flow chart is given in Figure II.2.6.

### Results of simulation studies

The following simulation results assume that the passive gains from the sun space are used prior to the use of the gains from the air collector. If the opposite assumption were made, the solar system would have performed better.

The setting point for auxiliary heating was 20°C. The higher level of comfort actually achieved by the solar air heating system has not been taken into account in the following simulations.

*Table II.2.5. Electricity consumption*

| | |
|---|---|
| Total demand | 21 kWh/m²a |
| Appliances and lighting | 18 kWh/m² a |
| Water heating | 2 kWh/m²a |
| Solar system fan | 0.7 kWh/m²a |

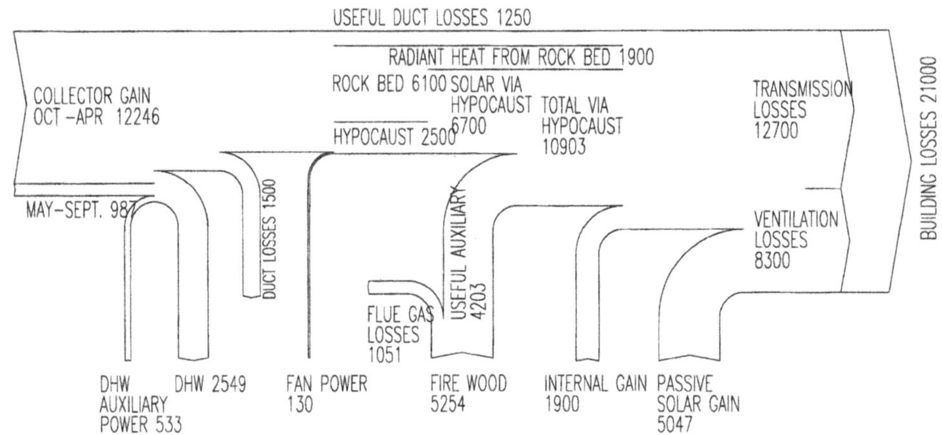


*Figure II.2.6. Energy flow chart (kWh) from October to April (DHW for the whole year)*

## Variation of the collector area

A constant rock bed volume of 30 m$^3$ was used to simulate the effect of collector areas from 0 to 60 m$^2$. The air flow rate was kept at 34 m$^3$/m$_{coll}^2$. An ideal flow rate is considered to be at 50 m$^3$/m$_{coll}^2$.

As a result of the passive solar gains of the sun space and the high standard of thermal insulation, the auxiliary heat load was 8310 kWh without any collector. With a 60 m$^2$ collector the auxiliary heat load was reduced to 4380 kWh (see Figures II.2.7 and II.2.8).

## Variation of the size of the rock bed

A constant collector area of 60 m$^3$ was used to simulate the effect of rock-bed volumes from 0 to 45 m$^3$. The hypocaust system was left constant. The option without a rock bed was thus handled as a single closed loop with a collector and a hypocaust. The setting points for turning off the fan for collector gain were 25°C for room temperature and 28°C for the hypocaust surface temperature.

The simulations show that the collector was able to deliver considerably more energy with rock bed than without (3300 kWh with 45 m$^3$, but also 2660 with 15 m$^3$ and 2080 kWh with only 7 m$^3$ – see Figures II.2.9 and II.2.10). A large part of this energy is used for a higher comfort level (above 20°C). The saved auxiliary heat needed to sustain only 20°C was 860 kWh.

The Frei House was intended to have a vertical-flow rock bed with 12 to 15 m$^3$. This required too much of the lower floor, leaving too little room for the bedsitter apartment. Consequently, this part of the rock bed was reduced to 8.6 m$^3$. A horizontal part with 25.8 m$^3$ was therefore placed under the lower floor. Because the changes in the construction of the foundations were minimal, the additional costs were low.

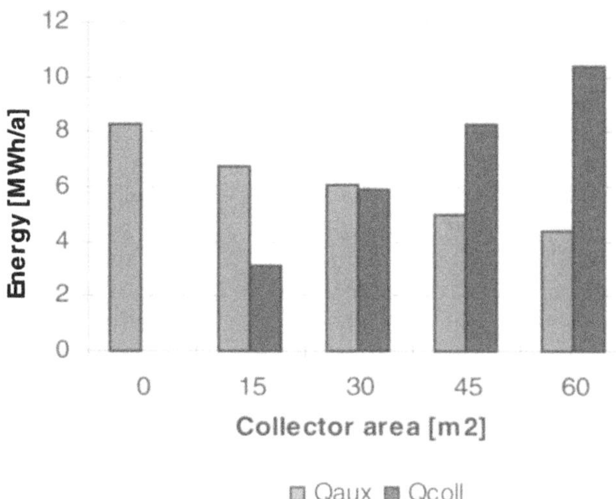


*Figure II.2.7. Auxiliary heating load and solar gain from the collector as a function of the collector area*

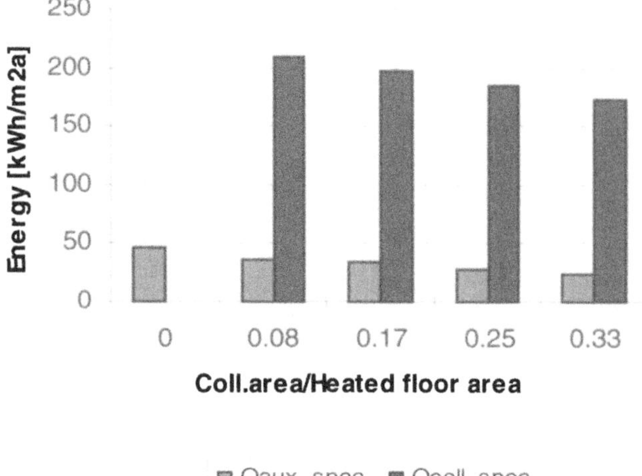


*Figure II.2.8. Specific auxiliary load and specific solar gain (per floor area) as a function of the ratio of collector area to floor area*

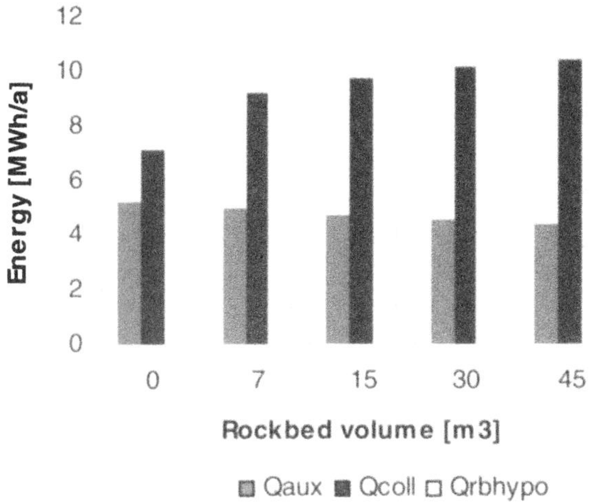


*Figure II.2.9. Auxiliary heat load, solar gain and heat transport from the rock bed to the hypocausts as a function of the storage volume*

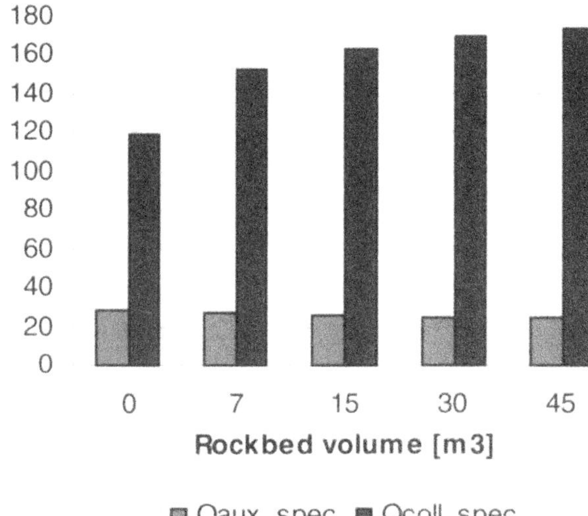


*Figure II.2.10. Specific auxiliary load and solar gain (per floor area) as a function of the storage volume*

## REMARKS

Auxiliary heat demand should be judged relative to the level of comfort provided. For economy and performance reasons, a reduction in the number of strategies leads to better results, but not always to improved comfort. The costs of constructional storage elements, rock beds, air collectors, sun spaces and, not least, thermal insulation depend just as much on practical building aspects as on the actual sizing. There is no sharp optimum for combining thermal-insulation and solar-gain strategies. This allows variations in the combination of strategies.

### The hierarchy of energy sources

The simulation results presented above assume that the passive gains from the sun space are used before the gains of the air collector. The assumption that gains from the air collector are used prior to those from the sun space would, however, give higher results for the solar air system. An attempt to apportion the usefulness between the two is purely arbitrary.

### The rock bed

The rock bed in this house saves energy and improves comfort since both the rock bed and the hypocausts help even out the temperatures. The size of the rock bed could be considerably smaller, but, given careful planning, this is not necessarily a major cost factor. The thermosiphon distribution loop worked without fan power.


## ACKNOWLEDGEMENTS

Architect, system designer:
Sture Larsen, Lindauerstrasse 33, A-6912 Hörbranz
Monitoring and simulations:
Manfred Bruck, Kanzlei Dr. Bruck, Prinz-Eugen-Strasse 66, A-1040 Wien, Austria. bruck@magnet.at
Christoph Muss, Vienna
Eckhard Drössler, Düns
Josef Burtscher, Energiesparverein Vorarlberg, Dornbirn
Manufacturers of the solar components:
Selective absorber surface: TeknoTerm Energi AB, Askim, Sweden

# II.3 Gwadt House

## St Gallenkappel, Switzerland

*St Gallen-kappel*

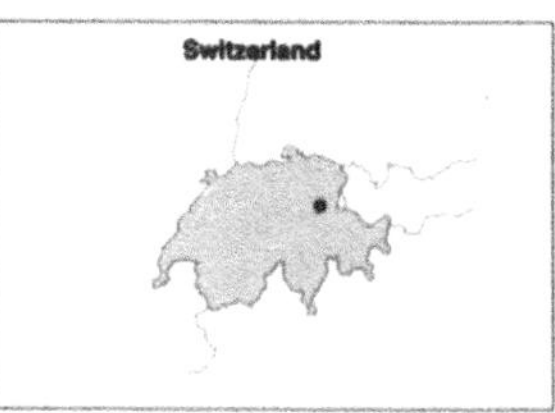


*System type 4*

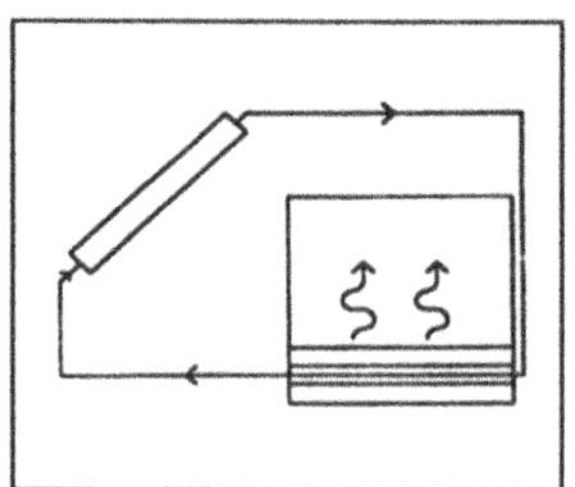

### PROJECT SUMMARY

In this solar building a window collector covers approximately 50% of the south facade. The airflow in the collector, duct system and storage elements occurs by natural convection only, and therefore no electric energy is needed for fans.

The air system delivers 1450 kWh of heat per year, saving approximately 150 kg of wood. The building structure provides heat storage and improves comfort. A wood stove provides auxiliary heat during periods with insufficient sunshine. The concept of letting the solar system and the wood stove deliver hot air to the same duct is promising, but still in its infancy.

#### Summary statistics

| | |
|---|---|
| System type: | type 4 |
| Collector type and area: | window collector, 26.8 m$^2$ |
| Storage type, capacity: | hypocaust, 9.8 kWh/K |
| Annual contribution of the solar air system: | 7 kWh/m$^2$a$_{floor}$ |
| Annual auxiliary heat consumption: | 30.6 kWh/m$^2$a$_{floor}$ |
| Basis: | monitored |
| Heated floor area: | 206 m$^2$ |
| Year solar air system installed: | 1993 |

### SITE DESCRIPTION

The Gwadt house is located in St Gallenkappel, a small village approximately 40 km east of Zurich:

| | |
|---|---|
| latitude | 47°N |
| longitude | 9°E |
| altitude | 550 m above sea level. |

The local climate is typical for a major part of Switzerland with average monthly temperatures ranging from –1.8°C to 18.2°C. The total annual horizontal solar irradiation is 1150 kWh/m$^2$. No major shading in the neighbourhood obstructs the sun.

### BUILDING PRESENTATION

The house is a simple rectangle of 7.8 × 11.6 m with the broad side oriented south. The room distribution is quite conventional, with living room, dining room and kitchen on the ground floor and bedrooms and bathroom on the first floor.

#### Construction

The basement is constructed of site-poured concrete with 100 mm of polystyrene insulation on the outside. The upper walls have an outer layer of exposed brick,

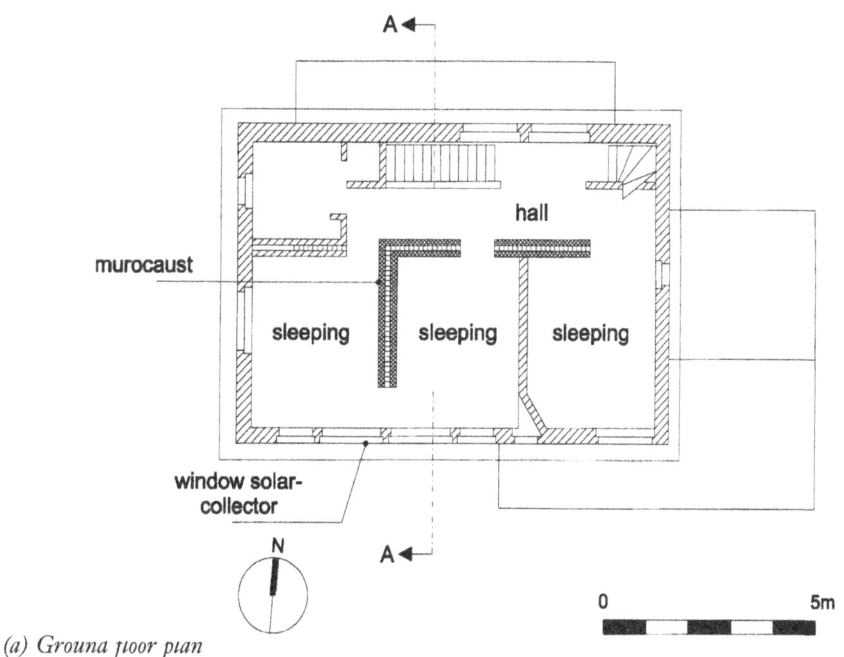


*(a) Grouna ƒloor pıan*

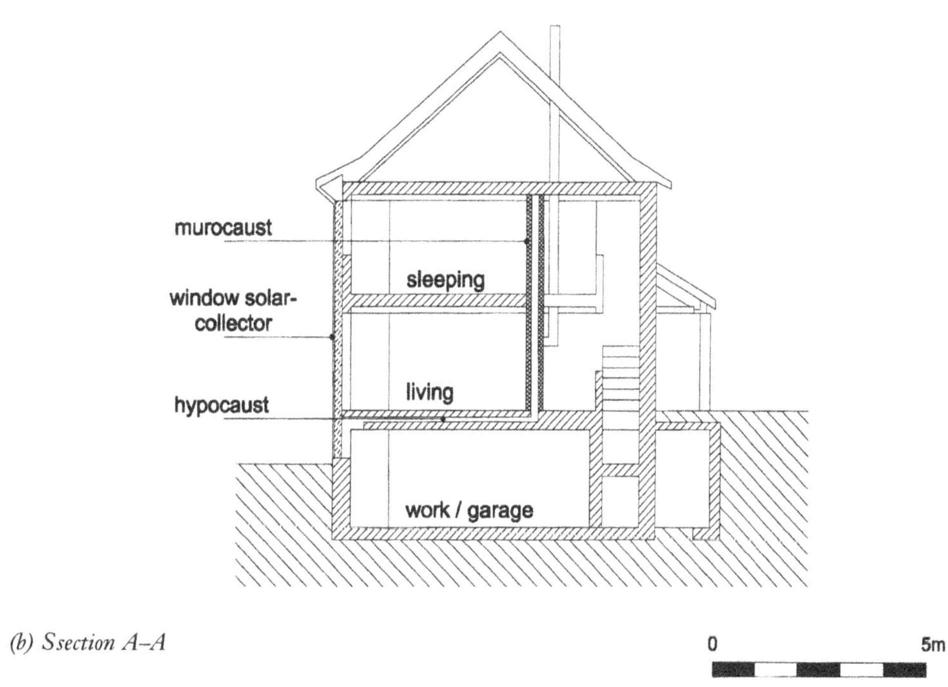


*(b) Ssection A–A*

*Figure II.3.1. (a) Ground floor plan and (b) section A–A*

followed by insulation, a vapour barrier and wooden panelling on the inside. The roof has 200 mm of mineral wool insulation placed between the rafters. All windows have wood frames with low-emissivity double glazing.

The inner walls are of light-weight wood-frame construction, except for the hollow concrete wall storage elements. Likewise the floors are wooden, except for the hollow concrete storage elements. The ground-floor plan and a section are shown in Figure II.3.1.

## Building statistics

The U-values reported in Table II.3.1 are typical for a low-energy house in Switzerland. The building heat losses are given in Table II.3.2.

*Table II.3.1. U-values for single elements*

| Element | *U*-value (W/m²K) |
|---|---|
| Outer wall | 0.24 |
| Roof | 0.22 |
| Cellar ceiling | 0.34 |
| Window | 1.60 |
| Window collector (including frame) | 0.97 |

*Table II.3.2. Building heat losses*

| | |
|---|---|
| Gross heated floor area | 206 m² |
| Conduction heat loss (estimated) | 138 W/K |
| Ventilation heat loss (estimated) | 40 W/K |
| Total building heating loss (estimated) | 178 W/K |
| Total building heating loss per m² | 0.86 W/m²K |

## SOLAR SYSTEM

The solar system is the primary space-heating system and a wood stove supplements it when required. The solar air system consists of a window air collector, the ductwork, the concrete hypocaust between floors and murocausts between all major rooms. All circulation is by gravity, i.e. without any fans. With the exception of a backdraft damper in the solar loop, there are no moving parts. Figure II.3.2 shows an isometric view of the solar system. The solar circuit is activated by the sun striking the collector, the auxiliary heating circuit by the occupants lighting a fire in the wood stove. The solar circuit and the auxiliary heating circuit are separated by means of a thermal trap analogous to a plumbing trap. The air paths are shown in Figure II.3.3.

*Table II.3.3. Properties of the collector and of the storage system*

| | |
|---|---|
| Collector area (glazing) | 26.8 m² |
| Storage volume (concrete) | 13.7 m³ |
| Storage capacity | 9.8 kWh/K |
| Maximum air velocity | 0.4 m/s |
| Maximum air flow rate | 500 m³/h |
| System height for buoyancy | 6.5 m |
| Flow cross-section | 0.36 m² |
| Maximum air flow rate per m² of collector | 19 $m^3/m^2_{coll}h$ |
| Storage capacity per m² of collector | 0.37 $kWh/m^2_{coll}K$ |

### The collector

The collector is a double window with a dark-coloured venetian blind in the airspace between the glass panes. The venetian blind is raised and lowered manually. The inner glass pane of the window collector is double-glazed ($U$ = 1.6 W/m²K), whereas the outer window has normal insulation glazing ($U$ = 3.0 W/m²K). The system properties are given in Table II.3.3. The air flow rates given are maximum values reached during sunny days. They are much smaller during the start-up of the collector and during cloudy weather periods.

### Storage and distribution

The characteristics of the hypocaust storage and distribution system can also be seen from Table II.3.3. The air is circulated through spiral-formed sheet metal ducts, 10 cm in diameter. This choice, together with careful design (e.g. avoiding abrupt changes in cross-section), minimizes the pressure drop and allows gravity circulation.

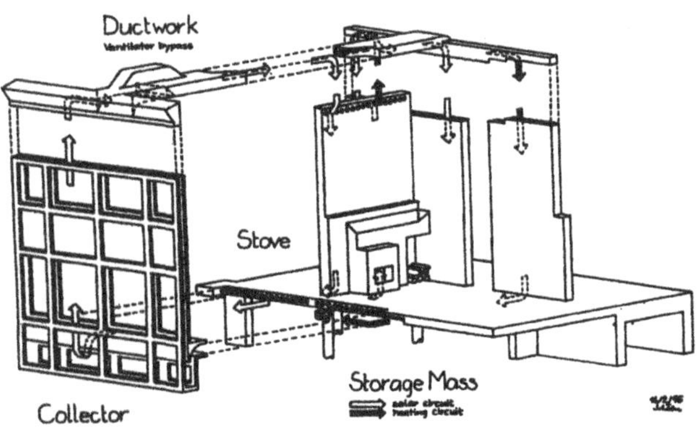


*Figure II.3.2. Isometric view of the solar system*

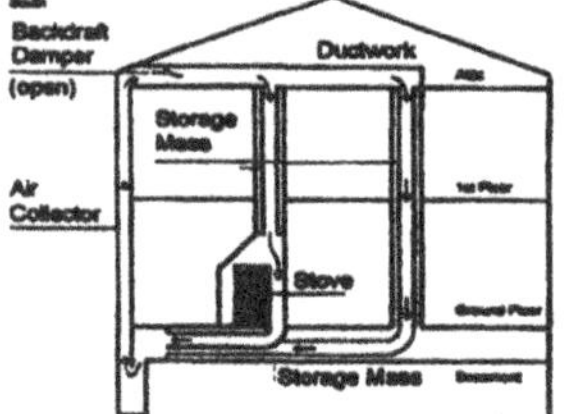


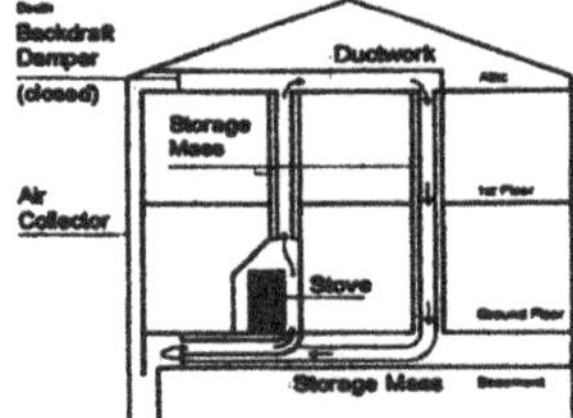


*Figure II.3.3. Air flow paths for collector (left) and the auxiliary heating circuit (right)*

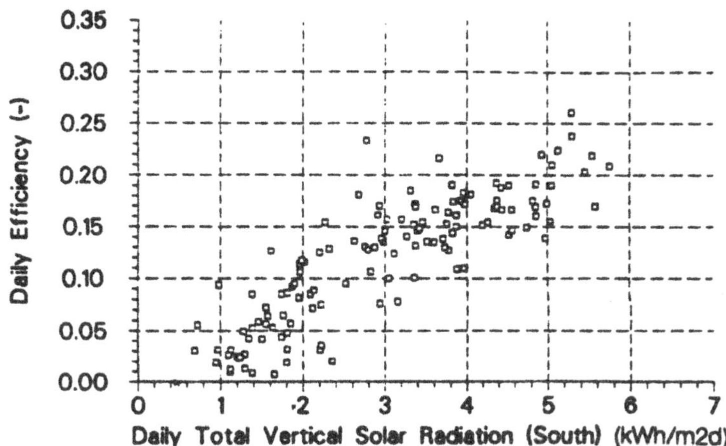


*Figure II.3.4. 'Daily efficiencies' of the collector in thermo-circulation air flow*

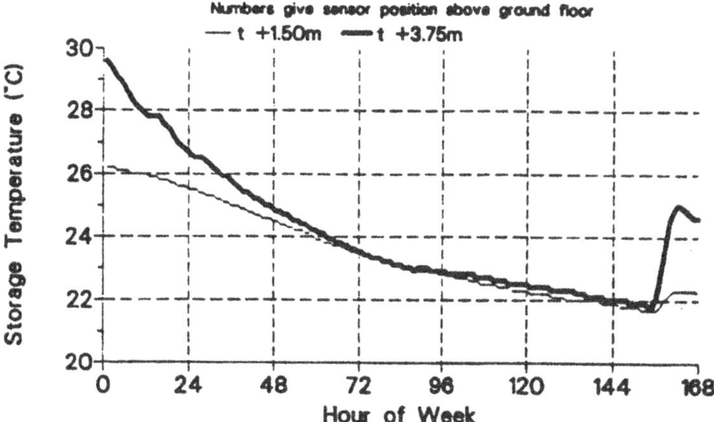


*Figure II.3.5. Temperature decay of the storage system during an overcast period*

The stored heat is released by radiation only. The insulation level of the storage system varies from room to room in order to achieve an optimal heat distribution throughout the house.

## PERFORMANCE

The performance of the heating system and the total energy consumption were monitored from August 1993 to April 1994.

### Collector

Standard collector efficiency curves for constant air flow rate do not apply to this system because thermo-circulation has variable flow rates. Instead 'daily efficiencies' are shown as a function of the total daily solar radiation in Figure II.3.4. They reach their highest values between 0.2 and 0.3 for very sunny days. Solar gains throughout the heating season are given in Table II.3.4.

*Table II.3.4. Solar gains of the collector during heating season 1993/94 (September through April)*

| | |
|---|---|
| Solar convective gains | 1450 kWh |
| Window collector area | 26.8 m$^2$ |
| Specific gains | 54 kWh/m$^2$ |
| Average daily collector efficiency | 13% |

The natural-convection solar air collector of the Gwadt house performs less well than comparable fan-forced collectors. For the Gwadt house the collector's annual gain in the heating season is 7 kWh per m$^2$ of heated floor area.

### Duct system

Because of the small difference in height between the collector and the storage, the air flows slowly. In the collecting circuit typical values were 0.4 m/s, whereas in the stove circuit values of 0.2 m/s occurred. The total flow rate of 130 m$^3$/h in the heating circuit is about one third of the value estimated from design. This suggests that the flow cross-section of 0.18 m$^2$ for the heating circuit, as compared to 0.36 m$^2$ for the collector circuit, is too small.

### Storage system

The storage temperatures reach 30°C in the uppermost section and 27°C at the top of the storage in the ground floor. Because of the long flow path, temperatures in the horizontal duct system seldom reach 20°C (see the

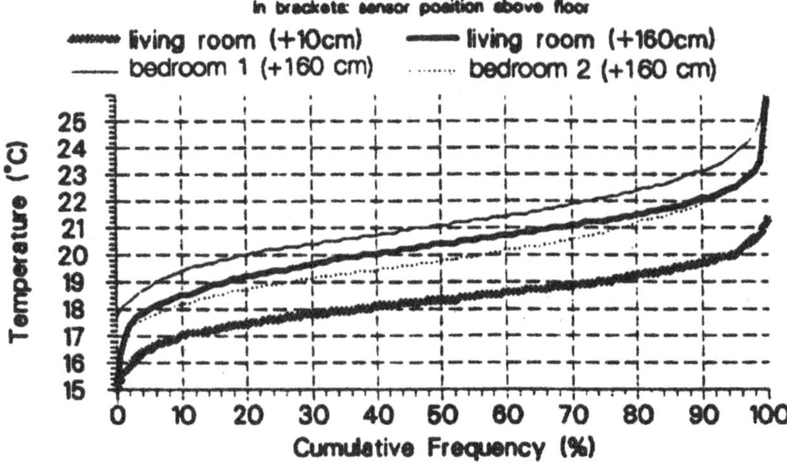


*Figure II.3.6. Cumulative frequency of room temperatures during the heating period*

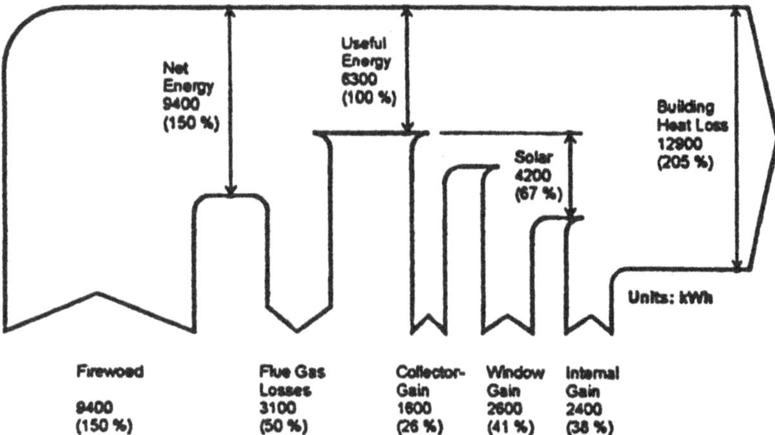


*Figure II.3.7. Energy flow chart for the heating season; all data in kWh*

next section, on *Comfort*). Figure II.3.5 shows the temperature decline in the storage during an overcast period, without there being any recharging through the solar or stove heating circuit. The storage releases half of its heat content within 2.5 days. Such a short discharge time was intended.

### Comfort

The combined solar/wood-stove system is able to maintain reasonable comfort in the house. This is illustrated in Figure II.3.6, which shows the cumulated frequency distribution of room air temperature in the living room during the heating period. Near the floor of the living room the temperature is too low. For 25% of the time the temperature falls below 18°C. This reduction in comfort was confirmed by the occupants.

The window collector is not vented in summer and continuous system operation, together with high passive solar energy gains, causes room air temperatures up to 30°C (~ 2 K higher than the ambient air temperatures). During the night the temperature can be lowered to an acceptable level by natural ventilation. In retrospect, openings should have been provided to vent the collector during summer.

### Energy consumption

The Sankey diagram (Figure II.3.7) shows that the auxiliary heat demand of 6300 kWh amounts to approximately 50% of the total building heat loss, which is well below the recommended target value for new buildings in Switzerland.

The amount of firewood consumed was determined by weighing and electrical power consumption was read from the meter. 2200 kg of wood was used during the heating period 1993/94, corresponding to an equivalent of about 800 kg of oil. The specific energy consumption for domestic hot water is 3 kWh/m$^2$a and the electrical consumption is 12 kWh/m$^2$a. Both also lie well below the target values.

## REMARKS

The Gwadt house demonstrates that window air collectors may be well integrated into the building envelope. Air flow sustained by natural convection (thermo-circulation) is possible in an air system with large cross-sections, but the energy output of the natural convection solar system is greatly reduced compared to air flow with forced ventilation.

It is possible to combine the duct systems for the collecting circuit and the wood-stove circuit, although much care must be taken to avoid backward circulation. When the collector and stove circuits are simultaneously operated, undesired interference may occur.

Temperatures in the floor hypocaust elements remain modest because thermo-circulation in the stove operating mode is too weak and the flow paths are too long. In addition, the ducts are inadequately insulated near the collector (causing backflow).

Summer venting of the collector should be provided.

## ACKNOWLEDGEMENTS

Chapter author: Charles Filleux, Basler & Hofmann, Forchstrasse 395, CH-8029 Zürich

Architect and design of solar system: Ueli Schäfer, Binz

Consultant and measurements: Daniel Brühwiler, Volketswil

## BIBLIOGRAPHY

*Messprojekt EFH Gwadt, St. Gallenkappel* (1995) (final report of the monitoring project, in German). HBT-Solararchitektur, ETH-Hönggerberg, CH-8093 Zürich, Switzerland.

# II.4 Lenherr House

Schwyz, Switzerland

*Schwyz*

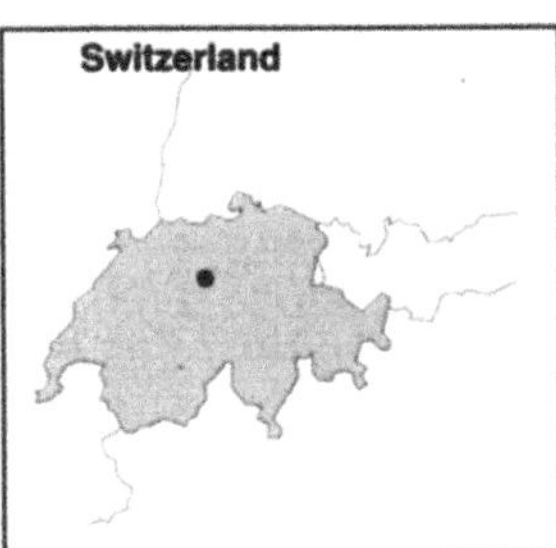


*System type 4*

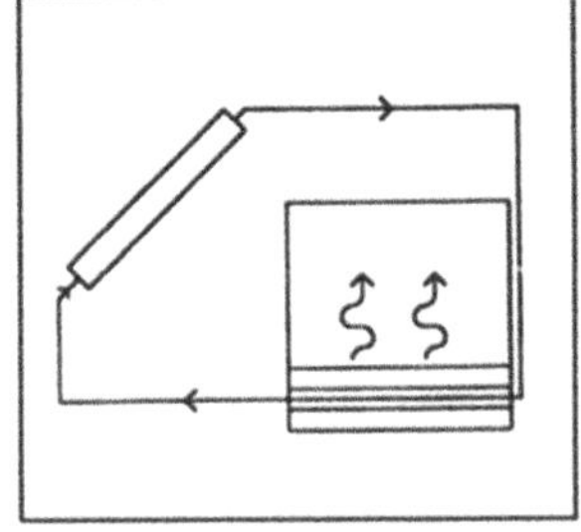

## PROJECT SUMMARY

This house includes an unusual window collector as the outer glazing of a sun space that covers the entire south facade. Thus, the living area behind the sun space is buffered from heat losses as well as sheltered from the overheating which can occur in rooms directly behind large areas of glazing. The combination of sun space and window collector reduces the heating load by approximately 30%.

The collector heated air is channelled through insulated hollow core floors and walls by fans. The stored heat is later passively released to the rooms. This closed-loop charging–radiant discharging solar system also reduces the heating load, by 30%, making a 60% overall reduction.

### General statistics

| | |
|---|---|
| System type: | type 4 |
| Collector type: | window collector |
| Collector area: | 33 m$^2$ |
| Storage type: | hypocaust |
| Storage capacity: | 44 kWh/K |
| Annual contribution of the solar air system: | 20 kWh/m$_{floor}^2$ |
| Annual auxiliary heating consumption: | 19.6 kWh/m$_{floor}^2$ |
| Heated floor area: | 230 m$^2$ |
| Year solar air system installed: | 1988 |

## SITE DESCRIPTION

The building is situated on a south-west slope in the residential area of the town of Schwyz:

| | |
|---|---|
| latitude | 47°N |
| longitude | 8°E |
| altitude | 610 m above sea level. |

The alpine peaks are close, but the main collecting facade is always in the sun even in December.

The climate is similar to the so-called 'central European climate', i.e. overcast in winter. Nevertheless, because of the altitude, the site is often just above the fog. Average monthly temperatures range from –1°C to 20°C. Relative to the base of 20°C/12°C there are 3900 heating degree-days. The total annual south-facing vertical solar irradiation is 3160 MJ/m$^2$.

## BUILDING PRESENTATION

- The building has the form of a cube and is earth-sheltered to the north (Figure II.4.1).
- The south side is generously glazed and covered by a sun space.
- There are few good-quality windows on the east and west sides and only two very small windows to the north.
- The envelope insulation (10 to 12 cm) has been carefully designed to avoid thermal bridges.

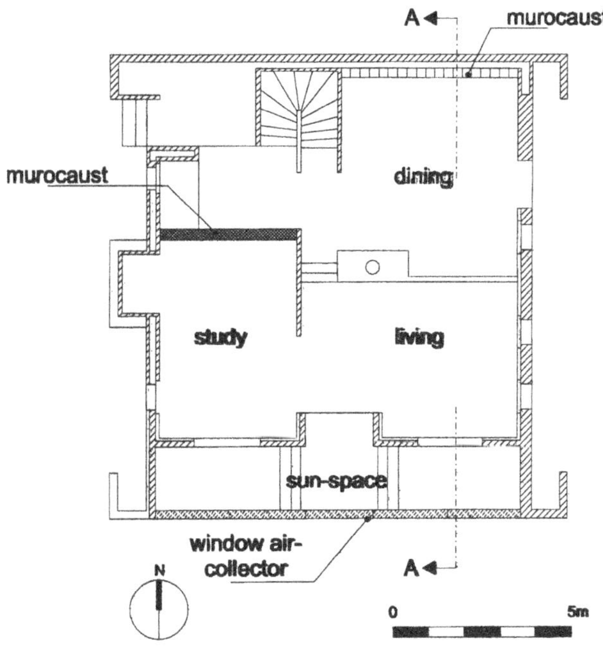


*(a) Ground floor plan*

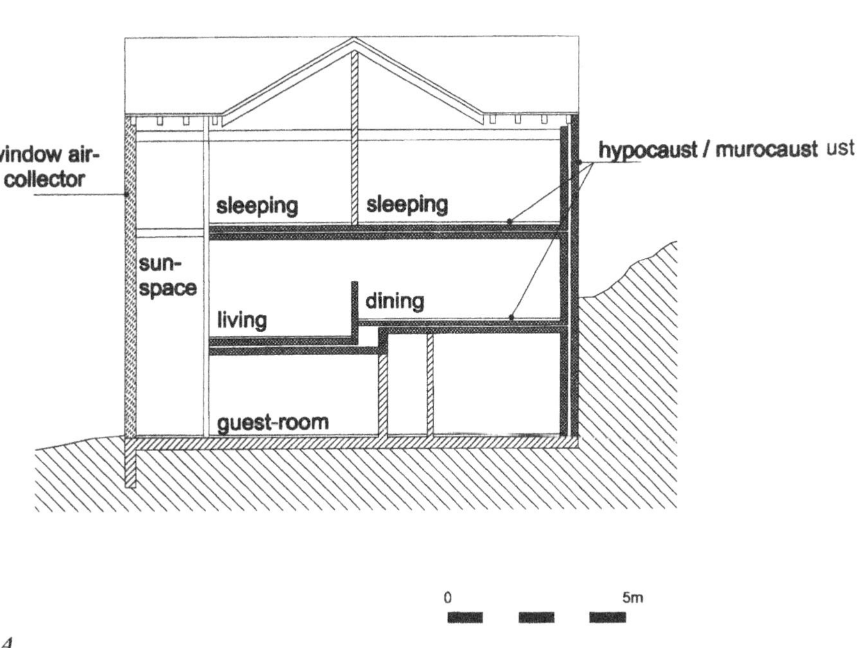


*(b) Section A–A*

*Figure II.4.1. (a) Ground floor plan and (b) cross-section A–A*

- The floors are built of 30 cm thick concrete slabs to accommodate the ventilation ducts of the solar system. The walls are mainly limestone brick.
- A central wood stove in the basement provides auxiliary heating. Hot water is provided by electricity.

*Table II.4.1.* U*-values of building elements*

| Element | *U*-values (W/m²K) |
|---|---|
| Walls | 0.3 |
| Roof | 0.3 |
| Windows east, west and north | 1.6 |
| Windows inner south (living room) | 2.8 |
| Windows outer south (sun space/collector) | 2.1 |

*Figure II.4.2. View into the sun space, where one fan and an electrically driven damper can be seen*

### Construction and building statistics

The *U*-values of the building elements are given in Table II.4.1 and Figure II.4.2 shows a view into the sun space.

## SOLAR SYSTEM

### The sun space

A sun space, 1.5 to 2 m deep, covers all three storeys of the south facade. Access to the sun space on the upper floors is via a large glazed door leading to a steel-grate balcony that can be moved on rails for maintenance of the sun space. The wall common to the house and the sun space is insulated (10 cm) on the sun-space side. The sun space therefore warms up quickly on sunny days.

### The collector

Except for a small strip of doors and windows that can be opened in the middle section, the whole outer facade of the sun space is built as a window air collector. Steel profiles provide an air gap for the dark-coloured venetian blinds, which are lowered and tilted by a manually controlled electrical operation. Black perforated steel sheets are located in the upper part of the facade. The solar facade has double glazing to the outside and single glazing to the inside. Air is fan-forced from the top downward, opposite to the direction of natural thermal convection. This permits the glazing of the lower sun space to be included in the loop and it ensures that the hottest air is delivered to the floors rather than to the ceiling (Figure II.4.3). There are two lower and two upper circuits (east and west), each equipped with an 'in-duct' axial type fan (Figure II.4.4) and an electrical damper to avoid backflow. Table II.4.2 gives the technical data for the collector and the storage system.

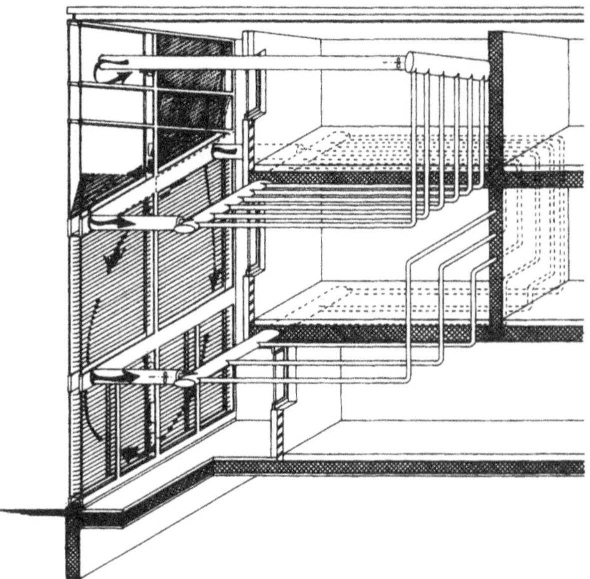

*Figure II.4.3. Isometric drawing of one side of the collector system with its small upper circuit (fixed perforated steel sheets, partly shown for visibility) and the large lower circuit (venetian blinds)*

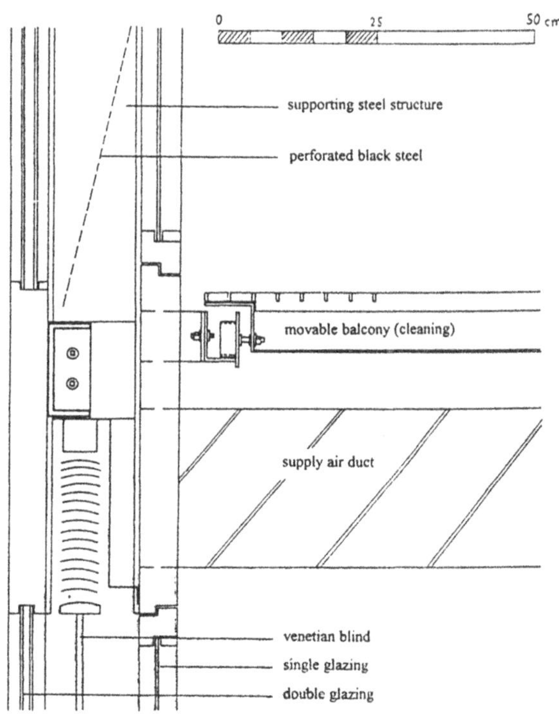


*Figure II.4.4. Details of collector construction*

*Table II.4.2. Technical data for the collector and the storage system*

| | |
|---|---|
| South facade – overall area | 83 $m^2$ |
| South facade – absorber area (perforated sheet and venetian blind) | 33 $m^2$ |
| Air volume rate/absorber area (maximum) | 52 $m^3/h\ m^2$ |
| Fan power/absorber area (maximum) | 25 $W/m^2$ |
| Storage volume/absorber area | 1.9 $m^3/m^2$ |
| Storage capacity/absorber area | 1.35 $kWh/Km^2$ |

### Storage and distribution

The sun-warmed air from each collector circuit is distributed evenly to ten flexible corrugated PVC drainage pipes that have been cast into the concrete floors and walls (Figure II.4.5). In the lower circuits the air is passed back to the collector through the ceiling, while in the upper circuit the air is collected and returned through a sheet metal duct running through the bedrooms.

To retard the release of the stored heat the floors are insulated with 2.5 cm of polystyrene on top and 5 cm below. The difference in thickness ensures that most of the heat is radiated up from the floor rather than down from the ceiling.

The automatic (two-speed) control of the fans is temperature regulated. The fans are switched to low speed as soon as the temperature of the blinds (sensor attached on the reverse side of a fixed 'dummy blind slat') is 7 K above the mean storage temperature. This low-speed operation also ensures that there is some air movement through the collector to give an accurate measurement of air temperature. The high-speed mode is engaged whenever the air temperature of the collector exceeds approximately 30°C (the storage never reaches this temperature, so there is no risk of the fan discharging the storage). The mean storage temperature is checked by four temperature sensors (one per circuit) located in the front part of each storage.

## PERFORMANCE

The system was monitored for 12 months.

### Comfort

Comfort measurements (air and globe temperature, humidity) and analysis (PMV values) were made according to ISO recommendations. The results indicate that the house was always sufficiently heated. Warm surface temperatures ensured comfort throughout the winter. In summer, even though the monitoring season was hotter than usual, opening the sun-space windows to the ambient air prevented overheating of the living areas. Being able to turn the bright (reverse) side of the collector blinds outwards had no effect at all on the temperatures in the collector. It did not affect the temperatures in the sun space either.

### System performance

The optimal air flow volume of 50 to 70 $m^3/h$ per $m^2$ of collector (experience and literature) was confirmed. But only 50 $m^3/h$ per $m^2$ of collector could be reached, even after replacing the fans of the lower circuits with a more powerful (and more electricity-consuming) model (450 W instead of 130 W) in February 1991. The corrugations in the flexible pipes caused so much drag that the original small fan moved only half of the required air volume. This led to very high collector temperatures and therefore high collector heat losses to the ambient. After the fan had been replaced, peak temperatures in the collector were more than 10 K lower and the efficiency of the collector performance increased from 0.27 to 0.42 (collector output/global daily solar incident radiation), or by more than 50% (Table II.4.3). It is very important to mention that the smaller fan would have done as well if standard smooth ventilation pipes had been used.

*Figure II.4.5. The reinforced floor construction before the concrete was poured*

*Table II.4.3. A comparison of measured system performance before and after fan replacement*

| Attributes<br>Day (date) | With original fans<br>6 November 1990 | With replacement fans<br>20 February 1991 |
|---|---|---|
| Mean daily outside air temperature (°C) | 1.1 | 1.9 |
| Total solar irradiation (kWh) | 75 | 78 |
| Collector output (kWh) | 20 | 33 |
| Fan running time (h): | | |
| low speed | – | 1.5 |
| high speed | 8 | 6 |
| Maximum collector air temperature (°C) | 60 | 50 |
| Change in storage temperature (K) | 0.4 | 1.8 |
| System efficiency | 0.27 | 0.42 |

The controls of the fan were reliable. The two-speed mode functioned well, but was not really necessary. Single-speed fans would have reduced first costs and not significantly increased the power consumption relative to the collector output; they would simply cycle on and off more frequently than a two-speed fan. Figure II.4.6 gives the collector efficiency curves for low and high specific air flow rates.

## Sun-space and ventilation strategies

The buffering effect of the sun space is important, not only to impede heat losses, but also to avoid overheating of the living space. Because the inner surface of the (single glazed) window collectors easily reaches 50°C, overheating would have been problematic, if the window collectors had been directly in front of the living spaces. On the other hand, the quick temperature rise in the sun space provides additional heated air, which can be moved into the house by opening the doors to the living area in the afternoon. The natural updraft through the three storeys is sufficient to provide strong air circulation. In summer, this updraft proves highly effective for venting the hot sun-space air to the outside. Windows on the outer middle part of the sun space are simply opened to the ambient air. With good manual management of the sun space, usability is not compromised in either winter or summer.

## Energy consumption

The auxiliary heating energy demand of the house was measured to be 4500 kWh or 19.6 kWh/m²a of heated floor area and was covered by burning wood (net energy 32 kWh/m²a).

The electricity demand was 27.3 kWh/m²a, including hot water. This is an average value for this kind of home and the number of occupants (four) and was composed of:

| | |
|---|---|
| Appliances and lighting | 14 kWh/m²a |
| Water heating | 12 kWh/m²a |
| Solar system fans | 1.3 kWh/m²a |

The energy balance is shown in Figure II.4.7.

## Results of simulation studies

Based on measured data, extrapolations were made by computer modelling with TRNSYS to answer additional questions. One key issue is how much energy the various solar systems built into this house save relative to a conventional house. The sun space taken alone reduces annual heating energy consumption by 35% compared to a similar direct gain house and the window air collectors taken alone save 40%. The two systems combined result in a saving of 60%.

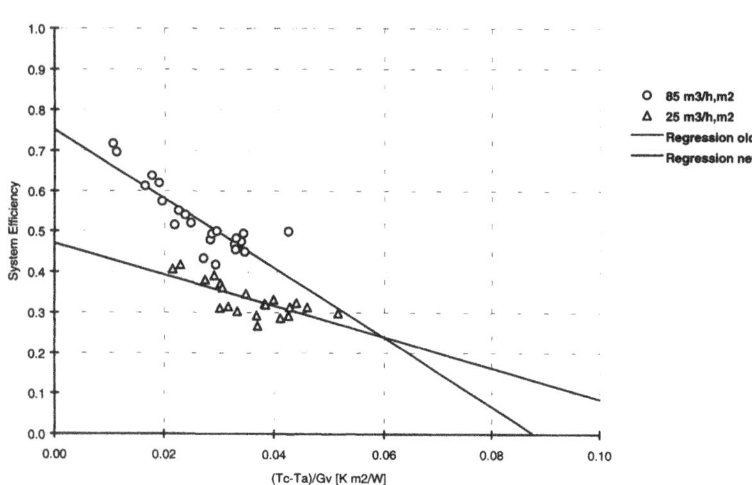


*Figure II.4.6. Collector efficiency curves for low and high specific air flow rates measured during operation; $T_c$ is the mean air-collector temperature, $T_a$ the mean outside temperature and $G_v$ the total global vertical solar radiation*

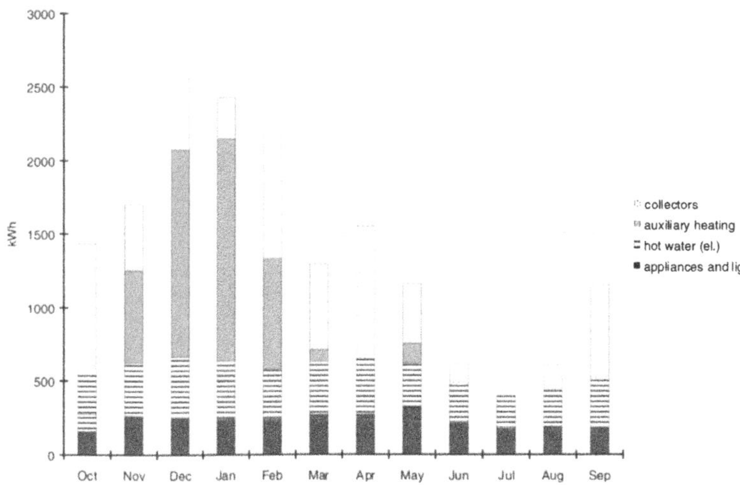


*Figure II.4.7. Measured monthly energy balance during period from October 1990 to September 1991*

Air flow rates exceeding 50 m$^3$/hm$^2$ in the window collector were investigated using simulation. An increase to 70 m$^3$/hm$^2$ lowered heating energy by only 300 kWh, but increased electricity demand by 150 kWh. The net benefit of this measure does not justify its implementation.

The thickness of the storage insulation was found to be optimal. Two-speed control for the fans is not necessary.

The ratio of storage volume to collector area (1.9 m$^3$/m$^2$) could be lowered to 0.8 m$^3$/m$^2$ without a major degradation in performance.

## REMARKS

The combination of window air collectors, sun space and hollow-core floor/wall storage proved to be very effective. The Lenherr house is a low-energy building providing excellent thermal comfort. With this system configuration, it is possible to use the whole south facade as the solar aperture with only one type of facade construction.

Window air collectors can achieve efficiencies exceeding 40% if the air flow rate is sufficient (> 50 m$^3$/hm$^2$). Single-speed fans are adequate for this purpose. With the absorber blinds in an almost horizontal position good visual contact with the outside is provided without any significant degradation of collector performance. Tilting the vanes of the blinds to a 'reflecting position' in summer (rather than keeping them horizontal) provides no advantage regarding collector air temperature and therefore avoidance of overheating.

Hollow-core storage in walls and floors offers good performance, low cost and high thermal comfort. Special attention has to be given to air flow resistance in order to keep ventilation power requirements reasonable. The surfaces of air-pipes, channels and ducts must be smooth and air speeds should be lower than 3 m/s to avoid excessive friction losses.

The sun space acts as a very effective thermal buffer against heat loss and overheating. Excess heat can easily be vented to the house in winter and to the outside in summer if someone is present to operate windows and doors.

Costs could be substantially reduced by omitting the upper two collector circuits (the absorber area is too small), selecting simplified controls and using a simpler, more standard facade construction.

## ACKNOWLEDGEMENTS

Architect: H. Oberholzer, Rapperswil Engineer for solar system design: K. Haas, Jona

Chapter co author: C. Filleux, Basler & Hofmann AG, Forchstrasse 395, CH-8029 Zürich

Chapter co author and measurement: A.Gütermann, Amena AG, Steinbergstrasse 2, 8402 Winterthur, Zürich

## BIBLIOGRAPHY

*Solar home Lenherr, Schwyz* (1992). (final report of the monitoring project, in German), HBT-Solararchitektur, ETH-Hönggerberg, CH-8093 Zürich, Switzerland.

# II.5 Schüpfen House

## Schüpfen, Switzerland

*Schüpfen*

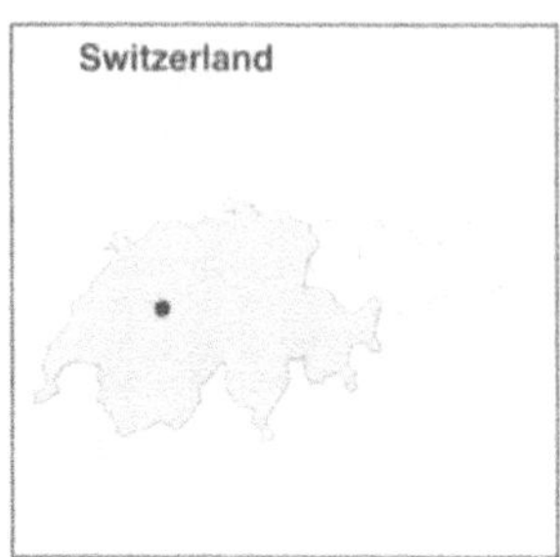


*System type 6*

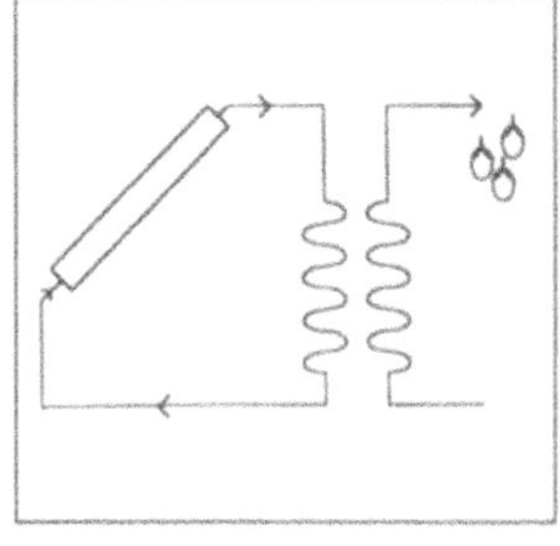

### PROJECT SUMMARY

This row (terrace) house, incorporates a PV-powered closed-loop solar air system constructed in 1985. It was rebuilt 10 years later to incorporate a water storage tank for domestic hot water production in the summer. The system delivers 2600 kWh of heat per year, saving approximately 450 kg of wood and 1200 kWh of electricity.

#### Summary statistics

| | |
|---|---|
| System type: | type 6 |
| Collector type and area: | double glazed, facade 15 $m^2$ |
| Storage type, volume: | water tank, 2000 litres |
| Annual contribution of the solar air system: | 14 $kWh/m^2a_{floor}$ |
| Annual auxiliary heating consumption: | 35 $kWh/m^2a_{floor}$ |
| Basis: | monitored/calculated |
| Heated floor area: | 183 $m^2$ |
| Year solar system rebuilt: | 1995/96 |

### SITE DESCRIPTION

The house is located in Schüpfen, a village about 10 km north-west of Berne:

| | |
|---|---|
| latitude | 47°N |
| longitude | 7°E |
| altitude | 520 m above sea level. |

The site is in hilly countryside, but there is no shading from the landscape.

The climate is typical for the middle of Switzerland, with average monthly temperatures between –1°C (January) and 18°C (July). The total annual horizontal radiation is 1150 $kWh/m^2$. Typical for the region is ground fog staying until noon on sunny autumn and winter days.

### BUILDING PRESENTATION

The house is situated in a row of four houses and forms part of a development with several rows (Figure II.5.1), all

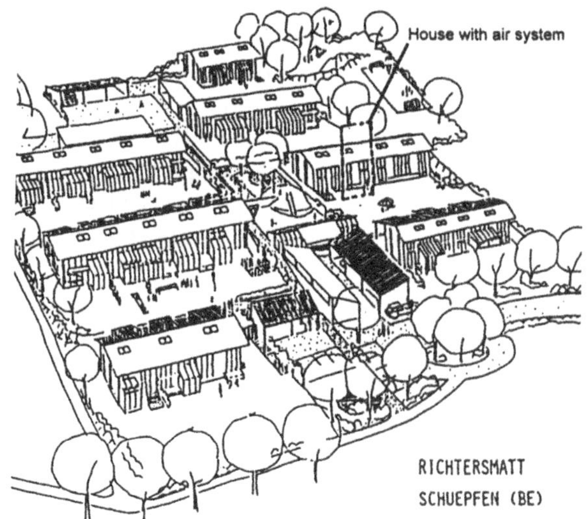


*Figure II.5.1. Development overview*

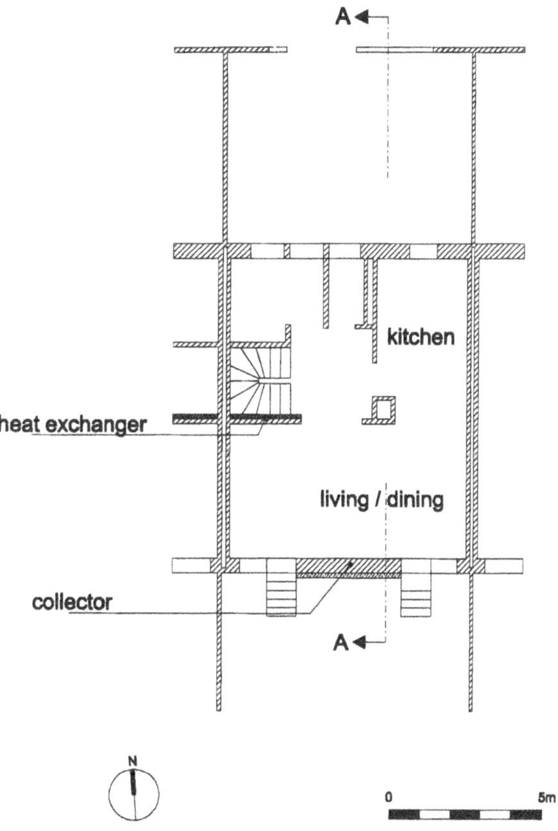


*(a) Ground floor plan*

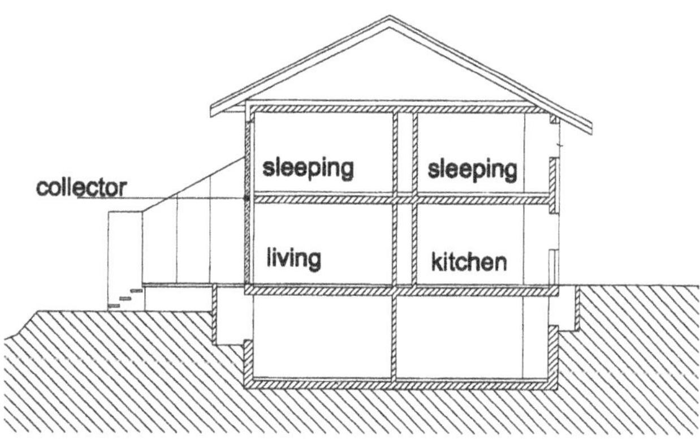


*Figure II.5.2. (a) Ground floor plan and (b) cross-section A-A*

oriented east–west, enabling all the houses to profit from the sun. The plan and section are shown in Figure II.5.2.

The house uses solar energy as follows:

- by direct gain through 13.9 m$^2$ of south-facing windows for the south zone of the house, supported by a heavy-weight construction;
- by use of a solar air collector (15.1 m$^2$), covering the rest of the south facade, providing heat to a 2000 litre water storage tank with an integrated domestic hot water heater via an air-to-water heat exchanger;
- by use of photovoltaic panels to power the fans and the pump.

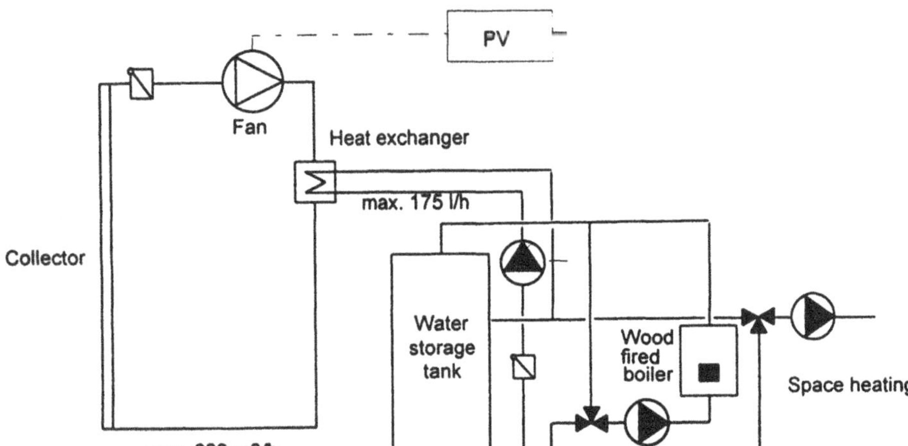


*Figure II.5.3. Schematic of the solar air system*

In addition, there is a wood-fired boiler feeding the same storage tank as the backup heating system

### Building statistics

Building heat losses are shown in Table II.5.1.

### Construction

The exterior walls are double-brick masonry with 12 cm of mineral wool insulation between. The roof is wood-frame construction with 12 cm of mineral wool insulation between the beams. The windows have triple glazing with a solar transmission of 0.60, wood frames and an inside moveable infrared reflection film for both solar and night heat-loss protection. The *U*-values are reported in Table II.5.2. They are not quite as good as they should be for a low-energy house in Switzerland, but are typical for the period of construction.

## SOLAR SYSTEM

The system was partly reconstructed in 1995/1996 because the original, phase-change storage walls were unsatisfactory. These were replaced by a heat exchanger linked to the existing water storage tank of the backup heating system. The system can thereby heat domestic hot water during summer. This nearly doubles the energy contribution.

*Table II.5.1. Building heat losses*

| | |
|---|---|
| Gross heated floor area | 183 m$^2$ |
| Heated volume | 510 m$^3$ |
| Estimated transmission heat loss | 96 W/K |
| Estimated ventilation heat loss | 85 W/K |
| Total building heat loss | 181 W/K |
| (Measured) total building heat loss per m$^2$ | 0.99 W/m$^2$K |

*Table II.5.2. U-values of the construction*

| Element | *U*-value (W/m$^2$K) |
|---|---|
| Opaque walls | 0.30 |
| Roof | 0.25 |
| Windows | 2.00 |

*Table II.5.3. Collector, storage and air system properties*

| | | |
|---|---|---|
| Collector | net collector area | 15.1 m$^2$ |
| | solar contribution | 2600 kWh |
| | airflow rate | 600 m$^3$/h maximum (radiation-dependent, PV-driven) |
| Storage | volume (water) | 2,000 litres |
| | capacity | 2325 kWh/K |
| Distribution | fan power | 200 W maximum |
| | fan power/collector area | 13 W/m$^2$ |
| | tension | 24 V DC |
| | consumption | none (photovoltaic) |

The collector was left unchanged and performs satisfactorily. It is a special construction with a wood frame, an absorber of perforated sheet metal and a double-glazing cover (Figure II.5.3, Table II.5.3). Figure II.5.4 shows an isometric view of the system.

The ductwork and the heat exchanger have been placed in the staircase (Figures II.5.5 and II.5.6) in place of one of the original phase-change material storage walls, which was removed.

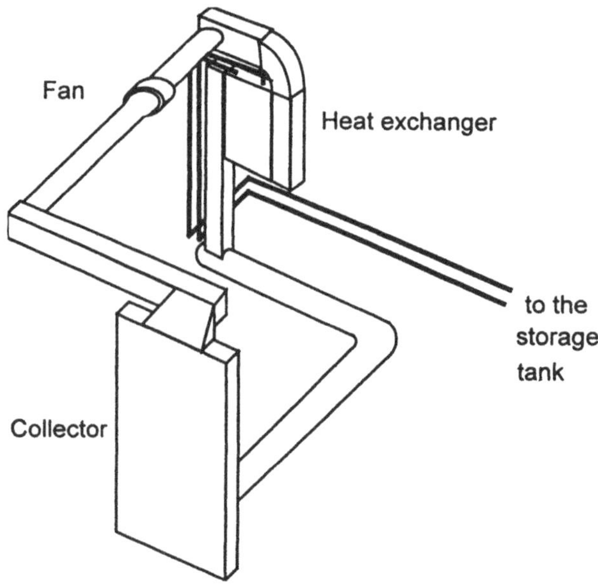


*Figure II.5.4. Isometric view of the solar air system*

*Figure II.5.5. Heat exchanger in staircase*

*Figure II.5.6. Duct and pipe connections to heat exchanger*

The air circuit is self-regulating, because the fans are photovoltaic powered (Figure II.5.7). No further control is necessary, except for a temperature sensor in the return air duct to stop the system when the storage tank is fully charged (summer operation).

*Figure II.5.7. The photovoltaic panel on south facade for fan and pump drives*

## PERFORMANCE

Upon reconstruction of the system in 1996, it was monitored and then further improvements were made, including:

- Addition of a second fan because the design flow rates had not been reached.
- Adjustment of the flow rates of the air and the water circuits to reach equal temperature differences. This was done by reducing the PV power (some of the existing PV panels were re-circuited to provide energy for some of the owner's appliances), and by changing the resistance of both the water circuit and the pump electricity circuit.
  - Figure II.5.8 shows the relationship between the solar radiation and the flow rates. The water flow rate needs more radiation to start than the air circuit, but shows a greater increase with radiation. This behaviour is better than the opposite, because the air can warm up the heat exchanger before the water starts circulating.
- The heat exchanger was designed for a maximum power of 5.1 kW with a difference of 5 K between air and water temperatures. This was exceeded (typically by 12 K) because of the poor distribution of air over the cross-section of the heat exchanger. Baffles were therefore added to the heat-exchanger inlet duct. Figure II.5.9 shows the considerable temperature loss between the collector and the storage tank. This occurs in spite of the fact that the heat exchanger was overdesigned and depends on the operation of the system.

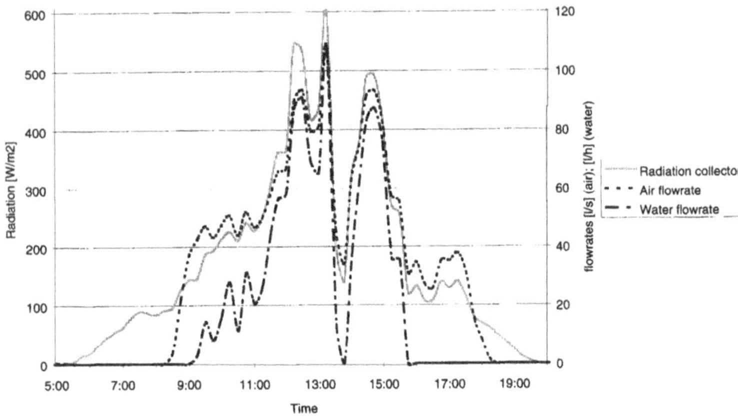


*Figure II.5.8. Air and water flow rates with solar radiation for one measured day*

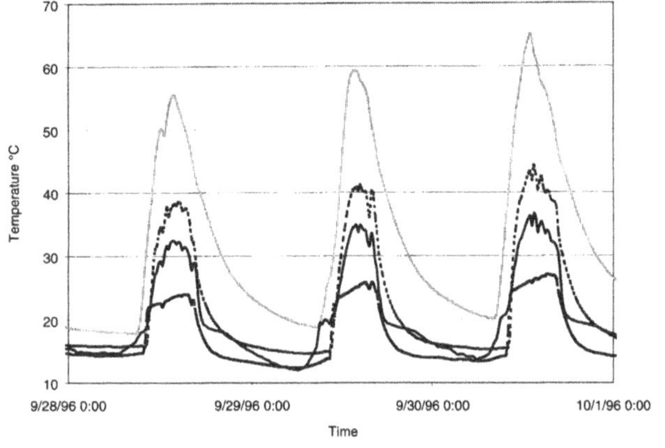


*Figure II.5.9. Air and water circuit temperatures for three measured days*

The storage tank was unfortunately completely mixed (except for the top layer) whenever the room heating pump was in operation. This led to heat exchanger (and thus collector) entry temperatures that were too high. The reason was that the connection of the return pipe from the room heating loop was at the bottom of the tank (common with solar circuits) and no volume was kept cold for the solar collector. The connection of the return pipe was therefore changed to a spare connection point 80 cm above the bottom of the tank.

The estimated seasonal energy performance of the system is shown in Table II.5.4. Because the system is also used for domestic hot water production (and therefore runs all year round), its energy contribution is almost doubled compared to a system that only provides space heating.

## REMARKS

The Schüpfen house demonstrates the operation of a solar air system in combination with a water storage tank and the control of the flow rates by photovoltaic driven fans and pumps. The photovoltaic control operates almost perfectly, if the components are correctly sized and tuned. This procedure, however, is not simple and requires observation under different operation conditions. The combination of a solar air system with a water storage tank has two advantages:

- Summer operation for domestic hot water production increases the solar usability.
- No frost protection is necessary for the solar circuit.

There are, on the other hand, also disadvantages. The most important is a relatively high temperature loss in the heat exchanger. The design of the heat exchanger is difficult. Even a generous sizing may not be sufficient under low part-load conditions, where, in combination with variable flow rates, the heat transfer may be inadequate. However, a large heat exchanger is not easy to integrate into an air system. Care has to be taken about the flow distribution, which may make flow-directing devices necessary.

*Table II.5.4. Seasonal energy contribution of the solar system*

| | Solar system energy contribution [MJ/a] | Solar gain per collector area [kWh/m²a] |
|---|---|---|
| Space heating | 5000 | 92 |
| Domestic hot water | 4365 | 80 |
| Total annual | 9365 | 172 |

## ACKNOWLEDGEMENTS

Sponsor:The Swiss Federal Office of Energy,
Owners: Roger and Ursi Spindler; thanks for their hospitality and all their help during the monitoring.
Chapter author: Gerhard Zweifel, Zentral Schweizerisches Technikum Luzern, Abt. HLK, 6048 Horw

# II.6 Morhenne Solar House

## Ennepetal Rüggeberg, Germany

*Ennepetal Rüggeberg*

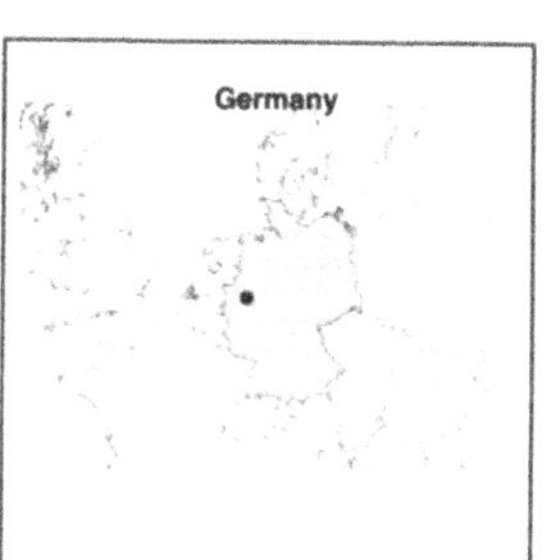


*System type 4*

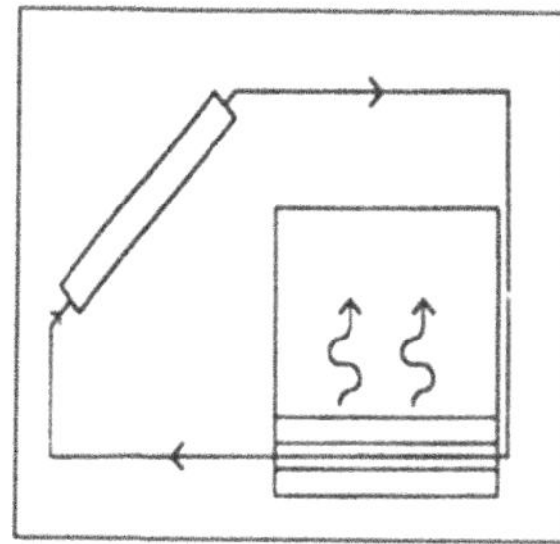

### PROJECT SUMMARY

This single family house, built in 1990 in a low mountain range, incorporates traditional building design with increased insulation and active solar systems. A hypocaust system, heated by solar air, with a PV-powered fan (DC), a solar domestic hot-water system, and a sun space that preheats the fresh air are the main solar installations. The hypocaust system installed in the ground floor is heated by 10 $m^2$ of solar air collectors.

The fan for the air collector system is powered by 212 $W_{peak}$ PV panels with a battery of 100 Ah. During the summer season thc PV system is used for lighting.

Warm air from the sun space is directly blown to the first floor. Important features of this building are the installed hypocaust system and the heavy mass construction, used to cover longer periods without sun (which occur even in early and late summer). This leads to higher performance of the solar system but carries the risk of overheating. Because of the small ratio of collector area to hypocaust mass, overheating has never happened, but the solar fraction is low (23%).

#### Summary statistics

System type, variation: type 4
Collector type and area: opaque, roof mounted, 10 $m^2$
Storage type, volume, capacity: hypocaust, 25.7 $m^3$, 8.2 kWh/K
Annual contribution of the solar air system: 12.3 $kWh/m_{floor}^2$
Annual auxiliary heat consumption: 42.2 $kWh/m_{floor}^2$
Basis: calculated
Heated floor area, volume: 226 $m^2$, 577 $m^3$
Year solar system built: 1991

### SITE DESCRIPTION

The building is located in Ennepetal Rüggeberg on the top of a hill:

latitude 51°N
longitude 7°E
altitude 360 m

The average monthly temperatures range from –27.7°C to 36.1°C, with a mean of 7.2°C. The total annual horizontal solar irradiation is 1150 $kWh/m^2$. No major shading in the neighbourhood obstructs the sun. Relative to the base of 20°C/12°C there are 3500 degree days. The mean wind speed is 5 m/s.

### BUILDING PRESENTATION

The plan and section of the building are shown in Figure II.6.1.

#### Building statistics

The orientation of this single-family house with attached sun space is such that the roof ridge runs east to west with azimuth of –10°. There are one and a half

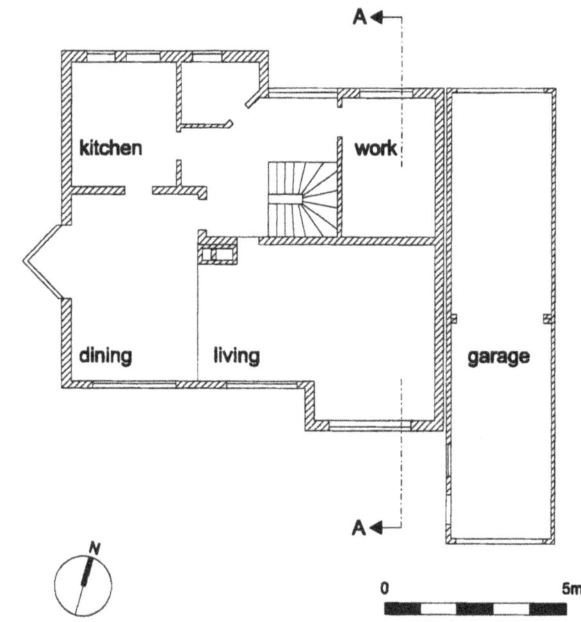


*(a) Typical floor plan*

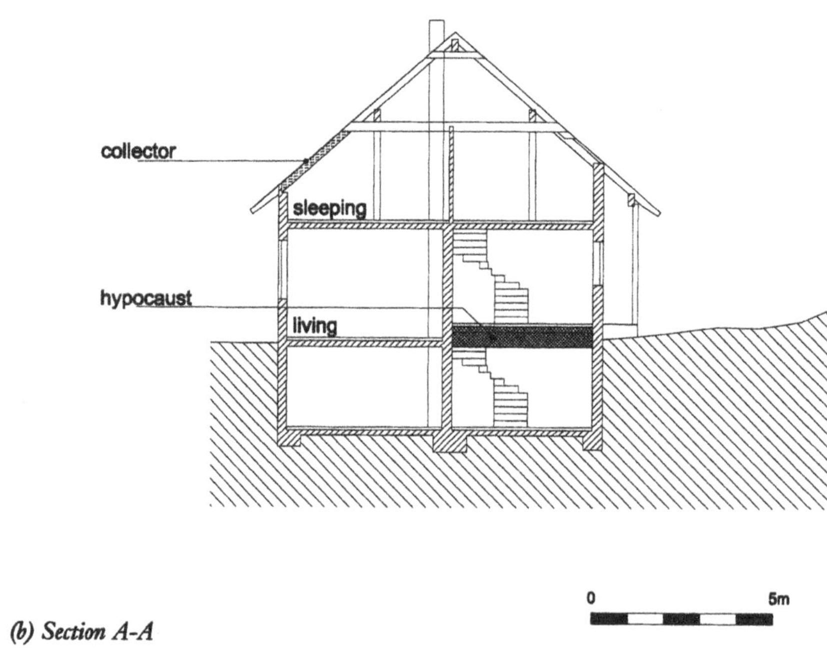


*(b) Section A-A*

*Figure II.6.1. (a) The floor plan of a typical floor and (b) a section of the house*

storeys and a basement. The gross heated floor area is 226 m$^2$, the heated volume 577 m$^3$ and the roof slope 40°.

The construction details are given in Table II.6.1 and the simulated and measured building heating coefficients in Table II.6.2.

## SOLAR SYSTEM

Long sunless periods made it a challenge to cover the heating demand from April to October without backup. A system with storage was therefore necessary. The solution, as built, includes:

- a hypocaust, solar air-heated by 10 m$^2$ of air collectors;
- direct air heating by a sun space;
- a solar DHW system (5.9 m$^2$ water collector, 500 litres of storage).

The sun space was integrated as an active air heating system. The hypocaust system in the ground floor is

*Tbale II.6.1. The construction details*

| | | | |
|---|---|---|---|
| Walls | $U$ = 0.28 W/m²K | lime sand bricks | 24 cm |
| | | mineral wool | 3 cm |
| | | Pu foam | 6 cm |
| | | wooden sheathing | |
| Windows | $U$ = 1.3 W/m²K | low-e glass | g =0.63 |
| | | wooden frames | |
| | | roller blinds | |
| Roof | $U$ = 0.25 W/m²K | plaster board | 1.2 cm |
| | | mineral wool | 10 cm |
| | | PU foam | 6 cm |
| | | between rafters | |
| Total | $U$ = 0.67 W/m²K | | |

*Table II.6.2. Simulated and measured building heating coefficients (per m² gross floor area)*

| | |
|---|---|
| Simulated | |
| Conduction load | 57 kWh/m²a |
| Air change load | 47 kWh/m²a |
| Passive gains | 32 kWh/m²a |
| Measured | |
| Auxiliary space heating including hot water in winter | 42 kWh/m²a |
| Appliances and lighting | 8 kWh/m²a |

designed like a rock-bed storage and is characterized by its heavy mass and insulation on the upper and lower surfaces to increase the floor mass temperature above temperatures produced by typical floor heating systems. An isometric view of the hypocaust and collector system is shown in Figure II.6.2 and the scheme of the system in Figure II.6.3. The auxiliary heating system is a condensing modulating gas furnace of 8–18 kW.

At the time of construction, no factory-manufactured roof-integrated collectors were available. Therefore, the collectors were mounted above the roof tiles. To avoid having ducts inside the living area and to keep them short, the air channels were integrated into the east and west facades. This arrangement required that the collectors run the full width of the roof. Collector modules of 2.5 m length each were connected to provide a maximum length of 10 m (10 m² in area). Only one row of collectors was installed, because an additional row would have increased costs and led to decreased solar gains per m² collector area. Properties of the collector and storage system are given in Table II.6.3.

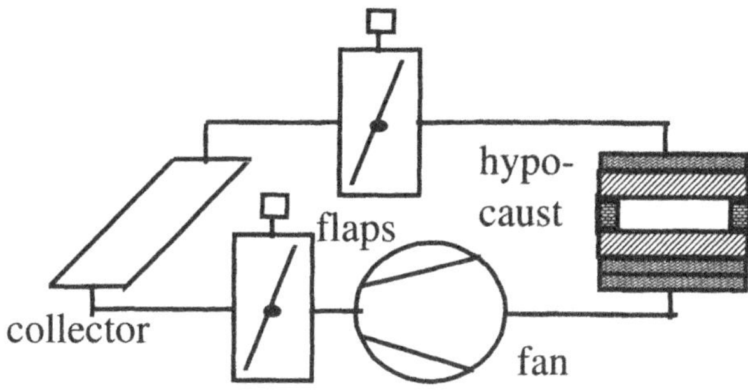


*Figure II.6.3. Scheme of the hypocaust and collector system*

The air inlet to the hypocaust is located in the northern part of the building in order to heat this part first. The ducts are connected in series to avoid inhomogeneous flow inside the hypocaust, which has often been a problem with such systems. The scheme of the air ducts is shown in Figure II.6.4 and the construction in the northern part of the building in Figure II.6.5.

## Distribution

The radial fan, which is belt driven by two d.c. motors (and therefore noisy), is mounted below the hypocaust to reduce noise in the living area. The fan location on the 'cold side' of the system means that there is a higher mass flow and allows the installation of the motors inside the system so that motor heat can be recovered and additionally the noise reduced. The wall-integrated ducts are made from plywood. During times of non-operation the inlet and outlet of the system are closed by motor-driven flaps to avoid thermosiphoning to the collector. The fan speed is adjustable by varying the DC voltage. The installed lead battery has a capacity of 100 Ah. The PV system consists of four Arco 53

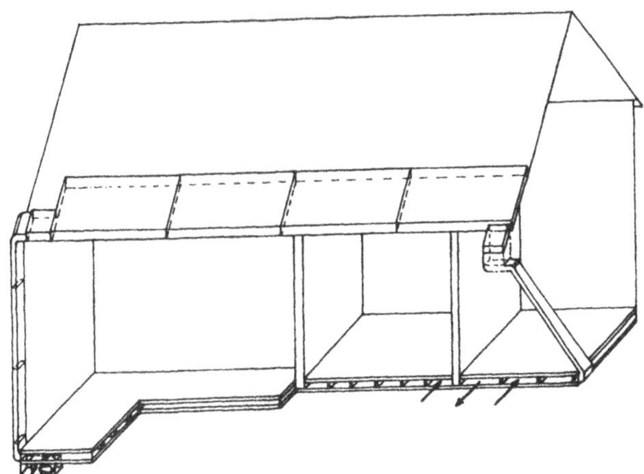

*Figure II.6.2. Isometric view of the hypocaust and collector system*

*Table II.6.3. Properties of the collector and storage system.*

| | |
|---|---|
| Collector | |
| Area | 10 m² solar air panel, net area 9.2 m² |
| Specific air flow | 50 $m^3/m_{coll}{}^2h$ |
| Solar gains | 280 $kWh/\ m_{coll}{}^2$ |
| Storage | 66 m² hypocaust in the ground floor<br>site built ducts, insulated on both sides<br>serial ducts, closed loop<br>29,600 kg mass<br>0.14 m³/s air flow |
| Storage | |
| Air speed | 1.94 m/s |
| Capacity | 8.2 kWh/K<br>0.88 $kWh/\ m_{coll}{}^2K$ |
| Air channels | width 0.4 m, height 0.18 m, length 46 m |
| Lower construction | 0.15 m concrete, 0.06–0.10 m insulation,<br>$U$ = 0.18 W/m²K |
| Upper construction | 0.1 m pumice stone, 0.06 m insulation<br>0.05 m light concrete $U$ = 0.5 W/m²K |

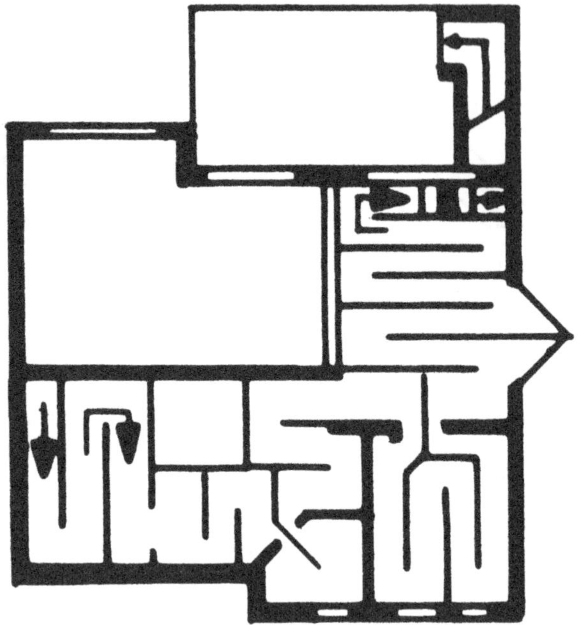

*Figure II.6.4. Scheme of the air ducts*

W panels with charge and discharge control. To avoid too deep a discharge of the batteries, a grid-connected charger comes on automatically.

### Control

The fan operation is controlled by a differential thermostat, which additionally opens the flaps at the inlet of the hypocaust. The set points are $DT = 5$ K for operation with a hysteresis of 2 K. The sensors are placed in the hypocaust and at the collector outlet. The control device has been enhanced by a relay, which switches the DC motors. An additional maximum thermostat, which switches the whole system off, is set to 35°C. All control devices can be switched off manually (for summer non-operation).

### Air heating in the sun space

Air heated in the sun space can be blown directly to the adjacent room by a small fan. The system includes a sound damper and a filter, is controlled by a thermostat set to 22°C and, because this system is often switched off by the inhabitants, does not make a significant energy contribution.

## PERFORMANCE

The energy performance of the air collectors is good, thanks to the large storage capacity. The collector inlet temperature never exceeds 24°C and, during the heating season, the collector is in operation whenever there is adequate solar irradiation. The temperature distribution in the hypocaust system is shown in Figure II.6.6.

The connecting ducts from the collector to the hypocaust were measured as quite leaky (20% of the air is exchanged). This is because one of the ducts was probably damaged during the installation of the prefabricated garages with a crane. Repair is nearly impossible because this part of the building is covered by the walls of the garage. The hypocaust therefore performs less well than expected. As a result of the reduced temperature of the inlet flow, the rise in temperature in the hypocaust is reduced and limited to the entrance area (see Figures II.6.6 and II.6.7). However, the temperature gradient is mostly due to the small ratio of the collector area to the hypocaust mass and the serial connection of the ducts.

The serial ducts cause a temperature gradient, which in this building is in accordance with energy demand, but the southern part of the hypocaust is ineffective for heating.

The low temperature of the hypocaust leads to an increased run time of the fan with a higher energy demand than that for which the PV system was designed. In times of low irradiation the supplied energy is

*Figure II.6.5. Construction of northern part of the building*

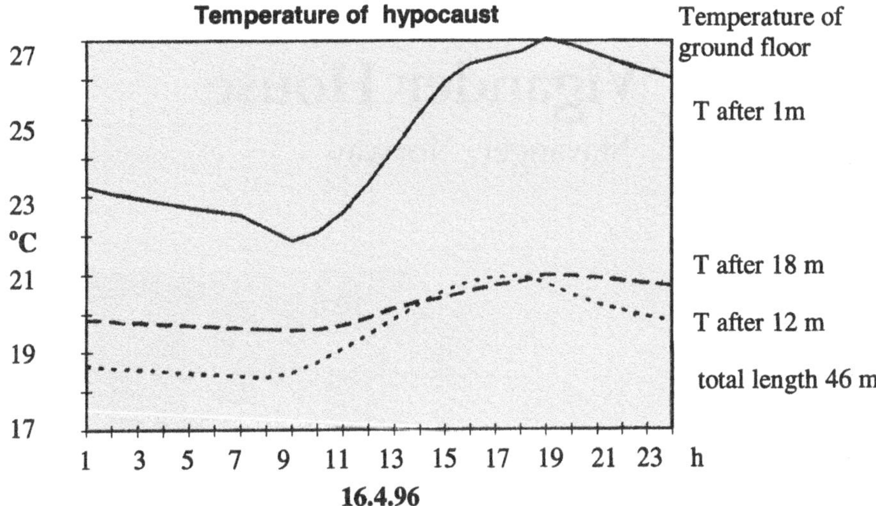


*Figure II.6.6. Temperature distribution in the hypocaust system*

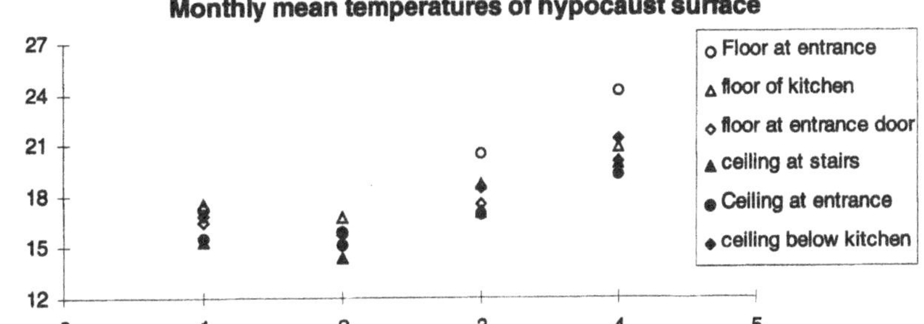


*Figure II.6.7. Monthly mean temperaature of different parts of the hypocaust (January–April)*

not sufficient to drive the fan. Consequently, the grid-connected charger operates more often than expected.

## REMARKS

The low ratio of collector area to hypocaust mass prevents a noticeable improvement in comfort. For new installations, therefore, it is recommended that a hypocaust with lower mass be installed, with separate additional storage if needed. An increase of the collector area increases the risk of overheating because of the high mass of the hypocaust.

The temperature gradient in the hypocaust, which is often not desirable, could be reduced by installing parallel ducts. The hydraulic design and precise construction in this case are important to avoid a non-uniform flow distribution.

Today, better planning tools are available and all design parameters can be predicted more accurately.

## ACKNOWLEDGEMENTS

### Manufacturers of solar components

Collector: Grammer SKT W. von. Braunstrasse 6, D-92224 Amberg

DC motor: Maxxon Motor GmbH, Wardeinstrasse 3, D-81825 München

Control: Resol GmbH, Fänkenstrasse 26, D-45549 Sprockhövel

### Building

Architect: W. Wülfing, D-58256 Ennepetal

Energy consultants: J. & J. Morhenne, Schülkestrasse 10, D-42277 Wuppertal

Owners: J. & J. Morhenne, Schülkestrasse 10, D-42277 Wuppertal

Chapter author: J. Morhenne, Schülkestrasse 10, D-42277 Wuppertal

# II.7 Vigander House

## Stavanger, Norway

*Stavanger*

*System type 4*

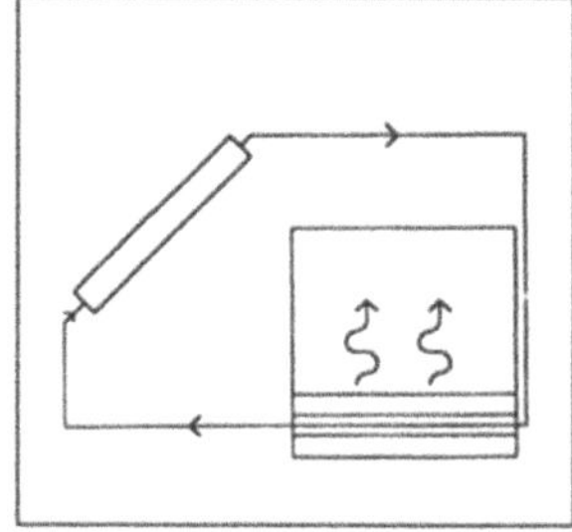

### PROJECT SUMMARY

Twenty solar buildings have been constructed in the Stavanger region, the majority with facade or roof-integrated solar air systems for space heating and/or heating of domestic hot water. The Vigander house is a typical example of architectural integration of a solar facade collector coupled to a hollow concrete floor heat distribution system. The concrete floor is 44 m$^2$. The system has been in uninterrupted service since 1989. The total additional costs for the whole system were US $2500 (ECU 2270). The occupants are extremely positive about the system.

#### Summary statistics:

| | |
|---|---|
| System type, variation: | type 4 |
| Collector type and area: | wall-integrated air collector |
| Storage type, volume, capacity.: | hypocaust, 5.7 m$^3$, (concrete), 4.0 kWh/K |
| Annual contribution of the solar air system: | no monitoring |
| Annual auxiliary heat consumption: | no monitoring |
| Basis: | calculated |
| Heated floor area: | 65 m$^2$ |
| Year solar system built: | 1989 |

### SITE DESCRIPTION

The house is located in Hafrsfjord, Stavanger on a steep hill facing west. The view to the west is unobstructed in all directions from south to north:

| | |
|---|---|
| latitude | 58°N |
| longitude | 6°E |
| altitude | 40 m above sea level |

Annually, there are 1727 sunshine hours, the mean outside temperature is 7.4°C and the annual solar radiation is 850 kWh/m$^2$, of which 50% comes from an overcast sky. The heating season is 230 days long (27 September to 14 May).

### BUILDING PRESENTATION

The plan and section of the building are given in Figure II.7.1 and the *U*-values in Table II.7.1.

### SOLAR SYSTEM

#### The system

This closed-loop system heats up a 44 m$^2$ concrete floor via 15 cm diameter air channels. The floor, which is 27 cm thick, radiates this heat to the rooms above and partly to those below. The whole prefabricated hollow

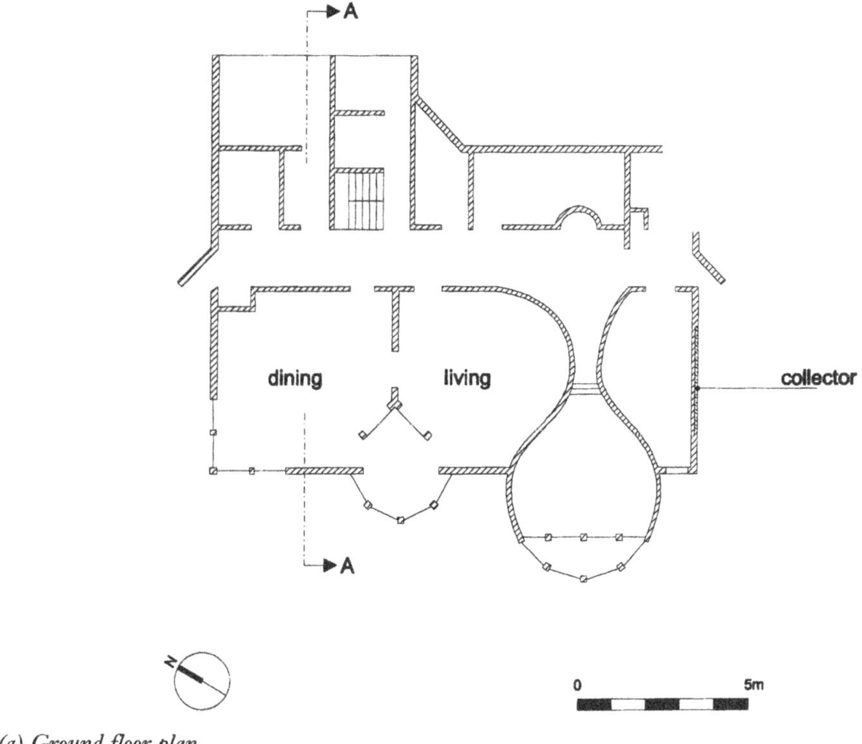


*(a) Ground floor plan*

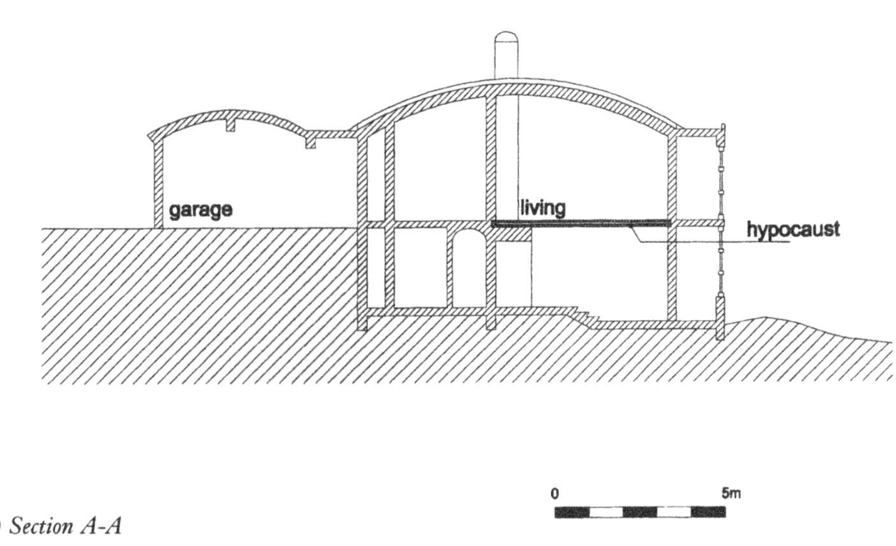


*(b) Section A-A*

*Figure II.7.1. (a) the ground floor plan and (b) a section through the house*

concrete floor was mounted with a crane in a matter of a few hours. Figure II.7.2 shows the total system.

*Table II.7.1.* U-*values*

| | *U*-value | |
|---|---|---|
| Walls | Insulated brick cavity | 0.30 W/m$^2$K |
| Windows | Triple glass | 1.30 W/m$^2$K |
| Roof: | Turf and glass wool | 0.12 W/m$^2$K |

## The collector

The 14 m$^2$ collector, facing 25° east of due south, is constructed from standard building materials. Window glass, 4 mm thick and mounted in wooden frames, covers a black non-selective aluminium sheet absorber, behind which there is a 10 cm space created by wooden battens against the back wall of pressed wooden fibre board.

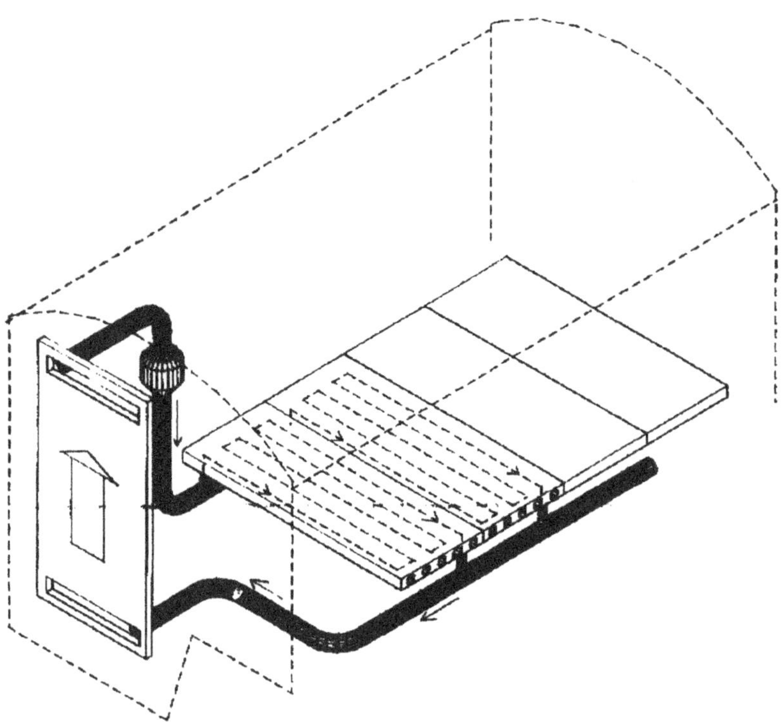

*Figure II.7.2. The total system showing collector and floor distribution/discharge*

### Storage and distribution

Solar-heated air passing through the concrete channels heats up the mass before returning to the collector in a closed loop.

### Controls

A temperature differential thermostat switches on the fan when the air temperature in the collector exceeds 10 K above the temperature sensed in the hollow concrete floor. The fan stops when the difference falls to less than the set temperature difference.

## PERFORMANCE

There is no monitoring. However, the owners/users are very satisfied with the system.

## REMARKS

The owner is very happy with the maintenance-free system operation, in uninterrupted service since 1989. The vertical collector does not overheat, as some roof collectors do, during hot summer months. The overall operation temperature is hence comfortable and does not push the capability of the materials to their limit.

The system even delivers hot air from low, weak winter sunshine. Visitors are impressed to feel the hot air from the fan in winter when use of solar energy does not seem possible. From a pedagogical point of view this project is thus a great success.

## ACKNOWLEDGEMENTS

Designer/energy consultant/author: Harald N. Rostvik, civil architect MNAL, Sunlab, Alexander Kiellandsgt 2, N 4009 Stavanger, Norway

System development: ABB MILJØ AS, Norway

# II.8 Summer-House Package

Slettestrand, Denmark

*Summer House*

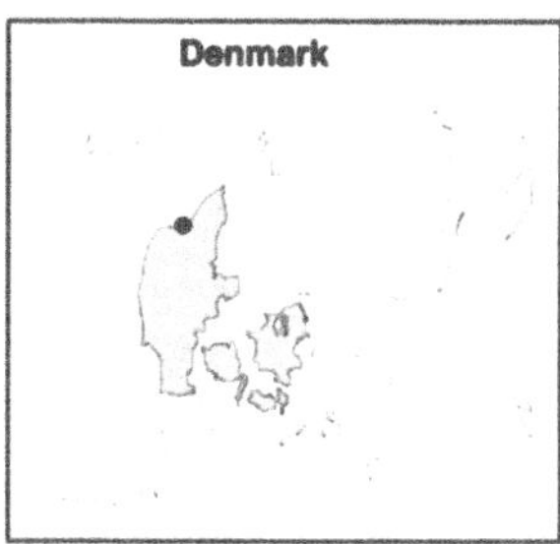


*System type 1*

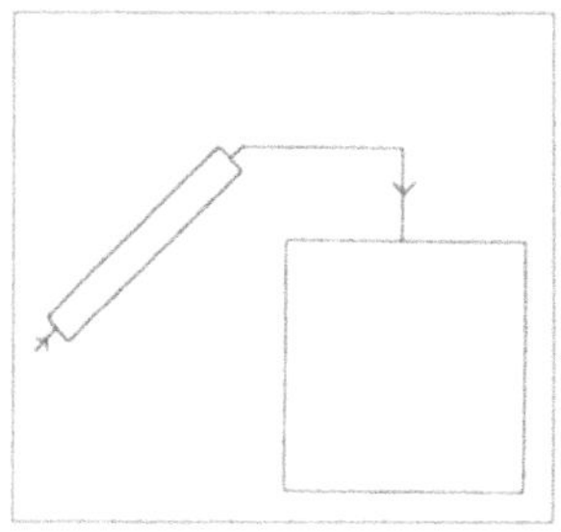

## PROJECT SUMMARY

The purpose of this system is to keep buildings dry by blowing solar-heated ambient air into the building during hours of sunshine. Several hundred of these mass-produced solar air systems have been sold, chiefly in Denmark and Spain. In Denmark, the Summer-House Package is mostly used to keep summer houses and week-end cottages dry during the winter, in order to prevent moisture damage to furniture, etc. In Spain it is used to dehumidify dwellings. For space heating gas is often burned in furnaces without chimneys, which generates much moisture in the building and consequently damages the walls. Yearly repainting of the house interior can be avoided by the installation of this system. Mass-production 'do-it-yourself' building kits make the system very inexpensive.

### Summary statistics

| | |
|---|---|
| System type, variation: | type 1 |
| Collector type and area: | opaque, facade integrated 1.28 $m^2$ |
| Storage type, volume, capacity: | no storage |
| Annual contribution of the solar air system: | 3.63 $kWh/m^2_{floor}$ |
| Annual contribution of the solar air system: | 225 $kWh/m^2_{collector}$ |
| Basis: | calculated |
| Heated floor area, volume: | 80 $m^2$, 250 $m^3$ |
| Year of development: | 1994 |

## SITE DESCRIPTION

The summer house used in this example is located in Slettestrand in the northern part of Jutland, 36 km west-north-west of Aalborg.

Latitude 57°N
Longitude 9°E
Altitude 10 m above sea level
There is a temperate coastal climate

## BUILDING PRESENTATION

The summer house is a three-room single-storey building with a floor area of 80 $m^2$. The floor plan and section of the house are shown in Figure II.8.1.

Figure II.8.2 shows the integration of the Summer-House Package on the south wall of the summer house. The air inlet of the solar collector is to one of the bedrooms of the summer house. In order to maximize the benefit of the Summer-House Package, the door

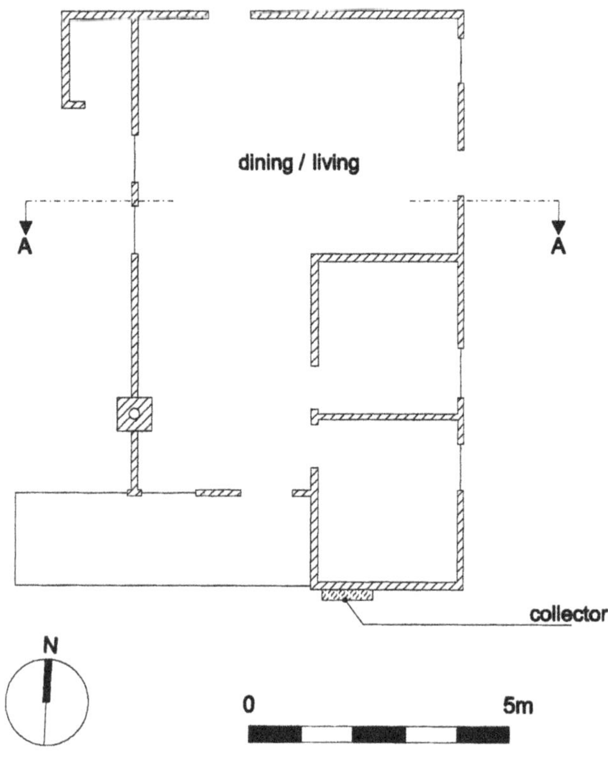


*(a) Floor plan*

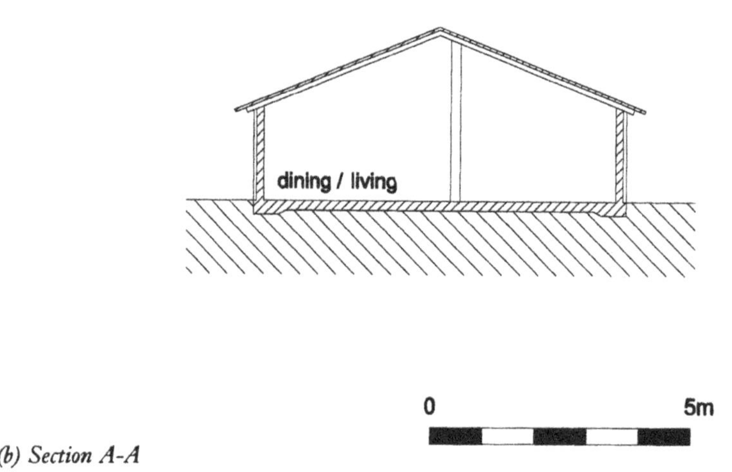


*(b) Section A-A*

*Figure II.8.1. (a) The floor plan and (b) a cross-section of the summer house*

between the living room and the bedrooms should therefore be left open, especially when the house is unoccupied.

## SOLAR SYSTEM

### The system

The Summer-House Package consists of a small 1.28 m² (glazing area) solar air collector with a small fan driven by a 0.28 m² (glazing area) solar cell panel. The main aim of installing the system is to prevent damage due to moisture; heating is only a second-order aim. The system is designed for buildings with a floor area of approximately 60 m². Larger buildings may require several systems.

The principle of the system is shown in Figure II.8.3. The air is sucked into the collector through holes on the reverse side of the collector. The air is then forced through a black fibre cloth heated by the sun. The air is finally directed through a duct at the bottom of the collector into the house near the floor of the room. This provides a better mixing of heated ambient air and room air.

*Figure II.8.2. The Summer-House Package installed on a summer house*

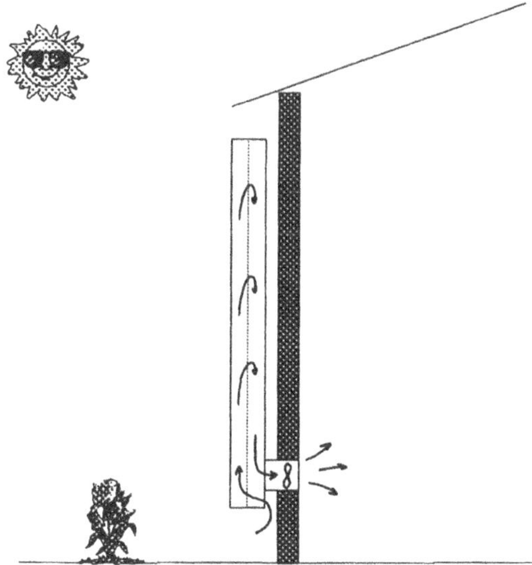

*Figure II.8.3. The principle of the Summer-House Package*

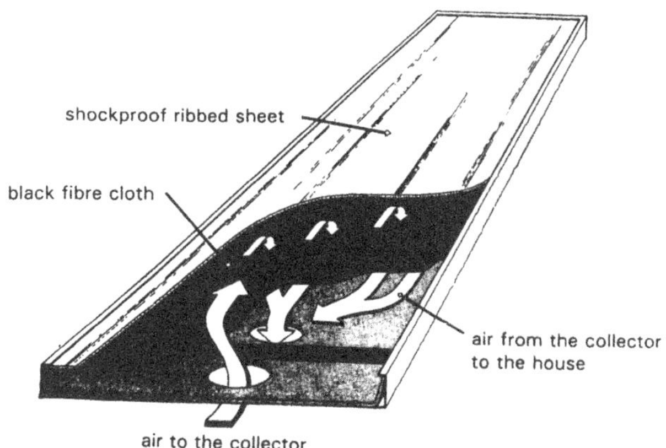


*Figure II.8.4. The principle of the solar air collector*

The system is easily installed by the owner of the building, with the help of an installation guide supplied by the manufacturer. No special tools are required.

## Collector

The principle of the solar air collector is shown in Figure II.8.4 and the characteristics of the collector are given in Table II.8.1.

## Storage

There is no storage in the system except for the storage capacity of the house itself.

## Distribution

The distribution is handled by a small fan driven by a solar cell panel, which also controls the system. The fan is connected to the solar air collector via a small flexible 100 mm duct. The fan is mounted right behind the air collector on the inside of the wall. The characteristics of the fan are given in Table II.8.2.

*Table II.8.1. Characteristics of the collector*

| | |
|---|---|
| Glazing area | 1.28 m$^2$ |
| Absorber | 2 mm temperature-resistant black porous felt mat, through which the air is forced twice |
| Cover | 10 mm ribbed sheet of UV-protected polycarbonate |
| Frame | extruded aluminium |
| Reverse-side insulation | 30 mm of mineral wool |
| Connection to the house | 100 mm flexible duct |

*Table II.8.2. Characteristics of the fan*

| | |
|---|---|
| Voltage | 12V DC |
| Power | 5 W |
| Dimensions | 175 × 175 mm |
| Start/stop | 100 W/m$^2$ solar radiation |

### Control

The fan is connected directly to the solar cell panel leading to a variable flow rate, which is dependent on the solar radiation, as seen in Figure II.8.5. The fan stops and starts at a usable radiation of approximately 100 W/m². The house is thus ventilated when the sun is shining. Characteristics of the solar panel are given in Table II.8.3.

## PERFORMANCE

The performance of the Summer-House Package has been measured under an artificial sun. The flow rate through the air collector depends on solar radiation, because the fan is driven by a solar-cell panel. The efficiency of the air collector depends on the flow rate. Relationships between radiation and the flow rate and between the flow rate and efficiency have been established and have been combined into the graph shown in Figure II.8.5. Although the dependency of the flow rate on solar radiation and the dependency of the efficiency of the collector on the flow rate are non-linear, the output of the collector is linearly dependent on the usable solar radiation.

In order to interpret Figure II.8.5 and obtain the usable radiation, it is necessary to correct the solar irradiation hitting the collector for the losses in the cover glass due to the incidence angle.

### Simulation of the performance

Using a simple simulation program, the performance of the Summer-House Package has been determined for Danish weather conditions using the Danish Test Reference Year. The solar air collector/solar cell panel is assumed to be oriented due south with a tilt of 90° in order to maximize the solar radiation on the system during the winter.

The yearly performance has been calculated to be 290 kWh (225 kWh/m² collector). It is assumed that

*Table II.8.3. Characteristics of the solar cell panel*

| | |
|---|---|
| Transparent area | 0.28 m² |
| Maximum power | 11 W |
| Rated voltage | 14.5 V |
| Rated current | 759 mA |

the solar cell panel is disconnected during the summer (from June to August), when Danish summer houses are usually occupied, in order to decrease the risk of overheating in this period. If the system were also allowed to run during June–August, the performance would be 430 kWh (335 kWh/m² collector).

The effectiveness of the Summer House Package in decreasing the humidity of a house during the winter is difficult to determine because the transport and accumulation of moisture in different materials is a very complex process. However, in order to give an impression of the capability of the Summer-House Package for dehumidifying a building, the maximum drying effect has been calculated.

If all of the energy gained were utilized for evaporation, the Summer House Package would evaporate 420 litres of water per year. How much water actually will be evaporated or prevented from penetrating into the construction and furniture of the house depends on the humidity of the air, the materials, the humidity of the materials, etc. According to statements from owners, the system is able to keep smaller buildings at an acceptable level of humidity.

## REMARKS

The Summer-House Package is inexpensive. The price in Denmark is approximately ECU 730 incl. VAT (August '98). The system may be purchased in 'do-it-yourself' building stores.

It is very difficult to discuss the pay-back time of the system. If it is installed in a summer cottage, the purpose is to increase the comfort and decrease damage

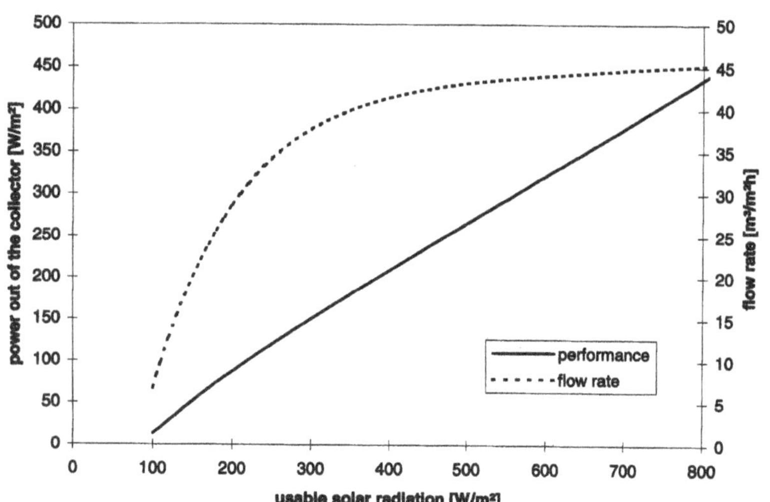


*Figure II.8.5. The energy gain from the Summer-House Package dependent on the (usable) solar radiation on the air collector/solar cell panel; the fan starts and stops at a usable radiation of approximately 100 W/m²*

due to moisture and thereby make it possible to leave, for example, bedclothes in the unheated house during the winter without their being damp-stained. If the yearly repainting is avoided in a Spanish house, then the pay-back period will, indeed, be very short.

The system saves energy during periods with sunshine in cases where the owners use the summer house off-season, because less energy is required to keep the house comfortable. The system may in some cases further act as protection against freezing, i.e. it will keep the temperature inside an unoccupied house above zero degrees, thereby preventing damage to plumbing. This protection against freezing would otherwise probably be achieved with electrical heaters having a set point around 5°C. In this case the installation of the system also saves energy.

In general, the owners of Danish summer houses with such systems are satisfied. They no longer have to take bedclothes, for example, with them when they leave the cottage for the winter and the table salt is no longer wet when they come to the cottage during the winter or in the spring.

## ACKNOWLEDGEMENTS

Manufacture of the solar system: Aidt Miljø A/S, Kongensbrovej, Aidt, DK-8881 Thorsø, Denmark
Chapter author: Søren Østergaard Jensen, DTI Energy, Denmark

## BIBLIOGRAPHY

Jensen SØ (1994). *Test of the Summer House Package from Aidt Miljø*. Report no. 94-1, Institute for Buildings and Energy, Technical University of Denmark. Building 118. DK-2800 Lyngby, Denmark.

# II.9 Humlum House

## Jutland, Denmark

*Humlum*

*System type 1*

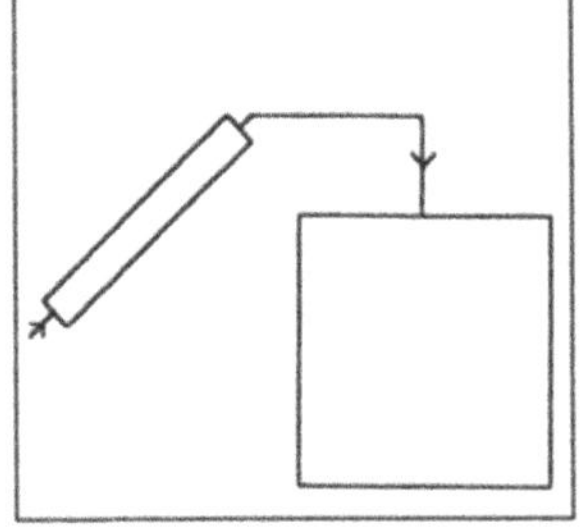

### PROJECT SUMMARY

This single-family dwelling has five small solar air collectors integrated into the south-east facade of the building. The solar air collectors are located under a window as an important feature of the building's architecture.

The collectors preheat fresh air for the living space, toilet/bathroom and office, replacing air that is exhausted. The solar air collectors are estimated to yield approximately 1075 kWh/year with a pay-back time of approximately nine years. The system has functioned efficiently and without problems since 1991.

#### Summary statistics:

| | |
|---|---|
| System type, variation: | type 1 |
| Collector type: | vertical, facade integrated |
| Collector area: | 5 $m^2$ |
| Storage type | No specific storage |
| Annual contribution of the solar air system: | 7.2 $kWh/m_{floor}^2$ |
| Annual auxiliary ventilation heat: | 47 $kWh/m_{floor}^2$ |
| Basis: | calculated |
| Heated floor area: | 150 $m^2$ |
| Year solar system built: | 1991 |

### SITE DESCRIPTION

The dwelling is situated at Duevej 15 in Humlum, north of Struer in Jutland, Denmark, which has a temperate coastal climate.

| | |
|---|---|
| latitude | 56°N |
| longitude | 8°E |
| altitude | 10 m above sea level |

West of the building site a ridge rises smoothly to the east, ending at the cliffs on the coast line to the Limfjorden, approximately 40 m from the house.

### BUILDING PRESENTATION

#### Plans and sections

The gross floor area is 215 $m^2$, of which the ground-floor area is 166 $m^2$ and the residential basement area 49 $m^2$, the total basement area being 93 $m^2$. The floor plan and section are shown in Figure II.9.1.

The building is east/west oriented and positioned so that there is no shading from the neighbouring house or any big trees. The geometry of the house ensures that it does not create shadow on itself.

The house is zoned with a kitchen/family room in the centre, open to the south with big studio windows for passive solar heating, and a smaller cooler

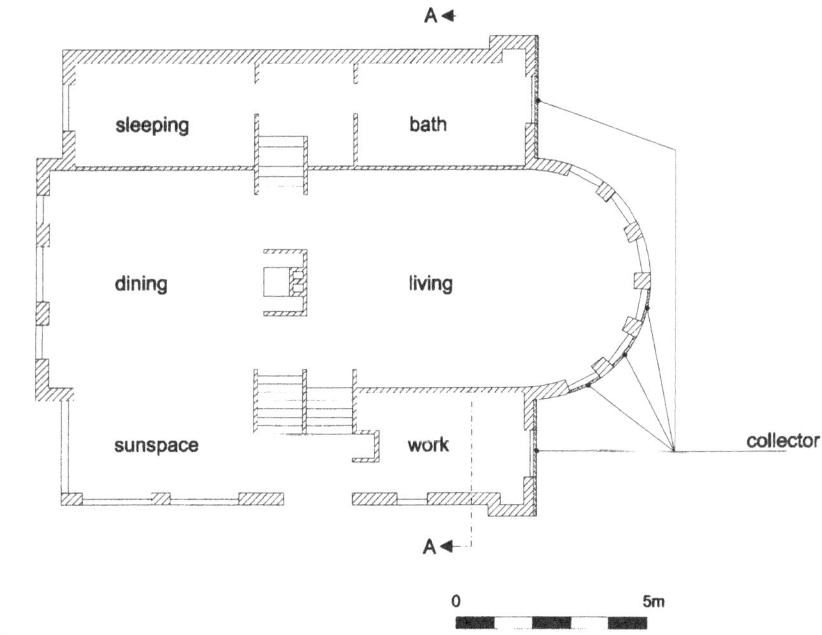


*(a) Floor plan*

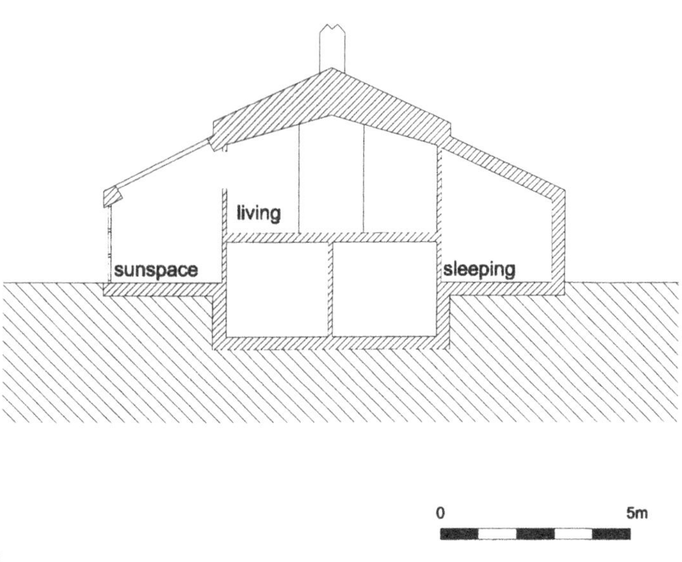


*(b) Section A-A*

*Figure II.9.1. (a) Floor plan and (b) section*

part with bedroom and toilet/bathroom to the north; this part has no windows to avoid unnecessary heat loss.

The outside walls consist of a 12 cm brick wall at the front and, at the back, 12.5 cm of mineral wool insulation and a back wall of 10 cm light concrete. The wood-framed roof has 25 cm insulation, the floor on the earth is insulated with 10 cm of mineral wool and 20 cm loose leca (light expanded clay aggregate).

All the windows have two layers of double glazing with low-emission coating and argon filling, providing a *U*-value of 1.5 W/m$^2$K.

**Ventilation**

The building shell is tightly constructed. The mechanical ventilation from the range hood in the kitchen and from the toilet/bathroom are demand-controlled. A supply of fresh air is provided either by opening windows and doors or by fresh-air intake through changeable ventilation openings from the solar panels.

## SOLAR SYSTEM

The location of the solar panels is shown in Figure II.9.2. Each solar panel (Figure II.9.3) consists of:

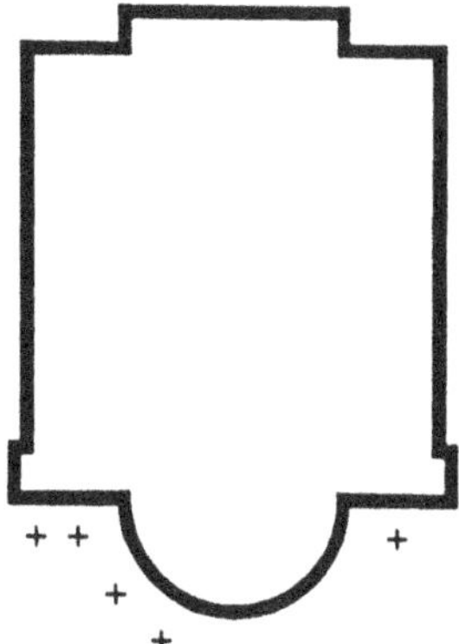

*Figure II.9.2. Location of solar panels*

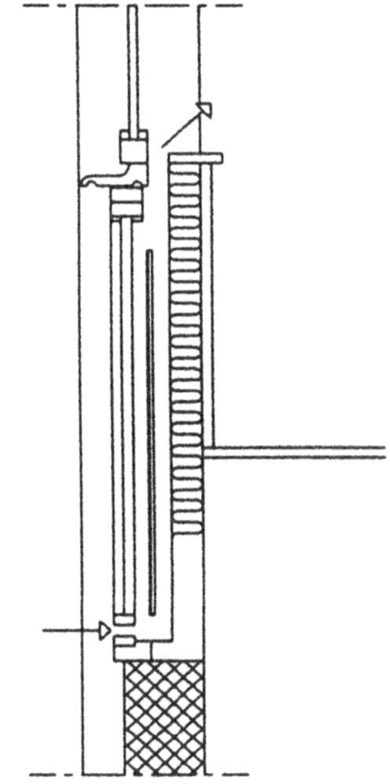

*Figure II.9.3. A solar panel element*

- one layer of glass in a standard window frame with a rift valve;
- an air space of 2–3 cm;
- a dark heat-absorbing surface;
- an insulation layer;
- a back wall;
- a manual damper in the top of the frame.

## The mode of operation of the solar panel

A manual damper in the solar panel controls the air flow. When it is open and the sun is shining, the cool fresh air slowly passes the heat absorbing surface and is warmed before it is drawn into the living room. When the sun is not shining the collector functions only as an ordinary fresh air inlet.

## Safety precautions against high temperatures

During the summer the fresh air supply is not taken through the solar panel. To prevent it from overheating, a built-in thermostat moves a sliding damper with openings to the outside. This allows natural thermal circulation of outside air through the solar panel. This is illustrated in Figure II.9.4.

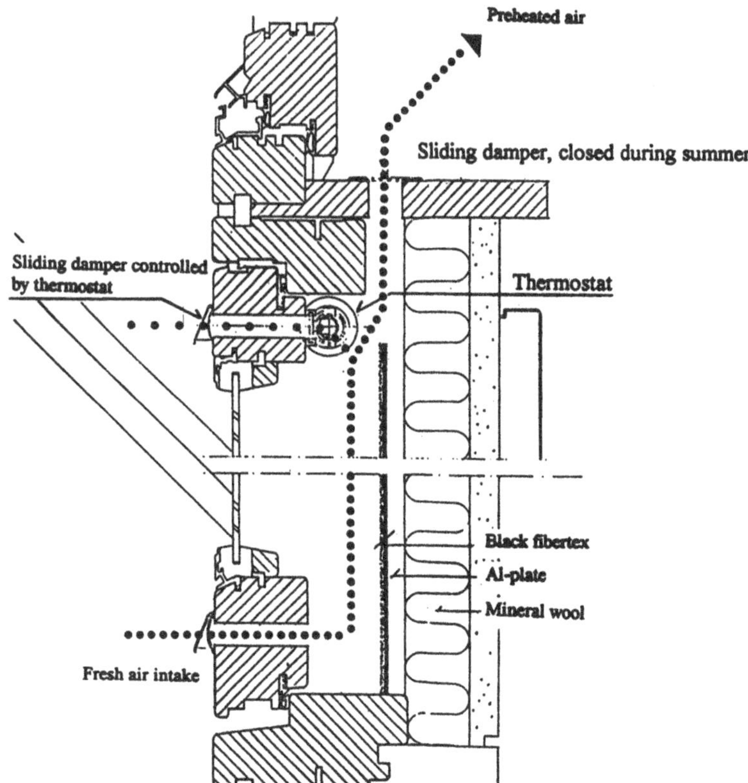


*Figure II.9.4. Section of the solar panel*

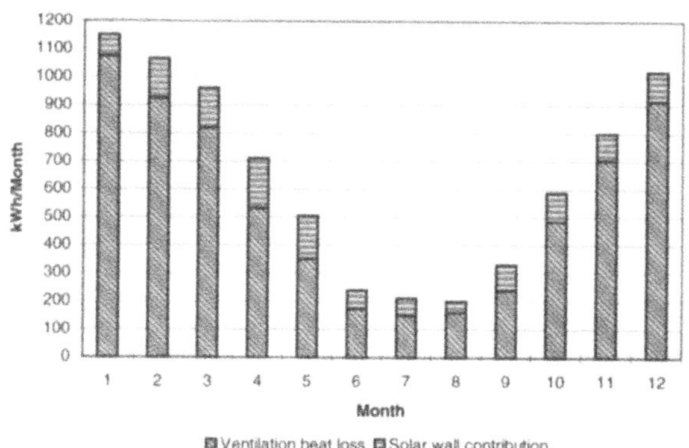


*Figure II.9.5. Solar panel output*

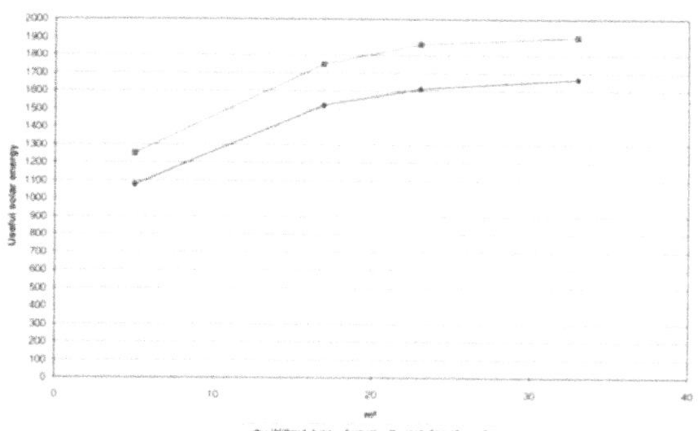

*Figure II.9.6. Calculated solar energy output as a function of solar-panel area*

The thermostat used was the Fiat 1300 C model 65 car radiator thermostat (Wahler, type 754) with an activation temperature of approximately 80°C. The sliding damper provides an opening area of 19 cm$^2$.

## PERFORMANCE

Figure II.9.5 shows the calculated output for the solar panels (5 m$^2$) at an assumed air change of 0.5 times in an hour with reference-year weather data.

The ventilation loss has been calculated from the difference between the inside temperature of 20°C and the outside temperature. Output in the summer is small because of the minimal need to heat fresh air. Excluding June to August, the yearly output is 1075 kWh from a total ventilation loss of approximately 7100 kWh.

Figure II.9.6 shows the solar output for the solar panels for the actual house. If an output of more than 100 kWh/m$^2$ is needed from the solar collectors, the area of the solar panels should not exceed approximately 16 m$^2$, corresponding to a ratio of solar-panel area to floor area of 0.1.

It should be noted that the performance has been calculated. The air exchange through the solar panel over the year was not measured. A lower air flow through the wall would result in lower output.

### Operation and maintenance

Cleaning of dust outside and behind the glass on the solar panel is a simple matter, as the frame holding the glass can be opened. Cleaning can therefore be done directly from the ground by the tenants.

The collector frame should be painted every six to eight years.

### Simple pay-back time

When a heating price of US $0.10 (ECU 0.09) per kWh is considered for the purchase those kilowatt hours that the solar panel saves, an additional price for the solar panel of US $140 (ECU 127) per m$^2$ gives a simple pay-back time for a 5 m$^2$ solar panel (215 kWh/m$^2$/year) of approximately nine years.

## REMARKS

This demonstration project, with its five collectors placed as window spandreils, has proved effective as a simple way to provide space heating as needed when both kitchen and bathroom are mechanically exhausted. The concept is also appropriate for retrofit projects.

## ACKNOWLEDGEMENTS

Manufacturers of the solar absorber: Fibertex: THHA 436, Fa. Helmut Oppenheim

Architect: Ulla Falck's Tegnestue, Architect Maa. Frederiksberg, Copenhagen, Denmark

Energy Consultant: Nick Bjørn Anderson & Thomas Genborg, DTI, Højetåstrup, Copenhagen, Denmark

Building owner: Søren & Hanne Smith Knudsen, Humlum, Struer, Denmark

Report: *Driftserfaringer med brystningssolvægge.* Danish Technological Institute, Energy Technology & Ulla Falck's Tegnestue, 1993. Journal no. 51181/91-007

Chapter author: Ove Mørck, Cenergia Energy Consultants

# II.10 OKA *Haus der Zukunft*

Schmiding, Austria

*Schmiding*

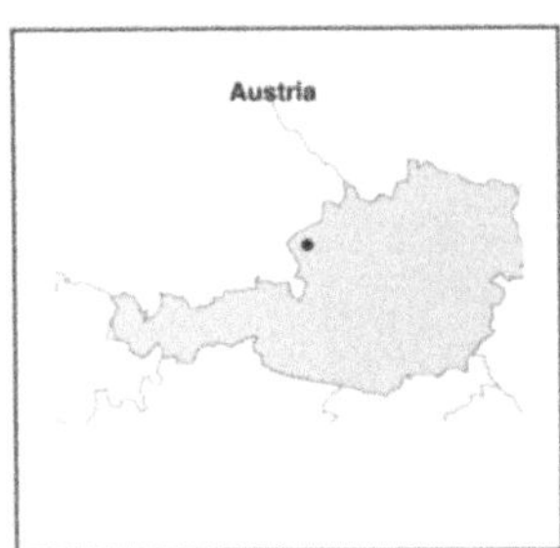


*System type 4*

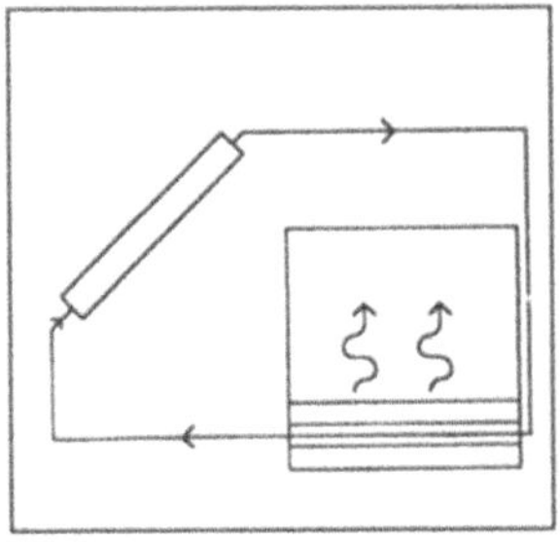

## PROJECT SUMMARY

The main feature of the *Haus der Zukunft* (House of the Future), a prefabricated wooden house, is an active solar system based on air collectors and thermal storage, together with an air–photovoltaic hybrid collector.

As an exhibition house at the 1997 Garden Exhibition near Linz, numerous other technologies were also demonstrated including:

- a sun space;
- preheating of ventilation air in an underground duct;
- air preheating with a heat exchanger;
- an air-to-air heat pump;
- a ground-coupled heat pump;
- a photovoltaic system.

It should be noted that the intention behind the planning of this building was to demonstrate different technologies taken rather than to demonstrate a well-balanced solution.

### Summary statistics

| | |
|---|---|
| System type: | type 4 |
| Collector type and area: | selective coating, single glazing, 83 m$^2$ |
| Storage type, vol., cap.: | concrete hypocaust, 100 m$^3$, 60 kWh/K |
| Annual contribution from the solar air system: | 18 kWh/m$_{floor}^2$ |
| Annual auxiliary heat consumption: | 30 kWh/m$_{floor}^2$ |
| Basis: | calculated |
| Heated floor area: | 362 m$^2$ |
| Year solar system built: | 1997 |

### Site description

Schmiding is situated 30 km west of Linz. The building is surrounded by gardens, fields and some low-rise buildings:

| | |
|---|---|
| latitude: | 48°N |
| longitude: | 14°E |
| altitude: | 310 m above sea level |

Ambient air-temperature data has been taken from the nearby meteorological station of Wels. The annual mean is 8.6°C and that for the heating period (October to April) 3.2°C. The heating degree days based on 20°C/12°C, October to April are 3520 (annual value 3775). The global solar insolation is summarized in Table II.10.1.

The climate is typical of the Danube valley, with prevailing foggy weather in November and December.

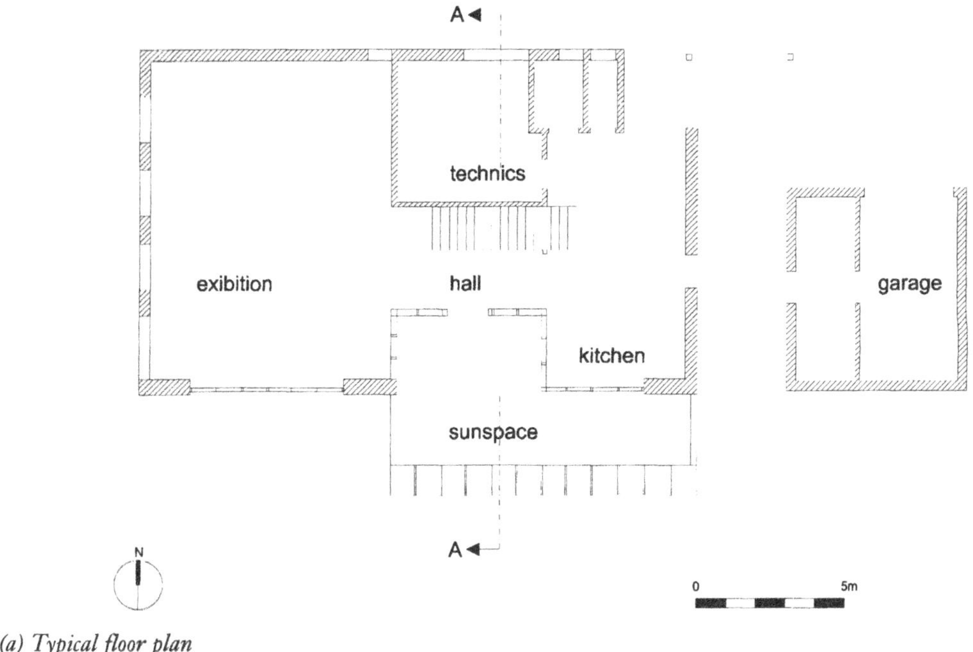


*(a) Typical floor plan*

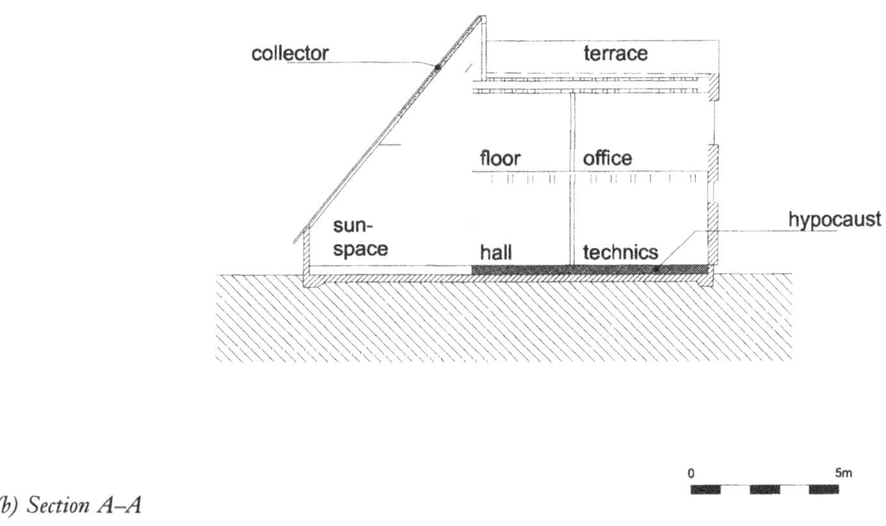


*(b) Section A–A*

*Figure II.10.1. (a) The floor plan of a typical floor and (b) section A–A*

*Table II.10.1. Global solar insolation*

| | |
|---|---|
| Horizontal | |
| annual mean | 1110 kWh/m$^2$ |
| heating period | 390 kWh/m$^2$ (35%) |
| At a tilt of a 45° | |
| annual mean: | 1220 kWh/m$^2$ |
| heating period: | 500 kWh/m$^2$ (41%) |
| On the south vertical facade | |
| annual mean: | 810 kWh/m$^2$ |
| heating period: | 380 kWh/m$^2$ (46%) |

## BUILDING PRESENTATION

A structural slab of concrete is the foundation of the building and this is insulated with 10 cm of expanded polystyrene (extruded styrodur foam, *U*-value = 0.34 W/m$^2$K). This concrete mass is also used for thermal storage (100 m$^3$). The gross heated floor area is 362 m$^2$ (excluding the sun space). The floor plan and section are shown in Figure II.10.1.

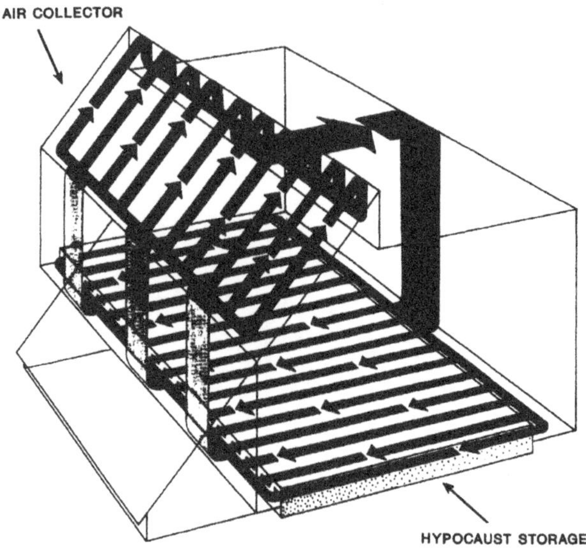


*Figure II.10.2. The air heating system (closed loop)*

The house is constructed of prefabricated wooden building elements insulated with 22 cm of rockwool (*U*-value: 0.2 W/m$^2$K). The flat roof structure is built of glued-laminated beams and the space between the beams is filled with 48 cm of rockwool (*U*-value: 0.1 W/m$^2$K). The double-glazed windows have a *U*-value of 1.3 W/m$^2$K ($g$ = 0.6). The design conditions were:

- inside temperature: 20°C for an ambient temperature of –16°C
- conduction heating load: 242 W/K
- ventilation heating load: 112 W/K

*Figure II.10.3. The design of the air collector*

## SOLAR SYSTEM

The solar air heating system is a closed loop (Figure II.10.2). Hot air from the roof-integrated collector (Figure II.10.3) is moved with a fan through the hypocausts in the floor and back to the collector again.

System data are as follows:

| | |
|---|---|
| Air collector area: | 83 m$^2$ aperture |
| Orientation, tilt: | south, 50° |
| Collector type: | selective coating of 'sun-strip' type, single glazing (Figure II.10.3) |
| Collector output: | ca. 9000 kWh/a (1 October–30 April) |
| Flow rate: | 50 m$^3$/m$^2$h |
| Fan: | axial type, $P_{mech}$ 550 W, 1420 rpm, $P_{electr}$ 736 W, 160 Pa, 4300 m$^3$/h |
| Electrical consumption of fan: | 528 kWh/a, 717 h/a |
| Storage type: | concrete hypocausts (26 hollow brick ducts) |
| Heat exchange area: | 480 m$^2$ |
| Storage capacity: | 60 kWh/K (100 m$^3$) |

### Control

When the temperature in the air collector rises more than 15 K above the hypocaust temperature, the fan is activated. For comfort reasons, the fan is stopped when the floor temperature reaches 25°C.

### Distribution

As can be seen in Figures II.10.2 and II.10.3, the 26 hypocaust ducts are connected in parallel by two channels with a very large cross-sectional area. It is very important to achieve a matched air flow distribution in the two hypocausts. The same also applies for the 18 air collector channels that are connected in parallel (Figure II.10.4).

*Figure II.10.4. Cross section of the hypocaust floor*

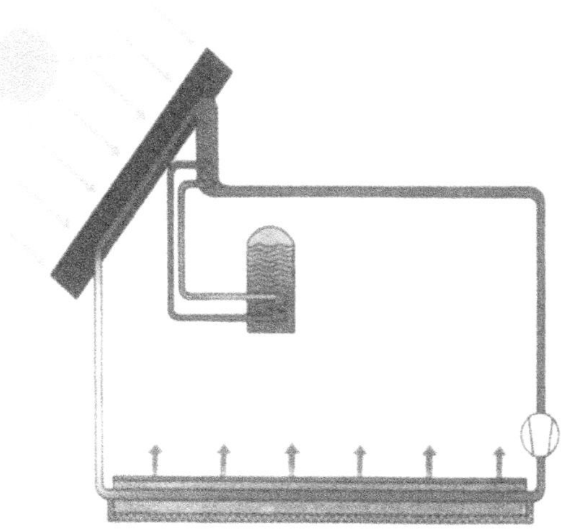

*Figure II.10.5. The domestic hot water system with air-to-water heat exchanger*

Flaps in the ducts leading to the air collector block the flow of cold air to the hypocaust at night. The flaps are opened by the air pressure caused by the fan and close by their own weight.

Air flow velocities in each segment of the loop are as follows:

Collector: 1.0 m/s
Hypocaust: 1.4 m/s
Air channel/fan: 3.2 m/s

In the summer the flaps are opened at both upper and lower ends of the roof-integrated collector to allow air flow by natural convection. This feature means that overheating of the structure is avoided. On a very sunny day natural air movement with a speed of 0.4 m/s has been observed in the collector. This corresponds to approximately 2200 $m^3/h$. On the same day a temperature of 95°C was measured at the upper end of the black absorber plate of the collector.

## Domestic hot water

An air-to-water heat exchanger has been installed in the collector air channel to heat the domestic hot-water

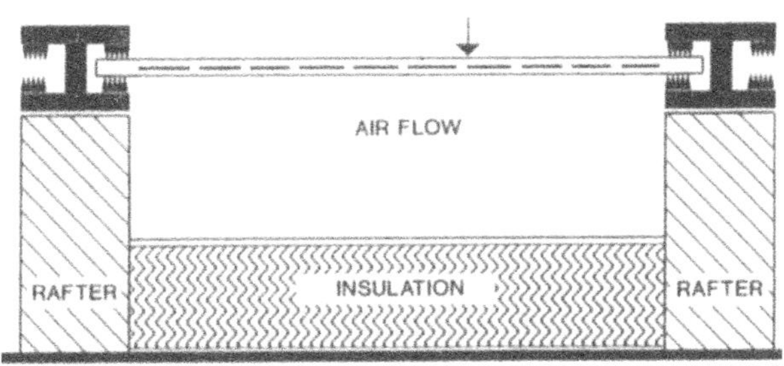


*Figure II.10.6. A cross-section of the PV-hybrid collector* (source: *Grammer/Germany*)

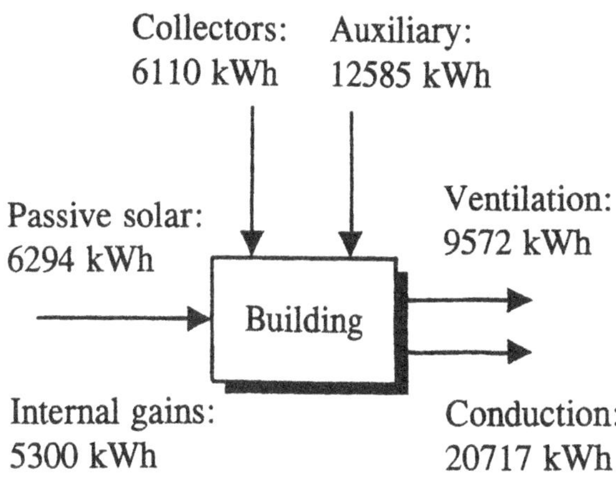


*Figure II.10.7. The system energy balance (with passive gains maximized). The specific heating energy demand is 35 kWh/m²a for the simulation period, 1 October to 30 April* (source: *Wilk, 1997*)

storage tank (500 litres). During bad weather conditions the upper part of the storage tank is heated by an electric heating element. The DHW system is shown in Figure II.10.5.

## Auxiliary heating system

When the solar collector output is less than the heat demand, two different auxiliary heat sources are used, an air-to-air heat pump situated in the forced ventilation system and a ground-coupled heat pump.

## The photovoltaic–air hybrid collector

The roof surface of the building is divided into 22 sections with three different types of solar energy systems:

- 18 air collector elements;
- two daylighting elements;
- two PV–air hybrid collector elements.

Four photovoltaic panels from Pilkington/Köln with a total peak power of 1072 W are used for the hybrid collector. They have been produced using a glass–glass technique. The efficiency of the mono-Si solar cells is 15% (panel efficiency is 12% with Siemens solar cells). Two panels are mounted like shingles to create a watertight roof element. The air channel below the panel is part of the air collector system. The PV-hybrid collector is illustrated in Figure II.10.6.

A sophisticated SPS control box controls the system. Priorities and strategies can easily be changed by reprogramming the unit (using a laptop computer).

An extension of the SPS system also monitors the data from the 100 sensors installed in the house. The temperature swing of the concrete hypocaust floor storage is limited to a span of 18°C to 25°C.

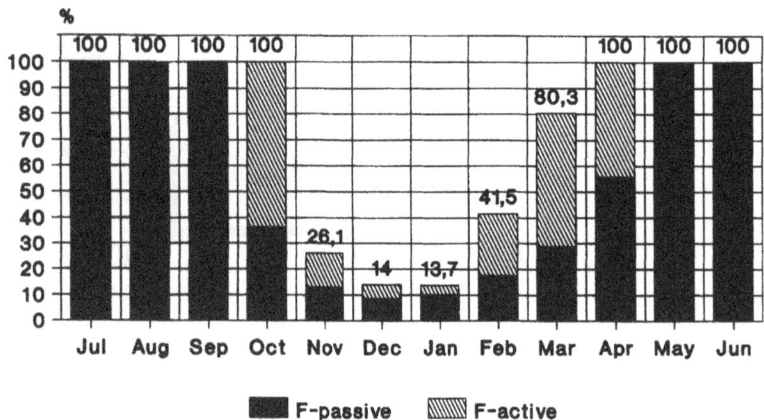


*Figure II.10.8. Active vs. passive solar*

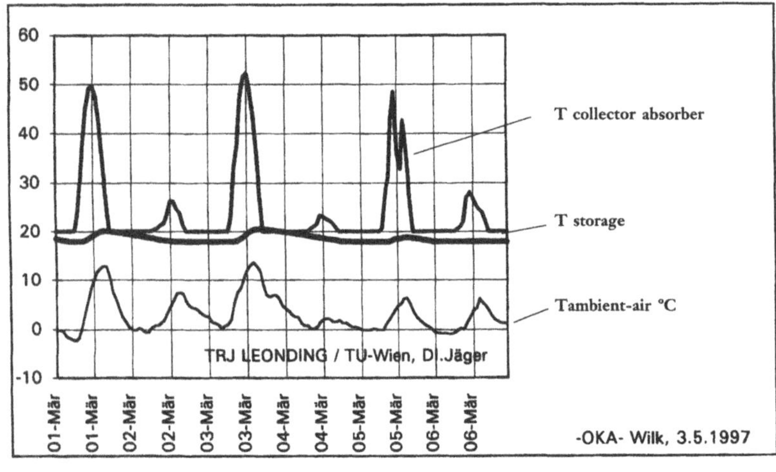


*Figure II.10.9. Simulation of collector/storage performance on an hourly basis (ambient air, storage and collector temperatures)*

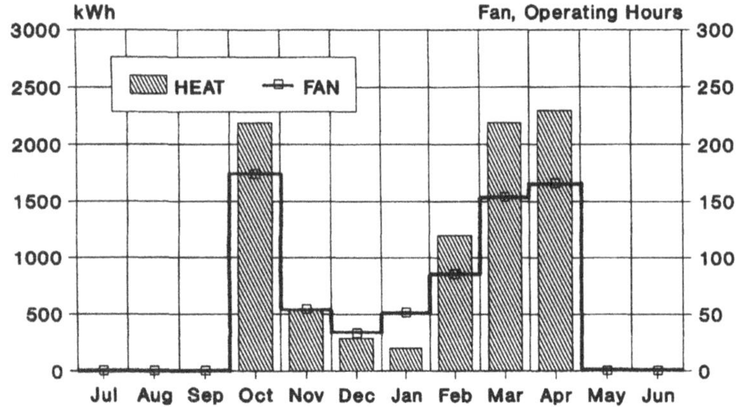


*Figure II.10.10. Solar collector system yield*

## PERFORMANCE

The system energy balance is shown in Figure II.10.7.

### Performance simulation

Simulations of the building behaviour for active and passive solar gains were done using the reference year Leonding/Linz (Figure II.10.8). Figure II.10.9 shows the slow temperature decay of the hypocaust mass after it is charged by the pulses of heat from the collector absorber.

## REMARKS

The limiting factor of this system is the small temperature swing allowed in the hypocaust storage, such that, in this case, solar energy can only be stored for approximately three days. A rock-bed storage would be more effective. The second drawback is that an air collector with single glazing is inefficient when ambient temperatures are low, so that most solar energy is gained in October, March and April (Figure II.10.10).

The roof integration of the different solar systems has been done very harmoniously. It is difficult to detect which roof sector serves which type of solar application. The hybrid system is more economical than PV alone since the hot air can be used for either space heating or domestic hot water.

## ACKNOWLEDGEMENTS

Chapter author: Dipl.-Ing. Heinrich Wilk, Oberösterreichische Kraftwerke AG, 4021 Linz, Böhmerwaldstrasse 3, Austria; Tel. +43 732 6593 3514; Fax +43 732 6593 3309; Email heinrich.wilk@oka.co.at

Architect: Sture Larsen, Lindauerstrasse 33, A-6912 Hörbranz, Austria

The author would like to thank Professor Gerhard Faninger, Dr Manfred Bruck, Dipl.-Ing. Fechner and Dipl.-Ing. Burtscher for their valuable advice.

## BIBLIOGRAPHY

Feist W (1996). Passivhäuser – Gebäude ohne Heizung. Institut Wohnen und Umwelt (Darmstadt), Zeitschrift '*Erneuerbare Energie*', **3**, 4–9, ARGE Erneuerbare Energie, Gleisdorf, Austria.

Feist W., Heidelberg 1997, *Das Niedrigenergiehaus* (4th edn). Verlag C.F. Müller.

Stahl W., Goetzberger A., Voss K., Heidelberg 1997. Das Energieautarke Solarhaus. Verlag C.F.Müller.

Strong S (1996). 1st International Conference on Solar Electric Buildings', Boston, USA.

Wilk H (1997). Haus der Zukunft, Systemtechnik und Auslegung. *VEÖ-Journal*, **6**, 56–61.

# II.11 CSU Solar House II, Fort Collins

Colorado, USA

*Fort Collins, Colorado*

*System type 1*

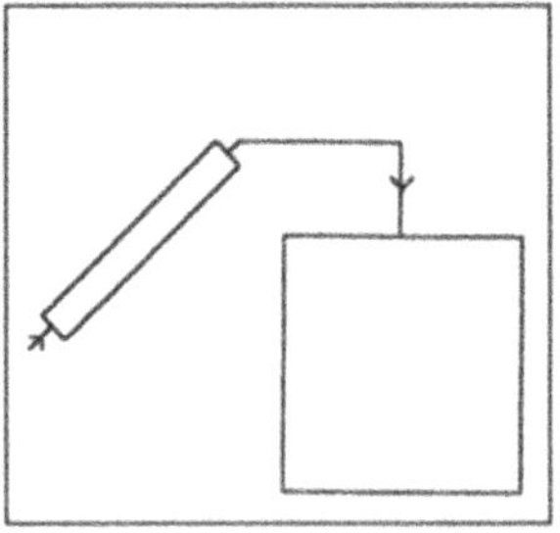

## PROJECT SUMMARY

This middle-sized single-family house, with its type 1 solar air system, has been in operation for almost a quarter of a century. The system consists of factory-built collectors, site-built rock-bed heat storage, an electric heat-pump for auxiliary heating and fully automatic temperature controls. Extensive measuring and testing facilities were provided for continuous monitoring of all energy sources and uses. The solar air system meets about three quarters of the annual heat requirements of the building.

### Summary statistics

| | |
|---|---|
| System type: | type 1 |
| Collector type: | single glazed, underflow |
| Collector area: | 58 m$^2$ |
| Storage type, volume | rock bed, 15 m$^3$ |
| Heat storage capacity: | 5.6 kWh/K |
| Annual contribution of the solar air system: | 91 kWh/m$^2$ (heated floor area) |
| Annual auxiliary heat consumption: | 44 kWh/m$^2$ |
| Basis: | monitored |
| Floor area, volume: | 129 m$^2$, 707m$^3$ |
| Years solar system installed: | 1976–1981 |

## SITE DESCRIPTION

Solar House II is located in Fort Collins, Colorado. It is on the south slope of a low hill, unshaded by trees or other buildings and the atmosphere is generally clear of local pollution:

| | |
|---|---|
| latitude: | 41°N |
| longitude: | 105°W |
| altitude: | 1585 m above sea level |

The climate is characterized by moderate temperatures, light precipitation and light winds, interrupted occasionally by strong north-westerly winds. The average monthly temperatures vary from –2.5°C in January to 21.8°C in July. The normal, total annual heating degree days (18.3°C base) are 3600. At the collector, monthly solar radiation averages vary from 125 to 160 kWh/m$^2$ and horizontal surfaces receive from 75 kWh/m$^2$ in winter to 200 kWh/m$^2$ in summer.

## BUILDING PRESENTATION

The building is a three-bedroom house, although the interior is used for offices. The plan of the ground floor and a cross-section are shown in Figure II.11.1. The 7.5 × 8 m space at the north-east corner of the building has

*(a) Ground floor plan*

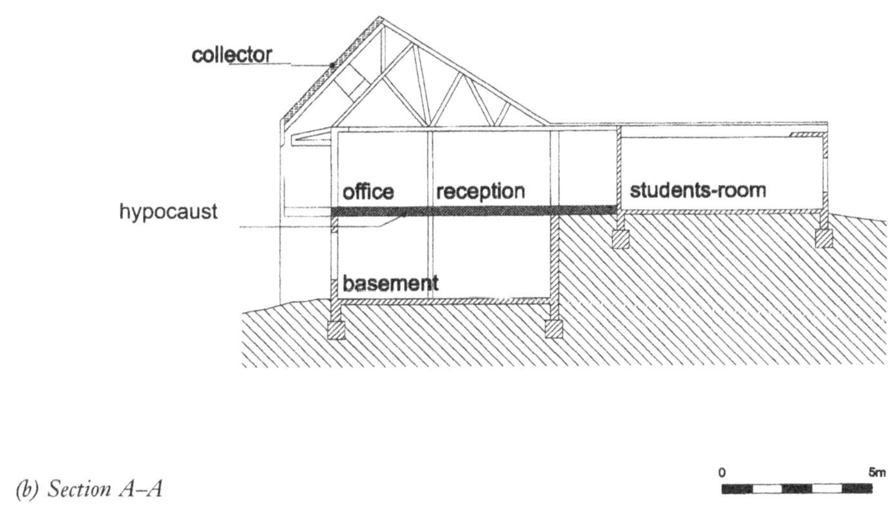


*(b) Section A–A*

*Figure II.11.1. (a) Plan of the ground floor and (b) cross-section A–A*

a separate heat supply not associated with the solar system.

The collectors occupy the entire south slope of the roof. As shown in Figure II.11.1(b), there is a second weatherproof roof below the collectors, to permit the replacement of collectors for experiments without exposing the building interior to the weather.

The walls are framed with 50 × 100 mm studs on 406 mm centres. Voids between the studs are filled with 89 mm fibreglass batt insulation. The exterior sheathing is 12.7 mm thick with 11 mm hardboard over the sheathing. The interior surface is 16 mm gypsum board. The roof has 165 mm blanket insulation. There are 28.25 m$^2$ of double-glazed windows set in

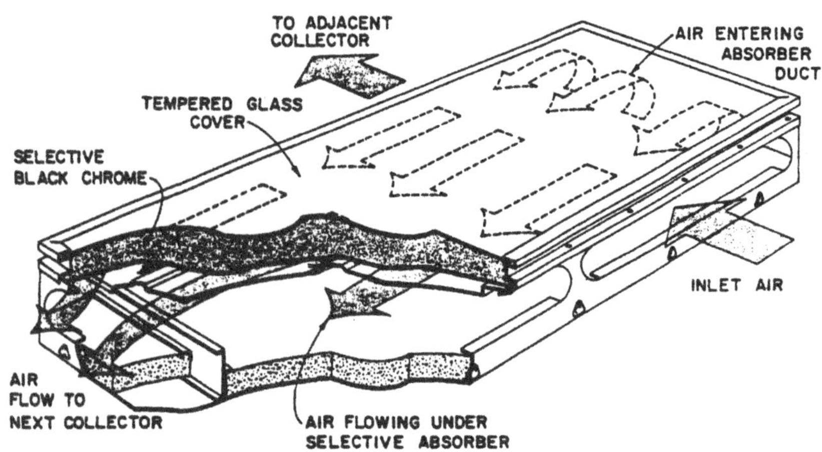


*Figure II.11.2. Cutaway view of the Series 3000 Solaron Collector (redrawn by permission of Solaron)*

wood frames. The building heat loss rate is calculated to be 419 W/K.

## SOLAR SYSTEM

### Solar collectors

A cutaway view of the collectors is shown in Figure II.11.2. Each collector module has a gross area of 1.74 $m^2$, with an aperture area of 1.59 $m^2$. Single-pane low-iron tempered glass covers the dead air space below which is the solar-heated air passage between two thin aluminium sheets. The absorber surface, which is the upper side of the air passage, is coated with a selective black chrome surface. Minimum absorbtance is 0.95 and maximum emittance is 0.15. The glass cover has a solar transmittance (normal incidence) of 0.89 and is lightly patterned on the top. A space between the air passage and the back insulation is designed for internal manifolding to minimize ducting for the array. The insulation has an overall heat conductance rate of 0.44 W/($m^2$K).

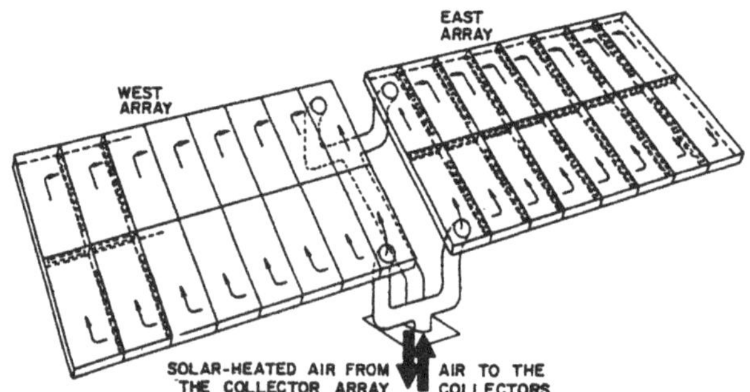


*Figure II.11.3. The collector array*

### Collector array

The collector array consists of 32 modules divided into two equal subarrays as shown in Figure II.11.3. Air passes through two collectors coupled in series, in 16 parallel circuits. Cold air is supplied to the lower collectors and heated air is delivered to the storage from the

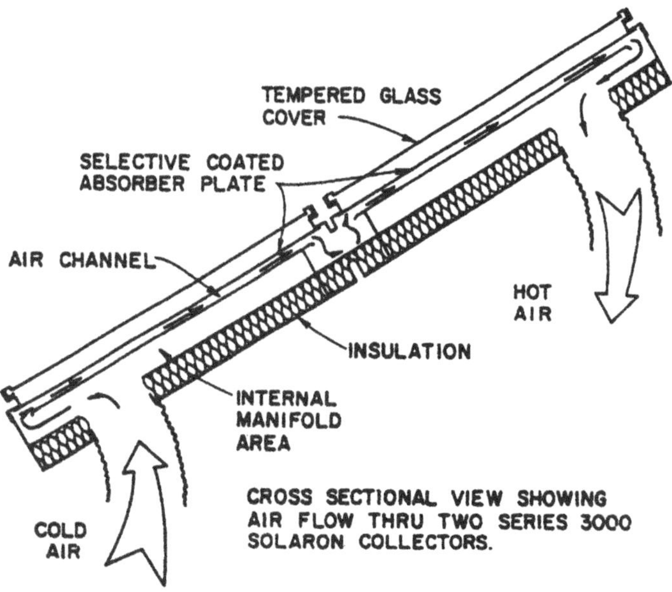


*Figure II.11.4. Cross-section of the Solaron Collectors in a two-high array*

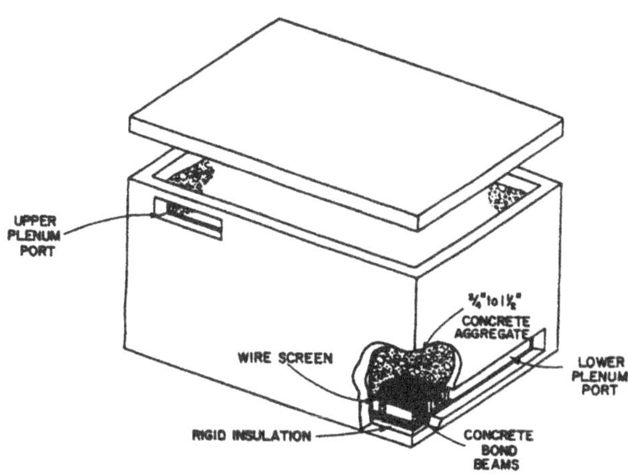


*Figure II.11.5. The pebble-bed heat storage unit*

upper collectors. Internal manifolds minimize ducting leading to and from the array. The connection of the top row and bottom row of collectors, together with the air-flow path through the collectors are shown in Figure II.11.4.

The gross area of the array, including the perimeter insulation strip is 57.93 m$^2$, which is used in calculating the thermal performance. Excluding the perimeter strip, the area is 55.77 m$^2$ and the aperture area is 50.9 m$^2$.

## Pebble-bed (rock-bin) storage

The heat storage bin (Figure II.11.5), located in the basement, is constructed of 50 x 100 mm wood framing, sheathed with 12.7 mm plywood, 89 mm fibreglass batt insulation and dry wall lining on the inside. The rock volume is 15 m$^3$. The corners of the box are sealed with silicone sealant to prevent air leaks.

## The solar air handler

An air handler supplied by the collector manufacturer circulates air from the collector to the rooms and to the heat storage unit. It originally contained a belt-driven centrifugal fan powered by a 750 W AC motor. To study the effects of flow variation, fan speed was changed from time to time. During one heating season, fan speed was manually reduced in steps (by pulley substitutions), providing flow rates of 750 litres/s, 550 litres/s and 475 litres/s. In a subsequent season, an automatically controlled four-speed fan provided solar-heated air at nearly constant temperature in spite of wide variation in solar energy intensity. Improved systems performance was obtained by use of the variable air flow rates.

## The heat pump

Because this house was also used for developing and testing space cooling, a heat pump was used as the principal source of auxiliary heat to supplement the solar supply. A 3 ton, air-to-air heat pump was selected and in extremely cold conditions this is supplemented by an electric resistance heater coil. The installed arrangement is shown schematically in Figure II.11.6. The fan is provided with a three-speed direct-drive, 560 W motor, but only the maximum speed was connected to the controls, primarily to match the single-speed air handler for the collectors. Return air from the rooms is always circulated either through the collector or through storage and then through the indoor coil of the heat pump before being returned to the rooms. The heat pump does not operate when the temperature of air from storage exceeds 35°C.

## Dampers

To facilitate testing and modification of the system, motorized dampers are used in several locations where they are not normally required in a conventional house. Two motorized dampers, MD4 and MD5, could be replaced with pressure-activated (back-draft) dampers. All damper blades have neoprene strips to minimize leakage. This design is reasonably effective in single-

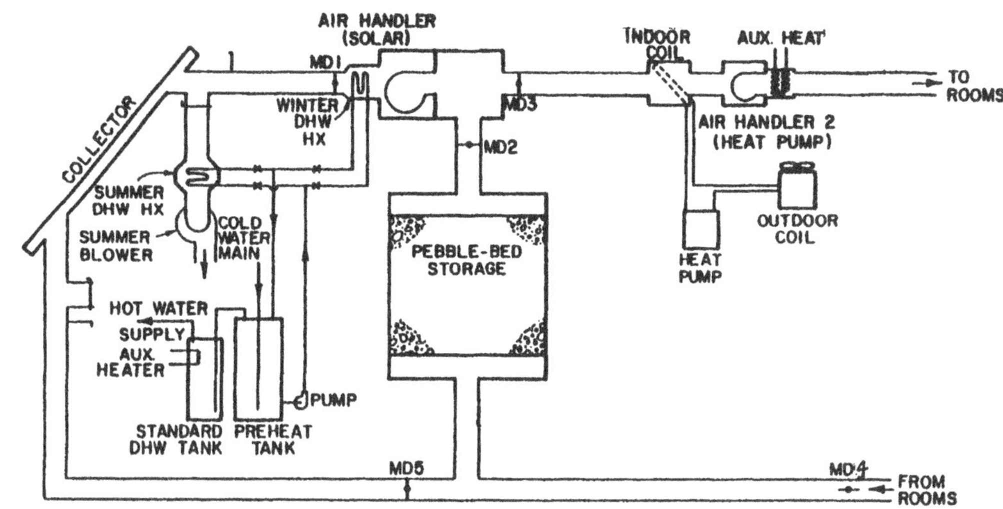


*Figure II.11.6. Schematic diagram of the solar heating and the heat-pump auxiliary system*

directional ducts, but when the dampers are pressurized from the reverse direction, there is a measurable amount of air leakage.

### Air-to-water heat exchangers

There are two heat exchangers for solar heating of domestic water: one is located in the basement, adjacent to the air handler for winter use and the other is located in the attic for summer use. Because the pebble bed and solar air handler were used for testing cool storage during the summer, a small alternative fan was used to deliver solar-heated air to the attic heat exchanger. With two heat exchangers and the duct design employed, summer cooling was tested without interference from solar-heated air inside the insulated building envelope. Only one water coil would normally be used. A by-pass duct would return the air to the collector.

### The preheat tank

The hot-water preheat tank is a 300 litre standard electric water heater. The electric heating elements are disconnected and the drain spigot has been replaced with a piping connection to enable the water to be pumped through the heat exchanger and the preheat tank.

### The auxiliary water heater

A 150 litre standard electric water heater stands adjacent to the preheater tank. Only the upper electric element in the tank is connected so that a minimum water temperature of 60°C can be maintained. Cold water from the mains is delivered to the bottom of the preheat tank and solar-heated water is delivered to the bottom of the auxiliary water tank. In a solar air heating system it is important to install a separate auxiliary water heater, rather than a one-tank solar-auxiliary unit, because the temperature of water delivered from a reasonably sized air-to-water heat exchanger will be lower than is desirable for domestic hot water.

### The preheater circulation pump

Water is circulated during the winter by a 25 W pump. In the summer, this pump is replaced by a pump with an adjustable flow rate. The recirculation rate is increased during the summer to improve the effectiveness of the air-to-water heat exchanger.

### Basic heating modes

*Heating from the collectors*
Solar heat is collected whenever there is sufficient solar radiation. If solar heat is being collected and the rooms require heating, as determined by the room thermostat, all five dampers except MD2 are opened by the controller to provide air circulation from the collectors to the rooms and back to the collector (Figure II.11.6). Damper MD2 is closed to prevent air flow through the storage. If the heat delivery rate from the collectors is insufficient, the heat pump is activated by a temperature sensor immediately upstream from the indoor coil.

*Storing solar heat*
When heating is not required in the rooms, the thermostat contact is open and the solar-heated air stream is circulated by the solar air handler through the storage. Dampers MD1, MD2 and MD5 are open, while MD3 and MD4 are closed. With the air handler positioned as shown in Figure II.11.6, the collectors are at slightly sub-atmospheric (negative) pressure, and the rock-bed container is at a slightly above-atmospheric (positive) pressure. Outdoor air therefore leaks *into* the collector, where it is heated and mixed with the warm recirculated air stream. An equivalent mass of warm air flows from the system into the rooms through cracks in dampers, ducts and the storage container. If the collector is pressurized, however, warm air leaks *out* of the collectors and cold air is drawn into the rooms (infiltration), necessitating additional room heating. Therefore, the collector fan should be on the outlet side of the collector.

*Heating from storage*
The rock bed supplies stored heat when solar energy is not available directly from the collectors. To extract heat from storage, air is circulated upward through the bed so that its delivery temperature approaches that in the hottest, upper zone in the bed, previously heated by hot air from the collectors. The circulation path for this mode is through open dampers MD2, MD3 and MD4. Dampers MD1 and MD5 are closed to prevent air flow through the collector. The room thermostat commands the controller to circulate room air through storage. When heat from storage can no longer meet the demand, the heat pump and electrical resistance heater add heat to the air stream. The heat pump is controlled by the air temperature ahead of the indoor coil and the resistance heater is activated by the second stage of the thermostat.

*Domestic water preheating*
Preheating of domestic water in winter is accomplished only during the heat-storing mode. Under those conditions, the pump circulates water from the preheat tank through the winter heat exchanger. When the house is heated from the collector, priority is given to space heating over domestic water heating. Experience has shown that the direct space-heating mode occurs principally early in the day when the temperatures of air from the collector are not high enough for effective water heating. Midday temperatures are more practical for heat storage and water heating.

In summer, solar-heated air passes to an exchanger coil in the house attic, through which the domestic water is pumped. The air is drawn through the collectors by a small fan and, after passing through the heat exchanger, is discharged into the attic space.

## PERFORMANCE

### Air flow rates

The most complete and representative solar air heating data were obtained during 1978–79 when collector air flow rates were maintained at constant levels. Three different constant rates were employed, during three portions of the heating season.

Analysis of system performance during that year showed that improved efficiency could be obtained by varying the air rate to provide solar-heated air at constant temperature. In the subsequent two years, variable flow was therefore employed by use of a four-speed air fan and a controller that selected fan speeds to match solar intensities, so that heated air could be delivered at nearly uniform temperature. Improved performance was observed, but detailed data were obtained for periods too short for seasonal evaluation. Projections of performance were made by computer simulations calibrated in terms of the monitored data.

*Table II.11.1. Summary of key monthly mean daily quantities and performance measures*

| No. | Item | Units | Season |
|---|---|---|---|
| 1 | Total incident solar insolation | MJ/m²d | 17.55 |
| 2 | Solar insolation during collection | MJ/m²d | 15.28 |
| 3 | Thermal energy collected | MJ/m²d | 5.77 |
| 3a | Thermal energy collected | MJ/d | 334.43 |
| 4 | Collector operating efficiency | | 0.38 |
| 5 | Collector daily efficiency | | 0.33 |
| 6 | Total heating load | MJ/d | 451.90 |
| 6a | Space heating load | MJ/d | 390.82 |
| 6b | DHW load | MJ/d | 61.08 |
| 7 | Solar energy utilized | MJ/d | 294.42 |
| 7a | Space heating | MJ/d | 261.16 |
| 7b | DHW preheating | MJ/d | 33.26 |
| 8 | Auxiliary thermal energy | MJ/d | 157.49 |
| 8a | Heat pump, space | MJ/d | 63.11 |
| 8b | Electrical resistance, space | MJ/d | 66.56 |
| 8c | Electrical resistance, DHW | MJ/d | 27.82 |
| 9 | Electrical energy to collector | MJ/d | 12.95 |
| 10 | Electrical energy to distribution/solar | MJ/d | 9.42 |
| 11 | Electrical energy for heat pump | MJ/d | 37.83 |
| 12 | Electrical energy for heat-pump blower | MJ/d | 5.75 |
| 13 | Solar system efficiency | | 0.29 |
| 14 | COP to collect solar | | 26.36 |
| 15 | COP to distribute solar | | 31.25 |
| 16 | COP overall solar | | 13.78 |
| 17 | COP heat pump | | 1.67 |
| 18 | COP auxiliary heating | | 1.18 |
| 19 | COP overall heating | | 2.82 |
| 20 | Number of days of measurement | | 160 |
| 21 | Solar fraction total | | 0.65 |
| 22 | Solar fraction, space | | 0.67 |
| 23 | Solar fraction, DHW | | 0.54 |
| 24 | F-chart fraction | | 0.64 |
| 25 | Difference (%) | | –1.5 |

COP = Coefficient of Performance, energy delivered/energy purchased

### Space and water heating

Monthly mean values of the daily totals of energy flows through the heating system from 1 December 1978 to 11 May 1979 are listed in Table II.11.1.

Monthly mean solar insolation, insolation while the system was running and heat collected are shown graphically in Figure II.11.7. For the season, 87% of the measured insolation was incident on the collectors while the system was operating with a mean daily collection efficiency of 33%, based upon total insolation, and 38%, based on insolation while collecting. During 160 days when data were taken, 45,200 kWh of solar radiation was incident on the collectors, 39,300 kWh of which was recorded during collector operation, and 14,900 kWh of thermal energy was collected. Collection efficiency was high in December because of the high rate of air flow through the collectors (13.1 litres/ms²) but reduced by 3% to 4% in January and February when air flow was reduced to 9.6 litres/ms². Further reductions in collector efficiency in March and April are due to warm air returning to the collector from the bottom of storage, at temperatures between 60 and 70°C, on many days when storage was 'full' following successive sunny days and the heating requirements in the house were low. The unbalanced collector operation during the last four days in April and 11 days in May did not noticeably reduce collection efficiency.

The mean daily solar and auxiliary energy supplied to the load are shown in Figure II.11.8. As expected, the total load is highest in January and decreases each month into spring. The mean daily seasonal space and water heating loads totalled 125 kWh/d, of which 65%, 80 kWh/d, was supplied from the solar system. Since a total of 14,800 kWh was collected, 1800 kWh (12%) was lost from the system through duct losses in the attic and through heat losses from storage to the ground. The monthly solar contribution to the load varied from a low of 48% in January to a high of 85% in April. The first 11 days in May were dominated by cold cloudy days and the solar contribution was low. (The mean daily solar insolation was only 61% of that in April).

The monthly average daily space-heating load is shown in Figure II.11.9, separated in terms of the solar heat pump and electrical resistance heating contributions. The seasonal space-heating load was 17,400 kWh, of which 67% was supplied by the solar system. The heat pump accounted for 16% (2730 kWh) and the balance, 2900 kWh (17%) was supplied by electrical resistance back-up provided with the heat pump.

The overall seasonal performance of the heat pump is low, with a mean daily COP of 1.67, and varies from 1.47 in December to a high of 2.17 in March. The COP is low because the heat pump operates under adverse conditions when solar heat has been depleted and ambient air temperatures are low. The measured performance of the heat pump as a function of ambient air temperature is shown in Figure II.11.10. The COP

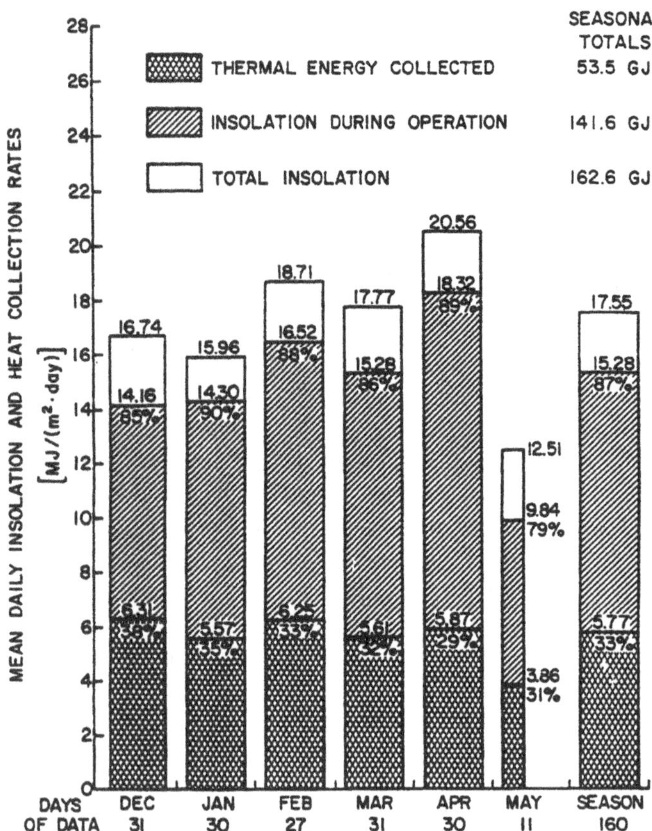


*Figure II.11.7. Solar insolation and thermal energy collected*

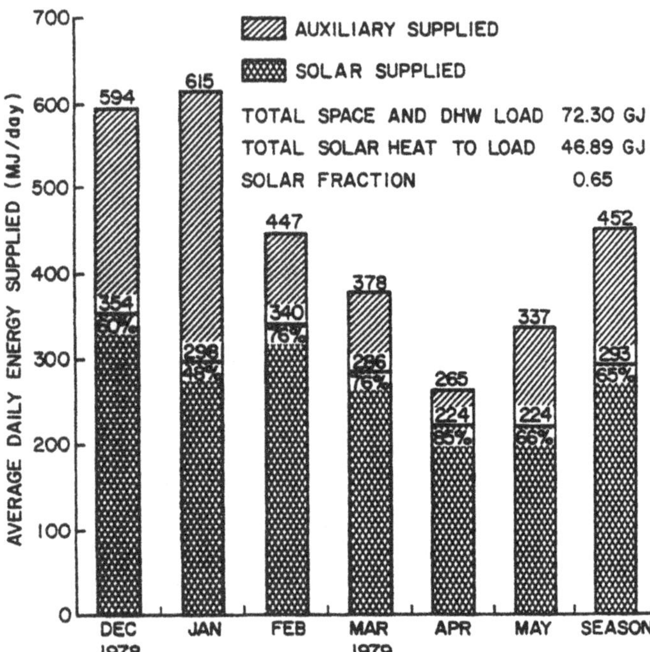


*Figure II.11.8. Mean daily energy supplied to the load*

varies from slightly greater than one at –20°C to about 2.5 at 3°C. The seasonal COP for auxiliary heating, including both heat-pump and electrical resistance heating, was 1.18.

The monthly mean daily domestic hot-water load and the solar contributions are shown in Figure II.11.11. Domestic water preheating is limited because preheating occurs only during the day when solar energy is being collected and because priority is given to space heating. When solar heat is being delivered from the collectors to the rooms, preheating of water is not permitted according to the control program adopted. The energy required for DHW increased from December through April, as also did the solar contribution. The seasonal

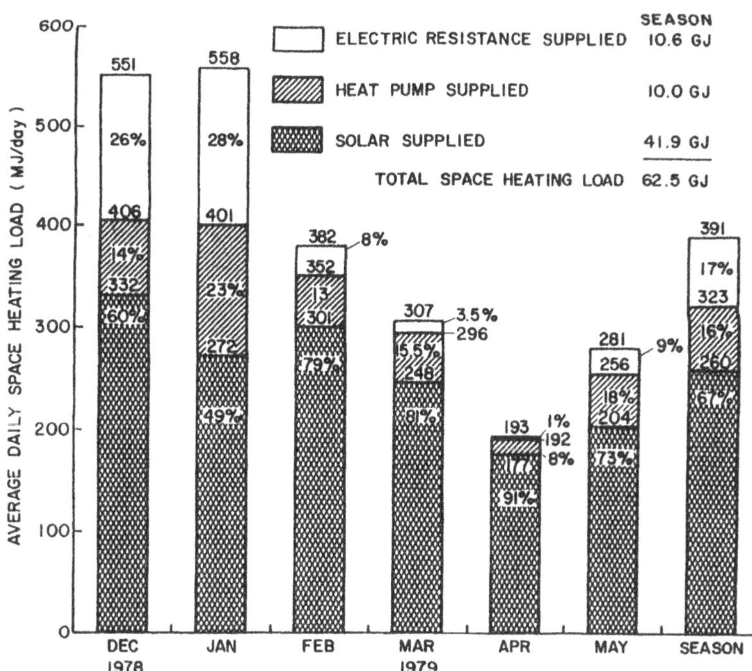


*Figure II.11.9. Monthly average daily space-heating load*

solar contribution to the DHW load was 1500 kWh (54%) and the DHW load was 13.5% of the total load.

The net system efficiency for the season was 29%, varying from 36% in December to 19% in April. During the cold winter months, collection efficiency is high, despite cold ambient air conditions, because the inlet air temperature to the collector is nearly always at room temperature. During the warmer months when storage is 'full' for a significant number of days, the temperature of the return air to the collector is high and collection efficiency decreases. Heat losses from ducts in the attic and from storage also increase during the warmer months.

The ratio of solar energy collection to electrical energy required by the fan (i.e. collection COP) averages 26 for the season, which means that 3.8% (in fan energy) is required for collection. Another 3.2% is required during the season to distribute the stored solar heat. For the total solar energy used for space and water heating, 7.3% electrical energy was required for an overall solar COP of 13.78. The range in COP was from 11 in May to 15 in February. It will be noted in Table II.11.1 that the COP to distribute solar heat to the rooms varied from 19 in May to 63 in April. Since April was a warm month and the heating load was small while solar radiation levels were high, a large portion of the load was satisfied by heat losses from the ducts and from storage during collection, so that only small amounts of air circulation through storage were necessary.

## Solar preheating of domestic water during summer

In summer (daytime only), attic air is drawn through the collectors, is heated and flows across an air-to-water heat exchanger. The air, which is still hot, is discharged out of the attic space. The heat exchanger is a 600 × 600 mm single-loop cross-flow unit with fins 100 mm wide.

The 1979 summer season has been divided into 3 experimental periods: in Period 1, from 11 to 29 June, the air flow rate through the collector is high (12.45 litres/sm$^2$), and water circulation rate is low (about 4 litres/min). In Period 2, from 30 June to 3 August,

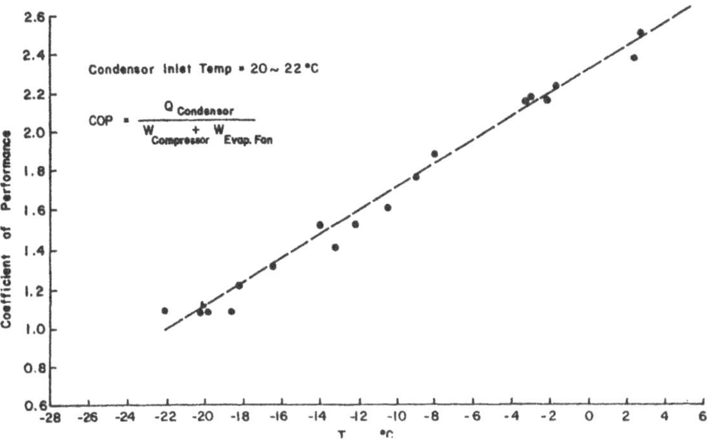


*Figure II.11.10. Heat pump COP as a function of ambient air temperature*

*Table II.11.2. Domestic water heating, summer 1979*

| No. | Item | Units | Period 1 (11–29 June) | Period 2 (30 June–3 August) | Period 3 (4 August–9 September) | Season (11 June–9 September) |
|---|---|---|---|---|---|---|
| 1 | Number of days of data | | 19 | 35 | 37 | 91 |
| 2 | Total solar insolation | MJ/m²d | 18.85 | 19.72 | 19.00 | 19.24 |
| 3 | Solar insolation during collection | MJ/m²d | 15.57 | 16.74 | 16.40 | 16.36 |
| 4 | Total thermal energy collected | MJ/m²d | 9.45 | 8.62 | 8.01 | 8.54 |
| a | East array | MJ/m²d | 9.36 | 8.63 | 8.03 | 8.54 |
| b | West array | MJ/m²d | 9.54 | 8.60 | 7.99 | 8.55 |
| 4c | Total thermal energy collected | MJ/d | 547.88 | 499.1 | 464.09 | 494.96 |
| 5 | Collection operating efficiency | No. 4/No. 3 | 0.61 | 0.51 | 0.49 | 0.52 |
| a | East array | No. 4a/No. 3 | 0.60 | 0.52 | 0.49 | 0.52 |
| b | West array | No. 4b/No. 3 | 0.61 | 0.51 | 0.49 | 0.52 |
| 6 | Daily collector efficiency | No. 4/No. 2 | 0.50 | 0.44 | 0.42 | 0.44 |
| 7 | Total DHW load | MJ/d | 80.70 | 85.84 | 84.68 | 84.30 |
| a | Solar preheating | MJ/d | 61.36 | 73.72 | 70.85 | 69.97 |
| b | Auxiliary heating | MJ/d | 19.34 | 12.12 | 13.83 | 14.33 |
| c | Solar fraction | | 0.76 | 0.86 | 0.84 | 0.83 |
| 8 | Electrical energy for collector blower | MJ/d | 18.98 | 12.05 | 8.14 | 11.91 |
| 9 | Electrical energy for water circulation pump | MJ/d | 5.04 | 4.84 | 4.44 | 4.72 |
| 10 | Solar COP for preheating | No. 7a/(Nos 8 + 9) | 3.36 | 4.36 | 5.63 | 5.07 |
| 11 | Overall COP for DHW | No. 7/(Nos 7b + 8 + 9) | 1.86 | 2.96 | 3.21 | 2.72 |

the air flow rate was reduced to 8.61 litres/sm² and the water circulation rate remained the same. In Period 3, from 4 August to 9 September, the air flow rate changed slightly to 7.97 litres/sm² and the water circulation rate was increased to about 6 litres/min.

The performance results are summarized in Table II.11.2. As a consequence of reducing the rate of air flow through the collector, its temperature increased and the effectiveness of the heat exchanger increased from 0.24 during Period 1 to 0.32 during Period 2. The collector efficiency decreased from 60% to 50%, but the reduction is not important because the entire 57.9 m² of collector area is available for water heating. The electrical energy requirement for preheating water decreased as a consequence of the change in flow rate and the solar COP increased from 3.4 in Period 1 to 4.4 in Period 2.

There was little change from Period 2 to Period 3, except that further reduction in air flow rate increased the solar COP to 5.6. The change in water flow rate did not affect performance of the heat exchanger because the capacitance rate of the air and physical size of the heat exchanger are the limiting factors. The solar system covered 83% of the seasonal water heating load with a COP of 5.1. Overall COP for domestic water heating for the summer was 2.7.

## ACKNOWLEDGEMENTS

The planning and extensive testing of the solar air heating system in Solar House II at Colorado State University were supervised by Dr Susumu Karaki, principal author of the technical report on which this case study is based. His contributions and those of participating graduate students under his direction are gratefully acknowledged.

In addition to sponsorship by Colorado State University, construction and testing of the building and its heating facilities were supported by the US Department of Energy, the sheet metal and heating industry and solar equipment suppliers.

# II.12 Wannseebahn Row Houses

Berlin, Germany

*Wannsee-bahn, Berlin*

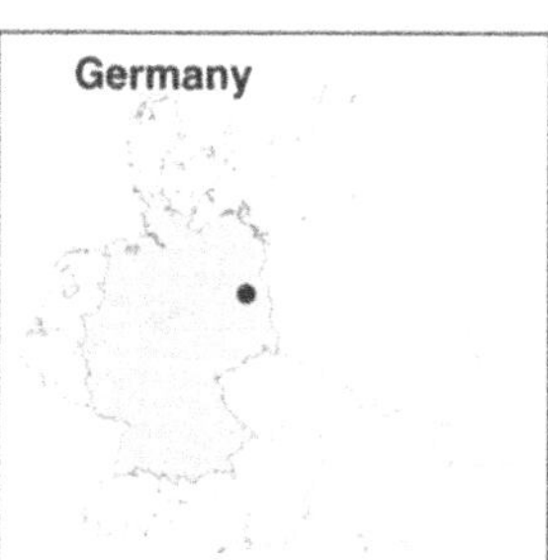


*System type 5*

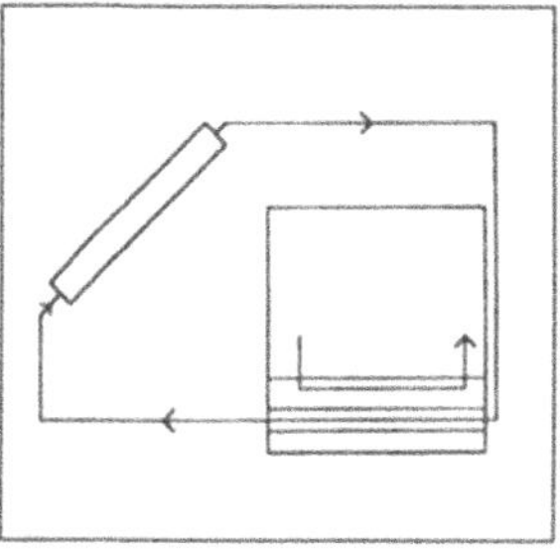

## PROJECT SUMMARY

The energy-saving concept of the four Wannseebahn row houses in Berlin comprises a compact construction method, good thermal insulation of the building envelope, deliberate use of solar energy with south-facing windows and sunspaces, and a solar air system. The energy gained by the air collectors is led through a closed cycle and released to the middle wall of the building. So that the stored energy is used in a controlled manner, there is a second closed air cycle with an active discharge.

### Summary statistics

| | |
|---|---|
| System type: | type 5 |
| Collector type: | air collector |
| Collector area: | 16 m$^2$ |
| Storage type: | concrete hypocaust |
| Storage capacity: | 1.8 kWh/K |
| Annual contribution from the solar air system: | 8.1 kWh/m$^2$a$_{floor}$ |
| Annual auxiliary heat consumption: | 70 kWh/m$^2$a$_{floor}$ |
| Basis: | monitored |
| Heated floor area: | 197 m$^2$ |
| Year solar system built: | 1994 |

## SITE DESCRIPTION

The row houses are located in Berlin-Zehlendorf:

| | |
|---|---|
| latitude | 52°N |
| longitude: | 13°E |
| altitude | 50 m above sea level. |

The mean ambient air temperature is 7.0°C (September to May) and 9.5°C annually.

There are 3750 annual heating degree days (base 20°C/15°C) and the horizontal global radiation is 566 kWh/m$^2$ (September to May) and 978 kWh/m$^2$ annually.

## BUILDING PRESENTATION

Four row houses have been built together in one three-storey building. This yields a good envelope surface-to-volume ratio of 0.49 m$^{-1}$ for the total row building, 0.60 m$^{-1}$ for the corner row house and 0.42 m$^{-1}$ for the middle row houses. The building faces 25° east. The westernmost middle row house was monitored during the heating periods 1994/95 and 1995/96 and is described below.

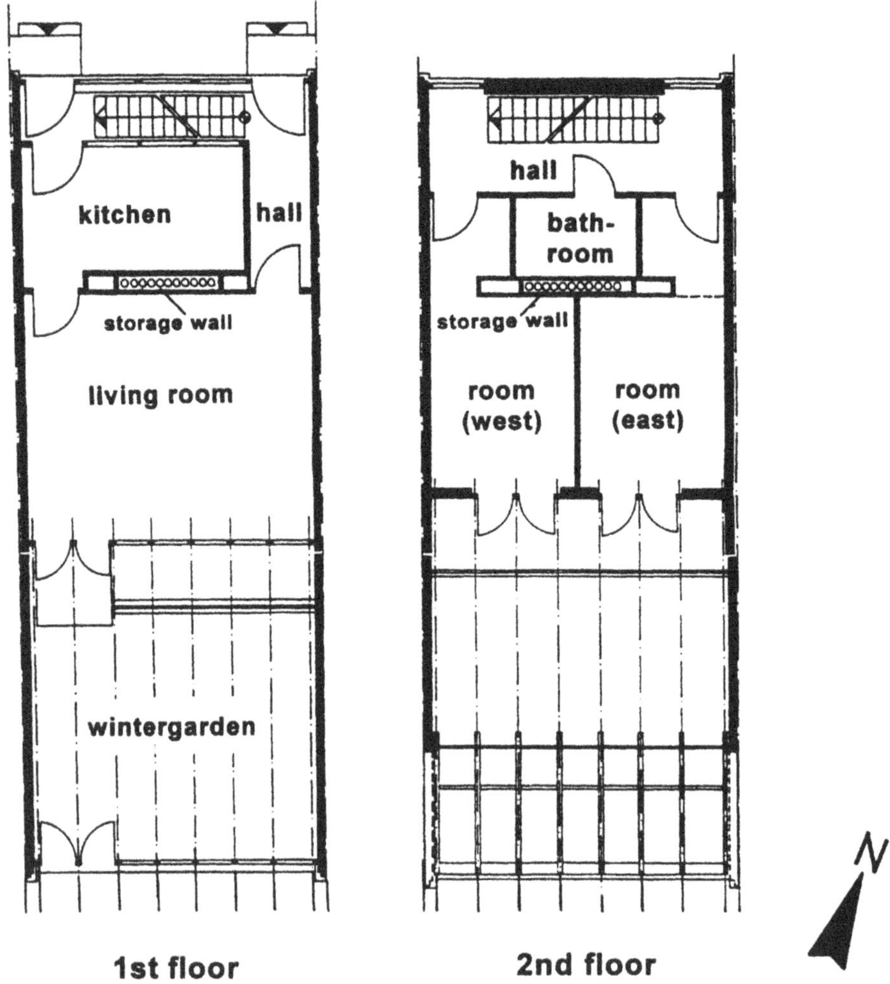


*Figure II.12.1. Plan of (a) the ground floor and (b) the first floor*

## Building statistics

The heated floor area of the middle row house is 197 m². The details of construction of the house are given in Table II.12.1. The plan of the first and second floors is shown in Figure II.12.1 and a section of the building in Figure II.12.2.

The energy required for space and domestic hot water heating is provided by district heating. The rooms are heated with radiant floor heating. The heating energy demand is given in Table II.12.2.

## SOLAR SYSTEM

The hybrid system consists of air collectors and thermal storage as a closed system. Sixteen square metres of air collectors are installed on the south-facing part of the roof. The operation of the system is illustrated in Figure II.12.3.

The air heated in the air collectors is fan-forced upwards in the middle wall of the building and then flows through vertical sheet metal ducts, which extend over two storeys. As the air rises, it releases heat into the

*Table II.12.1. Details of construction*

| | $U$ (W/m²K) | |
|---|---|---|
| Walls: | | |
| external wall | 0.3 | Gypsum plasterboard 1.5 cm; insulation 16 cm; ventilated exterior lining |
| wall to buffer zone | 0.45 | Interior plaster 1.5 cm; masonry 11.5 cm; insulation 6 cm; interior plaster 1.5 cm |
| Roof | 0.3 | Roof cladding; venting; insulation 14 cm; gypsum plasterboard 1.5 cm; |
| Floor | 0.31 | Screed 6 cm; insulation 10 cm; concrete 15 cm |
| Windows: | | |
| external window and window to buffer zone | | 1.6 ($g$ = 0.62) |
| window of buffer zone: | | 3.0 ($g$ = 0.8) |

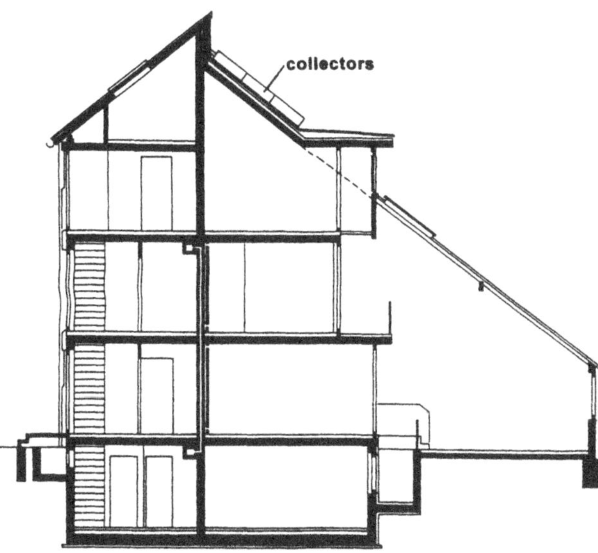


*Figure II.12.2. Section of the building*

*Table II.12.2. Heating energy demand (calculated values)*

| | |
|---|---|
| Overall building | 62 kWh/m²a |
| Middle row house: | 54 kWh/m²a |

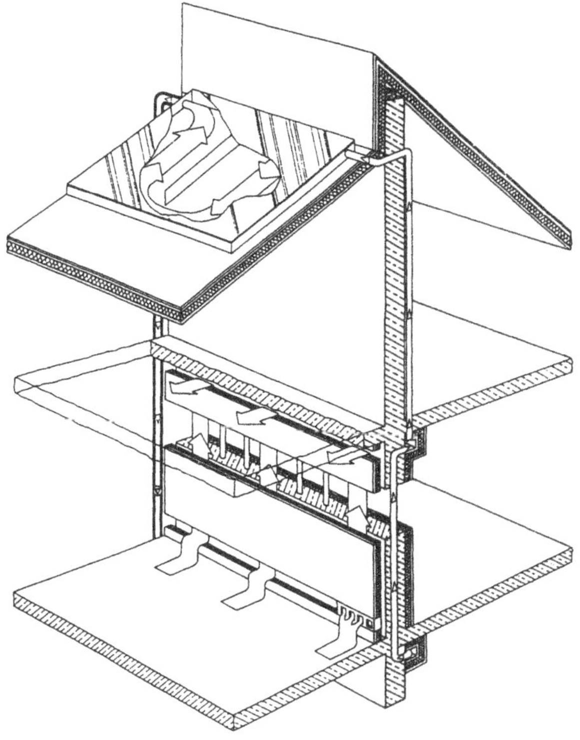

*Figure II.12.3. Isometric view of the solar system*

middle wall. The storage wall, which also extends over two storeys, is shown in Figure II.12.3 (for reasons of simplification here extending only over one storey). To ensure that the energy stored in the concrete wall is only released to the room when heating energy is actually required, there is an insulated facing shell on the room side of the concrete with an air space of 4 cm. This facing shell can be ventilated through openings at ceiling and floor levels. The concrete storage wall behind this has a 6 cm layer of insulation. The lower opening can be closed by a motor-driven sliding grating. The specific discharge of the wall is controlled so that this grating opens if heat is required in the room and if the surface temperature of the wall exceeds a specified value. The cooler air then flows through the lower opened air grating into the space and, as it rises, is warmed by the heat released from the storage wall. The heated air then flows through the upper air grating and back into the room.

## The collector

Nine air collectors are installed above the dormers on the sloping roof. Details are given in Table II.12.3.

## Storage

The middle wall on the ground and first floors is used as thermal storage. Eleven vertical sheet-metal ducts

*Table II.12.3. Details of the collectors*

| | |
|---|---|
| Area | 16 m² |
| Air volume rate/collector area | 22 m³/m²h |
| Collector manufacturer | Grammer |

*Table II.12.4. Air speed and storage heat capacity*

| | |
|---|---|
| Air speed in storage ducts | 1.8 m/s |
| Capacity | 1.77 kWh/K |
| Capacity/collector area | 0.11 kWh/m²K |

with a diameter of 8 cm run through the middle of this wall (L × W ×H 2.10 m × 0.24 m × 5.50 m). Details of air speed and heat capacity are given in Table II.12.4.

## Distribution

To transport the air, a 270 W radial fan was installed in the air duct, just in front of the entry into the storage wall. A mode with a reduced airflow rate was necessary to minimize noise. The electric power demand in this reduced operating mode is 120 W with an airflow rate of 350 m³/h (Table II.12.5)

## Control

If the air temperature in the collector is more than 5 K higher than the storage temperature, the fan is switched on automatically. Temperature sensors are integrated into the storage wall and the collector. Discharging is according to the heating energy demand of the room. If the indoor air temperature drops below the set value,

*Table II.12.5. Fan details*

| | |
|---|---|
| Manufacturer | Klein-Ventilatorbau-GmbH |
| Type | Radial standard blowers; ENG 4-14 |
| Pressure (max.) | 500 Pa |
| Fan power/absorber area | 7.5 W/m² |
| Fan power/air rate | 0.34 W/m³ |

the lower flap in the facing shell is opened by a control motor, allowing for the cool indoor air to flow through the lower opening behind the facing shell and through the upper opening back into the room again, as described above.

## PERFORMANCE

The westernmost middle row house was monitored from December 1994 through to May 1996. During the heating period 1995/96 (September to May) the solar energy incident on the 16 m$^2$ collector surface amounted to 10,630 kWh. 1,783 kWh was released to the storage wall. The fan consumed 89 kWh. As there is no flap installed between the collector and the storage wall, an unintended air circulation occurred during the time when the fan did not run. Thus, 196 kWh was transferred back from the storage wall to the collector. Table II.12.6 shows the energy usage (the total values and the values related to the collector area).

The heat produced by the fan motor is not released into the air flow because the motor is situated outside the air duct. Table II.12.7 summarizes the energy balances of the middle row house monitored during the heating period 1995/96, in relation to the floor area.

## CONCLUSION

The most important results gained from the evaluation of the measurements are summarized as follows:

- By using the hybrid system, the auxiliary heating consumption was reduced by 8 kWh/m$^2$a to 70 kWh/m$^2$a during the heating period 1995/96, which corresponds to 10% of the total heating energy consumption of the apartment.
- The active discharge of the storage wall proved to be efficient. The maximum amount of energy was released to the apartment, delayed relative to solar gains through the windows.
- As there are no flaps installed in the connecting ducts between collector and storage wall, an unintended discharge of the storage wall occurred during the time when the fan was not running. Thus, 11% of the energy gains was lost again.
- During the heating period, 15% of the solar energy that was incident on the collector surface was captured as usable heating energy in the storage wall. This corresponds to 100 kWh per square metre of collector surface.
- The electrical energy used by the fans costs approximately US $11 per annum and amounts to 5% of the energy transported.
- The airflow rate between the collector and the storage wall was set initially at 600 m$^3$/h. This had to be reduced to 350 m$^3$/h and it became necessary to decouple the rigid connection of the fan to the ceiling because the occupants could not tolerate the noise produced by the fans.

*Table II.12.6. Survey of the input and output values of the solar system and of the electric current consumption of the fans*

| Energy inputs/outputs | Energy usage (kWh) Total | Per square metre of collector area |
|---|---|---|
| Solar energy incident on the collector surface | 10,630 | 664 |
| Electrical power consumption of fans | 89 | 6 |
| Solar system output (including the power consumption of the fans) | 1,783 | 111 |
| Losses due to return flow | 196 | 12 |

*Table II.12.7. Survey of all energy balances measured during the heating period 1995/96, in relation to the floor area.*

| | Energy gains and losses (kWh/m$^2$a) |
|---|---|
| Gains: | |
| heating energy | 70 |
| passive solar gains | 20 |
| solar system | 9 |
| internal gains | 26 |
| Losses: | |
| discharge from the storage wall | 1 |
| transmission | 60 |
| ventilation heat losses | 64 |

## RECOMMENDATIONS

- Flaps should be installed in ducts between the collector and the storage wall to prevent discharge of the storage wall due to free-convection.
- In a vertical storage wall with active discharge the air should flow through the wall from top to bottom to ensure that air entering the bottom of the space between storage wall and facing shell is warmed constantly as it rises.
- The efficiency of the air systems should be improved. The determined operating ratio of 15% should be increased to achieve at least the same efficiency as hot-water collectors (40%).
- The total system costs are too high. They should be reduced to the level of those for solar water systems.
- More attention should be given at the design stage to reducing the noise due to airflow and fans, because a high noise level may cause the occupants to switch the system off. To prevent noise transmission, fans should never be rigidly fixed to the structure of the building.
- Hybrid systems should only be considered if a building is highly insulated, thus requiring only a minimum heating energy demand. In this case, reducing the heating energy by a further 8–10 kWh/m$^2$a by adding more insulation may be very expensive. Hybrid systems become economically competitive with other technical solutions once the heating demand falls to 20 kWh/m$^2$a.

## ACKNOWLEDGEMENTS

Architects: IBUS Architekten im Ingenieurbüro für Bau- u. Stadtplanung GbR, Caspar-Theyss-Strasse 14A, 14193 Berlin, Germany

Development and scientific consulting: Fraunhofer-Institut für Bauphysik, Nobelstrasse 12, 70569 Stuttgart, Germany

Builder: BBG Berliner Baugenossenschaft, Pacelliallee 3, 14193 Berlin, Germany

Research commissioned by: BMBF Bundesministerium für Bildung, Wissenschaft, Forschung und Technologie, Heinemannstrasse 2, 53175 Bonn, Germany.

Chapter authors: Johann Reiss and Hans Erhorn, Fraunhofer-Institut für Bauphysik, Nobelstrasse 12, 70569 Stuttgart, Germany

## BIBLIOGRAPHY

Reiss J, Erhorn H (1997). *Solare Hybridsysteme in einer Reihenhaus-Wohnanlage an der Wannseebahn in Berlin.* Report WB 91/97, Fraunhofer-Institut für Bauphysik, Postfach 800465, D70569 Stuttgart.

Hillmann G (1997). *Reihenhäuser an der Wannseebahn. Solare Energiesparhäuser der zweiten Generation im ökologischen Kontext.* Endbericht der IBUS GmbH, Caspar-theyss-strasse 14a, D14193 Berlin.

# III Apartment buildings

# III.1 Introduction

## BUILDING CHARACTERISTICS

Apartment buildings pose an interesting combination of attributes. Their internal gains, comfort requirements and occupancy profiles are those of a residence, but their geometry, unlike single-family houses is more like that of an office building. Namely, they are multi-storeyed with a small surface-to-volume ratio. What most differentiates this building type is that:

- Residents are typically not owners (even in the case of condominiums the facades and heating system are usually common property).
- Residents are not directly responsible for the heating.
- Investment tends to be more conservative and payback has a large importance.

Because heating demand per $m^2$ of floor is less than in the case of single family residences, a solar air system can cover a greater percentage of the heating load. Because typically the residents are not owners, operation and maintenance of a solar air system must be easy and trouble-free. Finally, because of the importance of payback, solar air systems must be as simple and as low-cost as possible. Ideally, they should be multi-functional, indeed often a secondary function, such as being a barrier to street noise, is of greater importance than the solar function.

## COMBINING DIFFERENT SYSTEMS

### Direct gain

Overheating is even more likely from direct gain in apartments than in houses because there are fewer heat-losing exterior surfaces. Accordingly, it is essential either to provide delayed release of solar air heating or to transport the heat to non-sunlit rooms.

### Glazed balconies

These can readily serve as the collector. Since typically kitchens, bathrooms and WCs are mechanically exhausted, glazed loggias can act as pre-heaters of intake air.

### Active solar water systems

An air collector can readily also serve for heating domestic hot water in the case of individual hot-water heater tanks.

### Auxiliary heating

This is typically a central system with heat distribution by a hot water system. Here a system of type 6 can readily be integrated.

### Central ventilation

Often corridors, lobbies and common rooms are mechanically ventilated. Here several types of solar air systems, such as the unglazed solar air collector, have proved very well suited.

## ACKNOWLEDGEMENTS

Author:
R. Hastings

# III.2 Marostica Passive Solar Dwelling

## Marostica, Italy

*Marostica*

*System type 2*

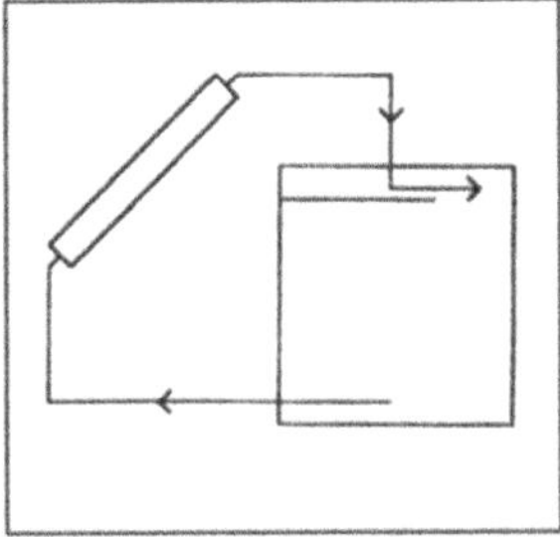

## PROJECT SUMMARY

This project consists of four separate buildings; three terraces comprising 24 dwellings in all, and one four-storey housing block containing 16 flats. The principal objective was to build low-cost housing in which innovative passive solar components could be incorporated at costs acceptable for public housing schemes (maximum 10% of the overall cost). An 'open-loop passive system', developed in Italy about 20 years ago by Barra-Costantini, was chosen. Warm air produced in the solar air panel circulates freely in the storage ceiling, into the rooms and back to the bottom of the air panel by gravity. The system supplies 30% of the net space-heating load.

### Summary statistics

| | |
|---|---|
| System type: | type 2 |
| Collector type, area: | thermo-circulating, 16 m$^2$ |
| Annual contribution of the solar air system: | 100 kWh / $m_{collector}^2$ |
| Basis: | monitored |
| Heated floor area: | 84 m$^2$, 230 m$^3$ |
| Year solar system built: | 1984 |

## SITE DESCRIPTION

The site is located in Marostica, a small town in northern Italy, some 20 km from Vicenza. The site is almost rectangular with the long axis running north–south and gently sloping at 5° towards the south. It has good solar access and is protected to the north by a hill:

| | |
|---|---|
| latitude | 46 °N |
| longitude | 12° E |
| altitude | 105 m above sea level. |

The climate is temperate continental, with an average horizontal winter solar radiation of 2 kWh/m$^2$d.

## BUILDING PRESENTATION

### Plans and sections

Correct separation between buildings avoids shadows on the solar panels during winter months. The tallest building has been placed at the north end of the site and roof pitches are designed to minimize solar obstruction. The terraced houses are two-storey dwellings; the plan and section of a typical dwelling unit are shown in Figure III.2.1. The south facade consists entirely of windows and solar air panels, which are unified by the use of the same red steel framing. The framing is used as a thin trimming applied to the glazing of the solar panels to soften the impact of their extensive surface area, which would otherwise appear entirely black. The results help to blend the windows and the passive components together in harmony.

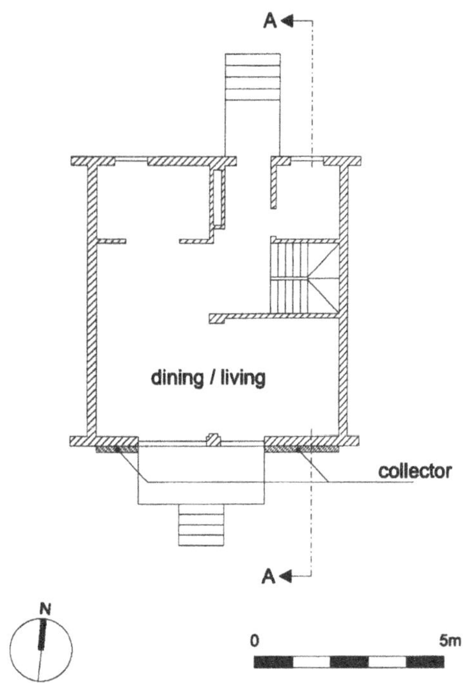


*(a) Typical floor*

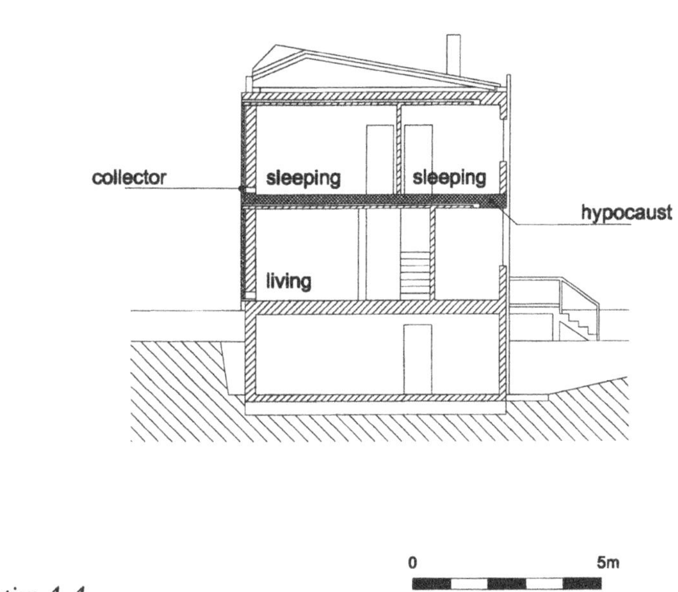


***(b) Section A-A***

*Figure III.2.1. (a) Floor plan and (b) section of a dwelling unit*

*Table III.2.1. Envelope specification*

| | *U*-value (W/m²K) |
|---|---|
| Floor: reinforced light concrete (30 cm) | 0.72 |
| Walls: reinforced concrete (20 cm) (5 cm) and insulation | 0.41 |
| Windows: double glazed steel frame | 2.80 |
| Roof: insulated, copper covered | 0.47 |

*Table III.2.2. Building heat loss coefficients*

| | |
|---|---|
| Transmission | 2.3 kWh/m²K |
| Ventilation | 0.48 kWh/m³K |

Active solar collectors for domestic hot water are integrated into the south slope of all the roof at an angle of 45°.

### Building construction

The envelope specification is given in Table III.2.1 and the building loss coefficients in Table III.2.2.

## SOLAR SYSTEM

Starting from a craft-made solar air panel, the team of designers and manufacturers built a prototype solar

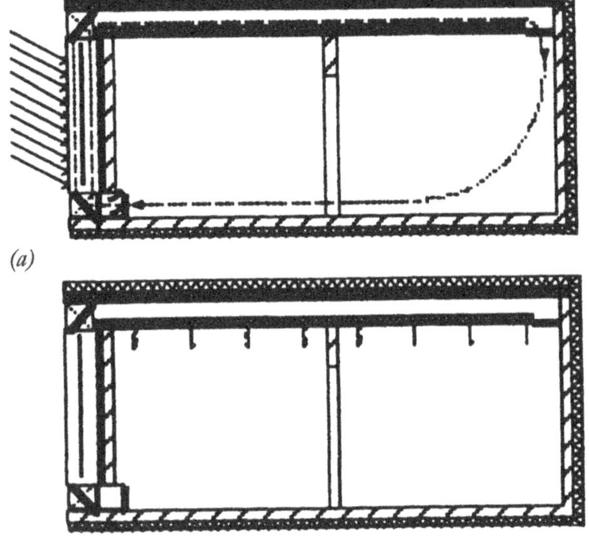

(a)

(b)

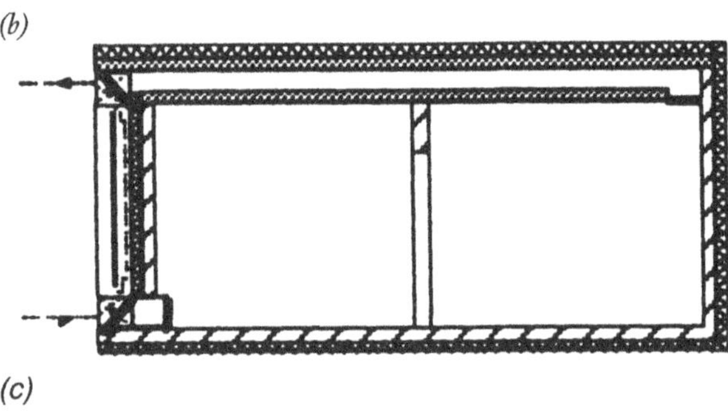

(c)

*Figure III.2.2. Diagram of the Barra-Costantini: (a) winter day operation mode; (b) winter night operation mode; (c) summer operation mode*

system, which was subsequently mass-produced at low cost (by Industry Secco, Treviso).

The integrated passive system operates as follows: on a winter day, the air heated in the solar panels rises and enters ducts in the concrete ceiling. The hot air warms the concrete structure and becomes cooler before entering the room. Here it mixes with the room air and returns to the panel through the low opening (Figure III.2.2a). At night, a backdraft damper closes the return opening, preventing reverse air circulation. The return air openings were reduced to one per solar panel, allowing greater freedom in furnishing the living space. The heat stored during the day is radiated from the ceiling to the room (Figure III.2.2b). During the summer, upper and lower seasonal butterfly dampers are manually shifted to vent the hot air to the ambient (Figure III.2.2c). A isometric view is given in Figure III.2.3.

### The collector

The collector is a thermo-circulating solar air panel integrated into the south facade; this consists of a single panel of glass, an air space, a dark coloured aluminium sheet, a second air space in front of 60 mm insulation and the load-bearing wall (Figures III.2.4 to III.2.6).

The net collector area is 16.5 $m^2$ and the yearly solar contribution is 100 $kWh/m^2$. The air flow is 30 $m^3/h$ per $m^2$ of air panel. Dimensions are based on a module 30 cm wide and a floor to ceiling height of 270 cm; the collector depth is 18 cm. Dimensions can be easily modified on demand.

### Storage

The storage is a passive charging and discharging thermal ceiling. It is a modification of a traditional prefabricated concrete slab. Galvanized steel ducts are substituted for polystyrene forms (Figure III.2.7).

The typical dimensions of the thermal ceiling are $120 \times 30 \times 700$ cm; duct dimensions are $40 \times 16$ cm. The thermal capacity of the ceiling is 4.17 kWh/K or 0.25 $kWh/Km^2$ net collector area. A special connection was designed to minimize the resistance to the air flow from the vertical collector to the horizontal ceiling ducts.

### Distribution

The distribution follows the following loop: through the thermal-ceiling inlet into the room, mix with room air, return via the opening in the solar air panel. Efficiency of distribution depends on the friction of the

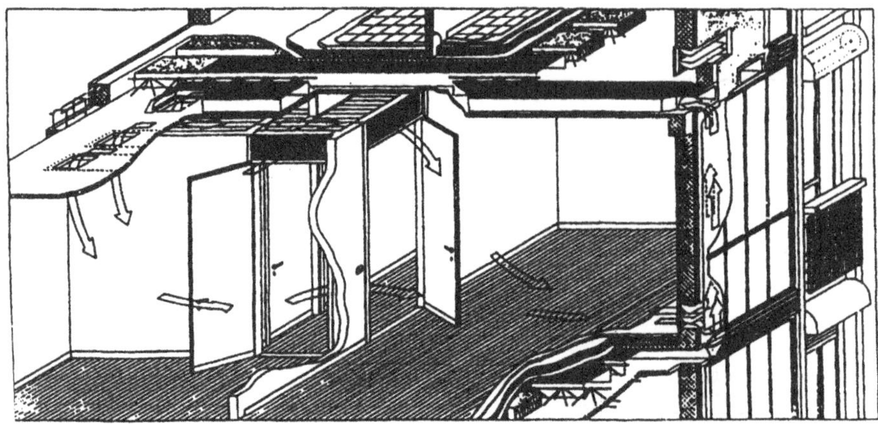

*Figure III.2.3. Isometric view of the system*

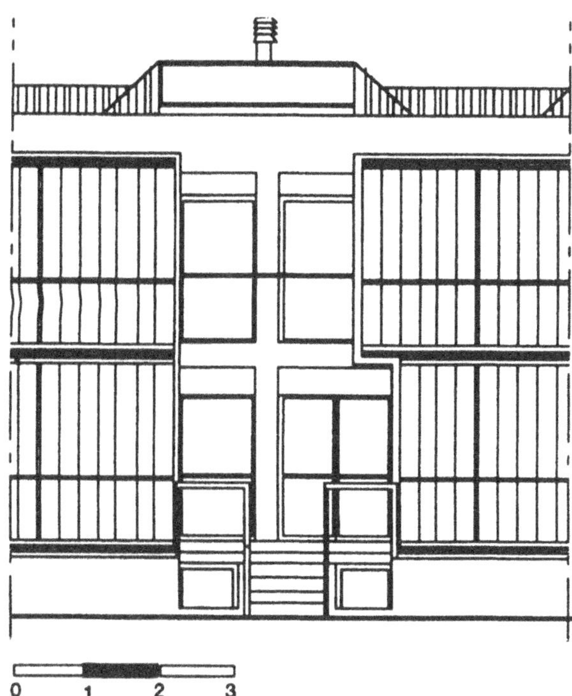


*Figure III.2.4. The south facade*

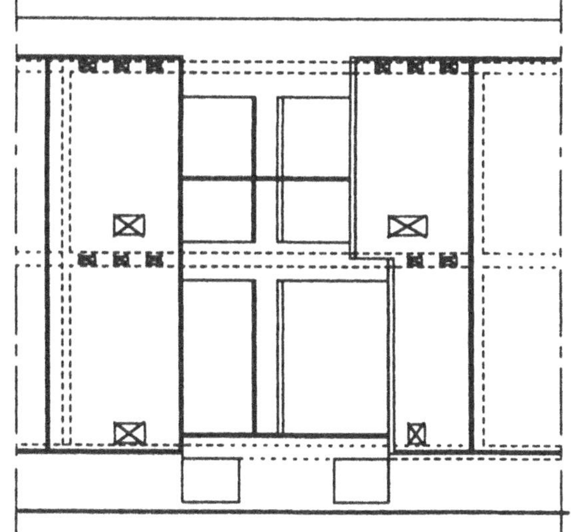

*Figure III.2.5. Layout of the building system behind the glazing and the absorber*

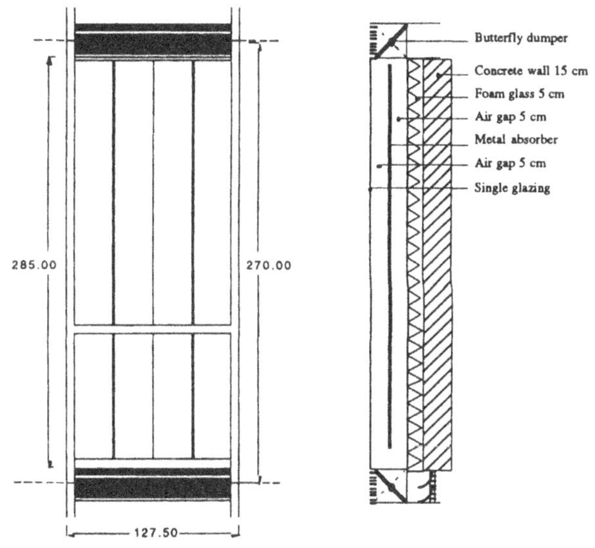


*Figure III.2.6. Front view and section of the solar air panel*

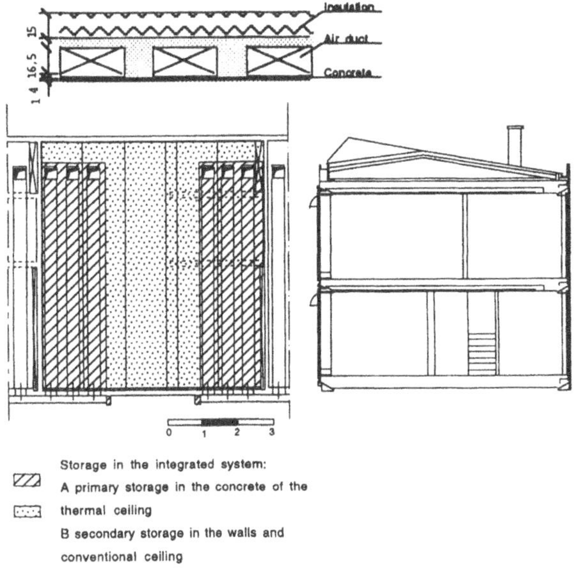


*Figure III.2.7. The thermal ceiling storage unit*

loop (Figures III.2.7 and III.2.8). It is advisable to have, as far as possible, constant cross-sections for the air ducts, the vent sections and all chimney sections.

## Controls

The system is modified seasonally and daily with dampers (Figure III.2.9).

In Marostica, controls are user-dependent: a lever connected to the dampers is pushed down for winter operation and up for summer operation. Reverse circulation at night is prevented by a plastic film damper, which opens by itself when the sun shines and warm air begins to flow. It is closed against a grid by the cold air of the solar panel when the sun is not shining.

## PERFORMANCE

Efficiency is defined by the ratio between the heat delivered by the system to the heated space (by convention through the inlet and by radiation from the thermal ceiling) and the solar radiation incident on the collector. Figure III.2.10 shows how the efficiency increases quickly during the morning, stabilizes around midday at 35% (on a sunny day) and then declines to zero in the evening. On a cloudy day the efficiency is around 15%. The ceiling fraction is the quantity of heat transferred to the heated space through the concrete ceiling, which is the ratio of the heat delivered by

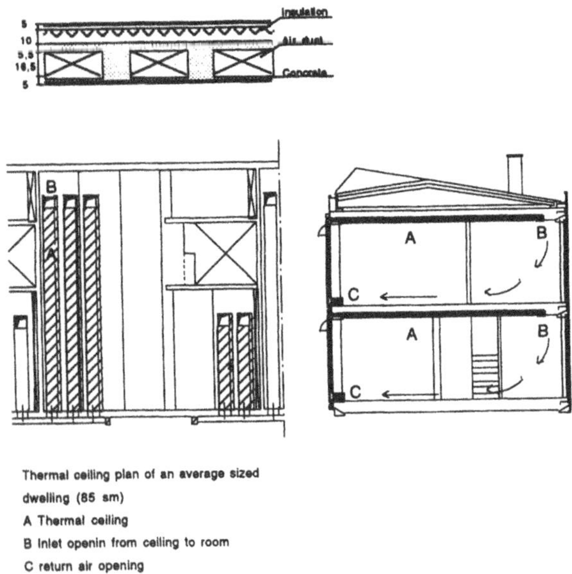


Figure III.2.8. Distribution system

ceiling to the total heat delivered by collector. The ceiling fraction is around 14% on sunny days.

## REMARKS

The overall performance of the Barra-Costantini system was 10–15% lower than the simulated values and the measured data from the Barra-Costantini experimental house in Salisano. This is due to the concentration of the air return openings (required by users) into a single return damper for each air panel; originally dampers were distributed along the full length of the wall, giving a better air distribution. Furthermore, storage efficiency was diminished by negative thermal flow during the night from the front part of the ceiling.

## ACKNOWLEDGEMENTS

Manufacturer of the solar collector: Industrie Secco, Preganziol (TV) Italy

Architects: COOPROGETTO s.c.r.l. (A. De Luce, M. Mamoli, G. Scudo), via Calderari 9, 36100 Vicenza, Italy

Energy consultant: L.I.F.E. s.c.r.l. (T. Costantini), via Palladio 2 B casa solare Salisano, 02040 Salisano (RI) Italy; ing. G.C. Rossi, Marghera (VE)

Building owner: Cooperativa Marostica, via Consolaro 39, 36063 Marostica (VI), Italy

Chapter author: G. Scudo DI.Tec, Politecnico di Milano, via Bonardi 3, 20133 Milano, Italy

## BIBLIOGRAPHY

Barra OA (1981). I sistemi passivi piani. *La conversione fototermica dell'energia solare*, Etas Libri, Milan.

Barra OA, Costantini T (1979).Un prototipo di sistema passivo a parametri modulabili per la climatizzazione degli ambienti. *La Termotecnica*, **8**.

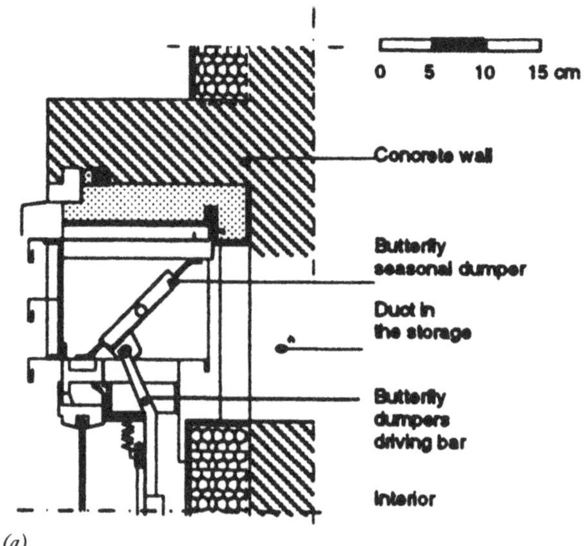


(a)

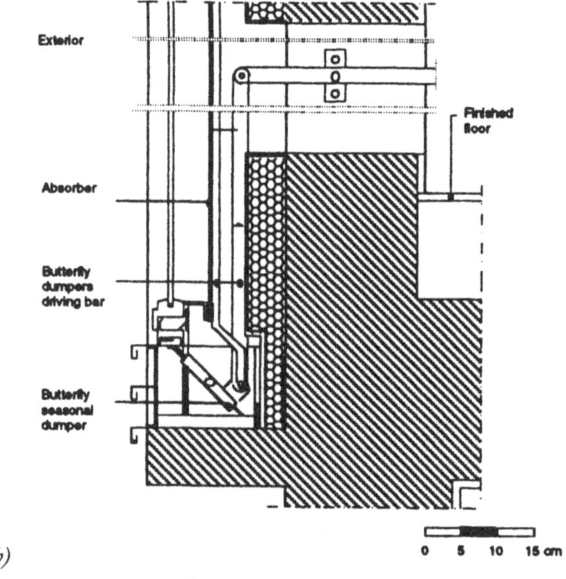


(b)

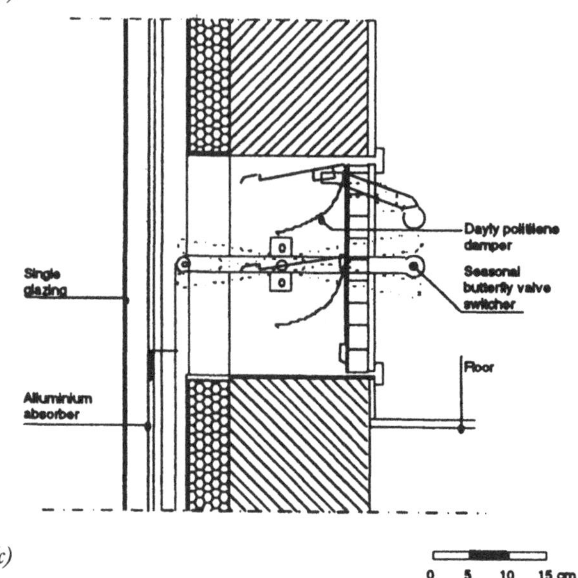


(c)

Figure III.2.9. (a) Upper seasonal damper; (b) seasonal butterfly damper; (c) daily butterfly damper

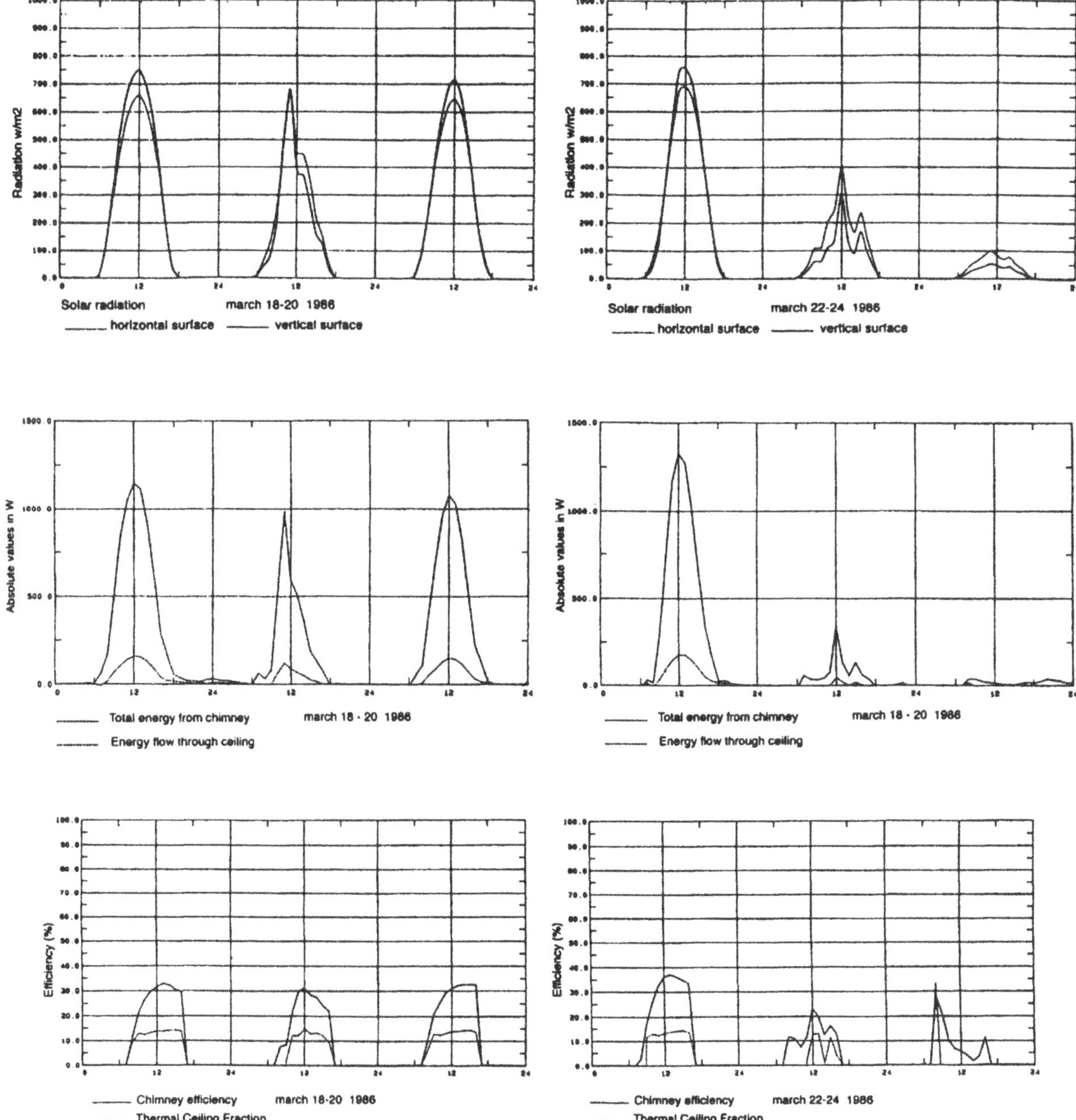


*Figure III.2.10. Energy performance of the system*

Barra OA, Pugliese Caratelli E (1979). A theoretical study of laminar free convection in I-D inducted flow. *Solar energy*, **23**(3), 211–215.

European Commission (1987). Marostica. *ProjectMonitor*, Issue **14**, December.

Izard JL (1982). Archi-Bio, *Architettura bioclimatiche*, Chapter 5, Section 4: Il sistema Barra-Costantini (a cura di G. Scudo). CLUP, Milan.

Scudo G (1984). Camino solare: la progettazione termoedilizia: quaranta alloggi solari passivia Marostica. *Modulo*, February 1984. BE-MA Editrice via Teocrito, 50 20128, Milan.

Scudo G (1988). *Costruire con il sole: progettazione e sperimentazione di tecnologie termoedilizie passive*. Rapporto finale di ricerca ENEA-CO.VE.CO. Milan, December. D. Malosti, ENEA-ERGSIRE, S.P.90, 00060 Santa Maria, Di Galeria, Rome.

Scudo G (1988). Tecnologia solare al controllo. *Modulo*, June 1980, BE-MA Editrice via Teocrito, 50 20128, Milan.

# III.3 Luino Apartments

## Motte, Italy

*Motte*

*System type 2*

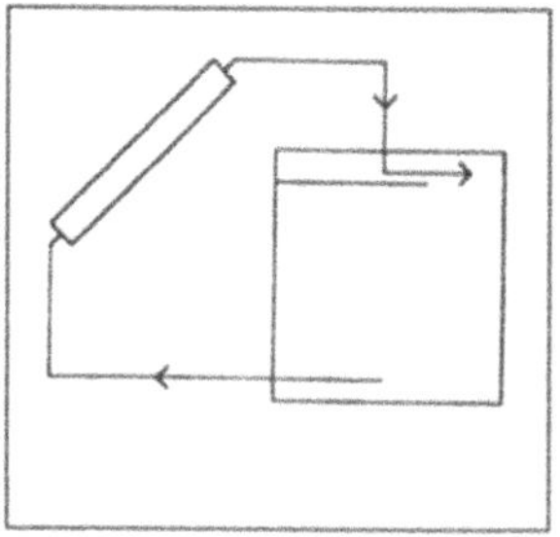

## PROJECT SUMMARY

These low-energy-consumption buildings are well insulated and incorporate a direct-gain solar air system, a greenhouse and a solar air system. The last of these is an open-loop system comprised of a backpass air collector, a concrete ceiling for storage and heat distribution and a new type of facade-integrated handling unit for ventilation and regulation.

The system has annual savings of 40% compared to a traditional building.

### Summary statistics

| | |
|---|---|
| System type, variation: | type 2 |
| Collector type: | opaque, facade-integrated |
| Collector area: | 5.4 m$^2$ |
| Storage type, capacity: | hypocaust, 4 kWh/K |
| Annual contribution of the solar air system: | 18.8 kWh/$m_{floor}^2$ |
| Basis: | monitored |
| Heated floor area: | 78 m$^2$ |
| Year solar system built: | 1994 |

## SITE DESCRIPTION

The site is located in Motte, a small village in northern Italy, about 2 km from Luino, a town close to Lago Maggiore. The site is fairly rectangular with the long axis running north-south and gently sloping at about 15° south-west. It has good solar access and is protected to the north by a hill:

| | |
|---|---|
| latitude | 46° N |
| longitude | 9° E |
| altitude | 200 m above sea level |

The area has favourable climatic characteristics (2340 degree days on the basis of 19°C and an average winter solar radiation of 1.7 KWh/M$^2$d).

## BUILDING PRESENTATION

The complex consists of four buildings: two buildings with two storeys side-by-side and two rows of single-family units, making a total of 20 units.

One of the apartments in the first two buildings was monitored during the 1995–96 season in order to establish the energy performance of the system. The plan of a typical apartment floor and a section of the building are given in Figure III.3.1.

Building statistics are given in Table III.3.1.

## SOLAR SYSTEM

This open-loop system incorporates 5.4 m$^2$ of solar air panel integrated into the south facade (Figure III.3.2). Air heated in the collector flows through the duct in the concrete ceiling and heats up the concrete structure. The hot air then enters the room, mixes with ambient

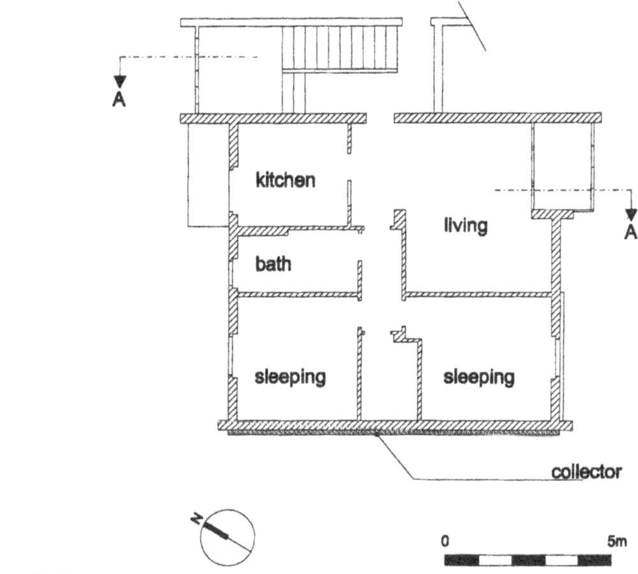


*(a) Typical floor plan*

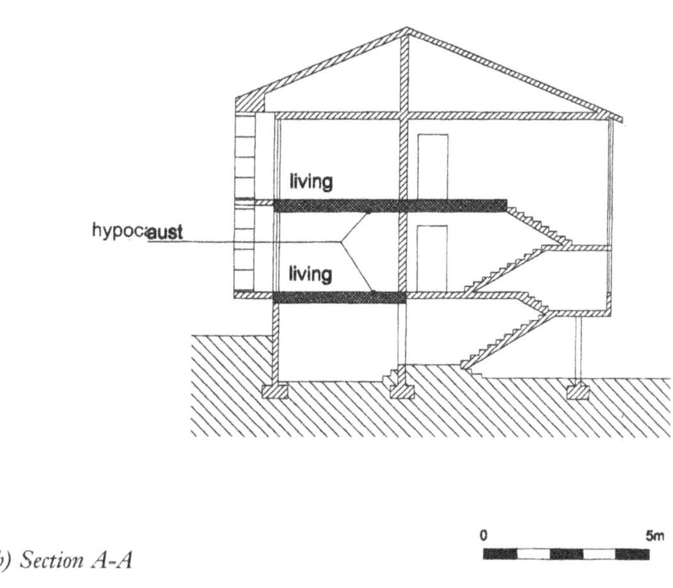


*(b) Section A-A*

*Figure III.3.1. (a) Plan of a typical apartment and (b) a section of the solar house*

*Table III.3.1. Building statistics*

| | |
|---|---|
| Gross heated floor area | 75 $m^2$ |
| Gross heated volume | 230 $m^3$ |
| Envelope specification | |
| floor: reinforced concrete with thermal insulation (30 cm) | $U$ = 0.72 W/$m^2$K |
| walls: mineral wood and concrete filled block (25 cm) | $U$ = 0.41 W/$m^2$ K |
| windows: double-glazed steel frame | $U$ = 3.35 W/$m^2$K |
| roof: insulated brick tile cover | $U$ = 0.37 W/$m^2$K |
| sun-space glazing | $U$ = 5.22 W/$m^2$K |
| Building global heat loss coefficient: | 0.76 W/$m^3$K |
| Calculated auxiliary space heating | 75 kWh/$m^2$a |
| Measured auxiliary space heating | 78 kWh/$m^2$a |
| Measured solar water heating | none |
| Appliances and lighting | 18 kWh/$m^2$a |

air and is convected back to the collector through return ducts in the floor (Figure III.3.3).

## The collector

Details of the collector are given in Table III.3.2. A single-glazed backpass collector has an absorber of

*Table III.3.2. The solar collector*

| | |
|---|---|
| Net collector area: | 5.4 $m^2$ |
| Annual solar contribution | 180 kWh/$m^2$ collector area |
| Air flow rate through collector: | 62 $m^3$/$m^2$h |
| Air flow rate through collector cross section: | 3240 $m^3$/$m^2$ |

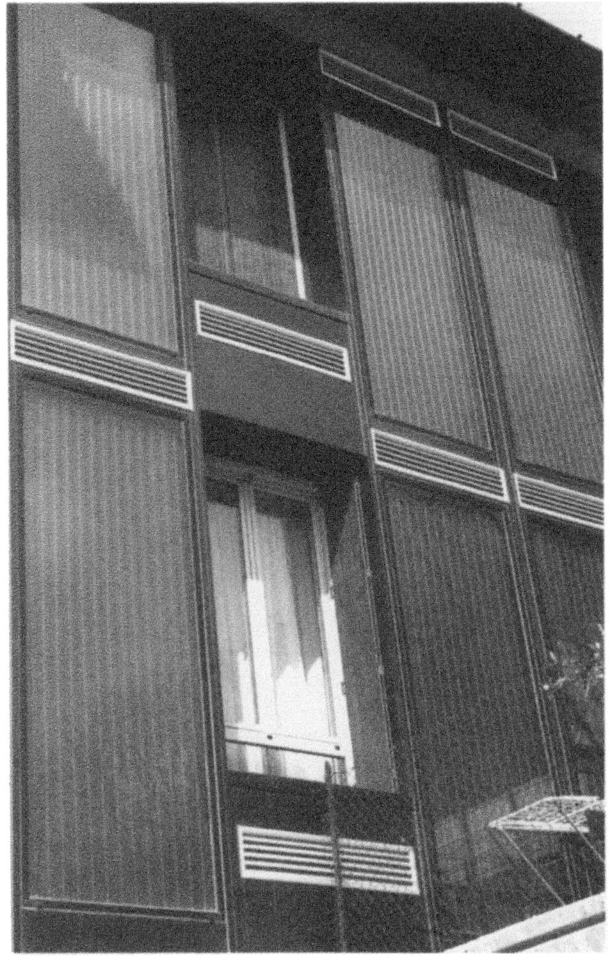

*Figure III.3.2. Solar collector integrated into the south facade*

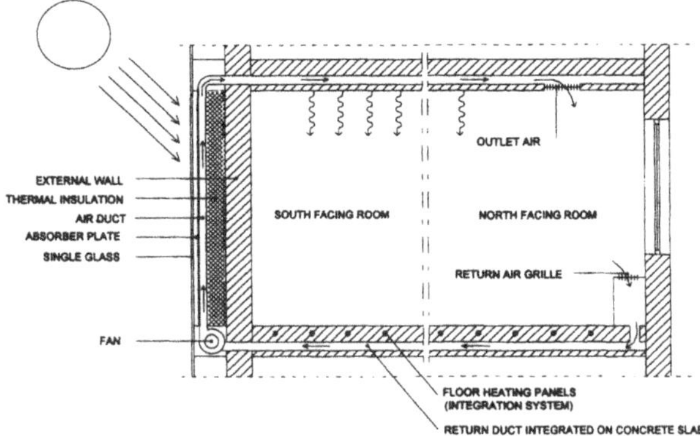


*Figure III.3.3. Schematic cross section of the solar system*

dark-green painted sheet steel. The collector (360 × 300 cm) is integrated into a solar facade and is composed of:

- a central part with window and handling unit;
- two lateral parts with solar air collectors each sized 1.20 m$^2$.

For the system design the following parameters were assumed:

| | |
|---|---|
| type | facade-integrated air collector; |
| glazing | single |
| absorber plate | dark-green painted aluminium sheet |
| $\alpha$, $\tau$ | 0.93, 0.88 |
| $U_L$ | 7 W/m$^2$ K |
| $F'$ | 0.72 |
| distance between the glass and the absorber: | 50 mm |
| thickness of the air-duct absorber plate: | 50 mm |
| thermal insulation material, thickness: | glass wool, 80 mm |

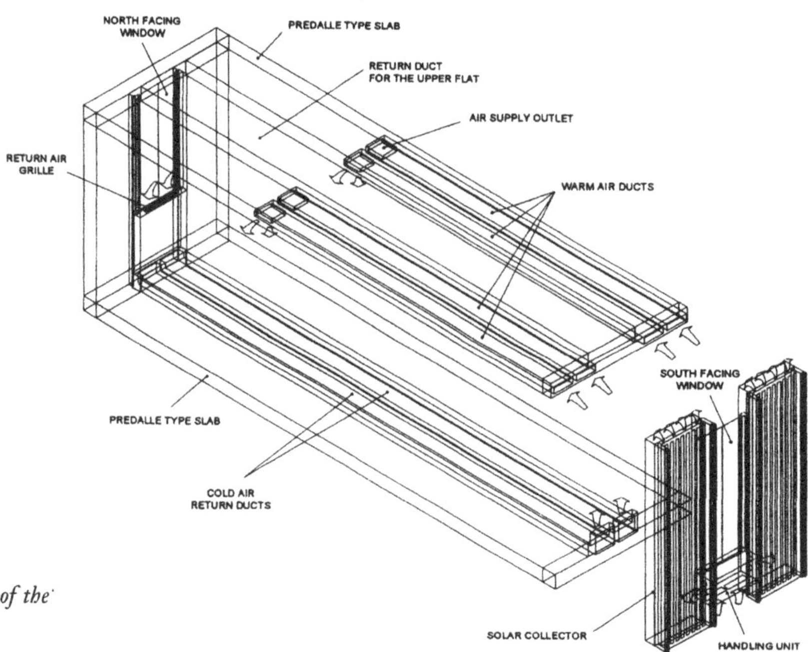


*Figure III.3.4. 'Open' axonometric view of the integrated solar system*

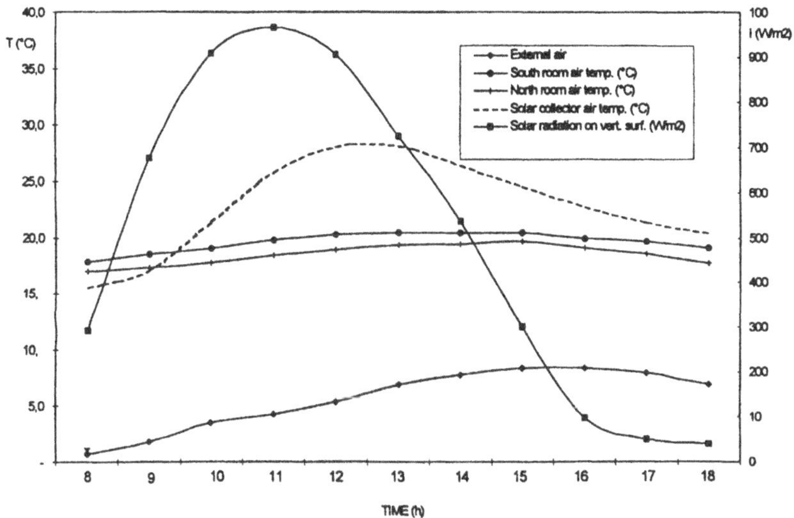


*Figure II.3.5. Solar radiation on the vertical plane and air temperatures recorded (12 March 1996)*

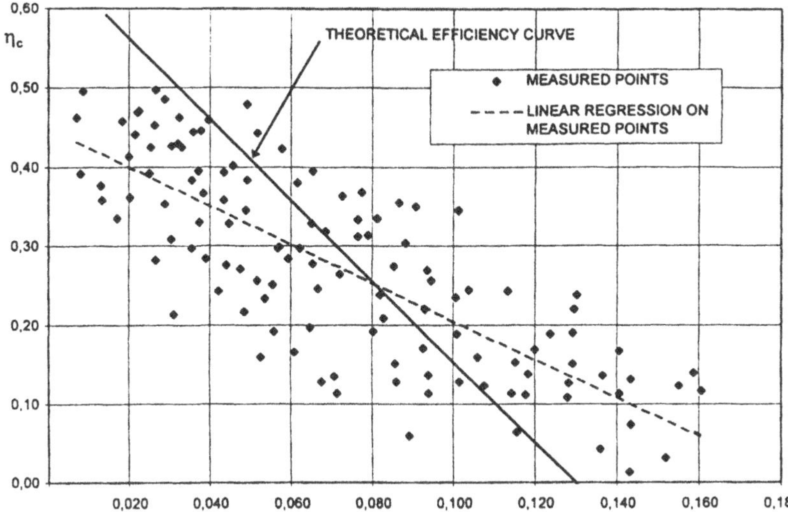


*Figure II.3.6. Diagram of the theoretical efficiency curve of the solar collector with the insertion of the monitored efficiency points*

*Table II.3.3. Fan characteristics*

| | |
|---|---|
| Fan power | 100 W |
| Consumption | 252 kWh/a |
| COP | 7.42 |
| Fan power/m² of solar air collector | 18.5 W/m² |

## Storage capacity and distribution

The primary storage is located in the concrete ceiling and is about 0.22 kWh/K per m² of solar collector, while the secondary storage, in the internal and perimeter walls, is about 0.13 kWh/K per m² of solar collectors. The total capacity is 3.64 kWh/K.

The collector-warmed air flows into the channels in the thermal ceiling or to the ambient air through vents. The air returns back to the handling units through two ducts integrated into the floor (Figure III.3.4). Fan characteristics are given in Table II.3.3.

## Controls

The solar system is regulated by an electronic control system specially designed for this project. During winter, the fans start if the temperature of the air in the solar collectors exceeds 25°C and stop when the ambient air temperature exceeds 25°C. The auxiliary heating system starts when the ambient air temperature is lower than 18°C.

## PERFORMANCE

The data obtained by the acquisition system were processed in order to determine, on an experimental basis, the performance characteristics of the system and of the solar collectors.

Figure II.3.5 shows the solar radiation on the vertical plane, as well as the external and internal temperature profiles recorded on 12 March 1996.

Examination of the internal air temperature curves shows that the system is effectively capable of maintaining internal acceptable comfort conditions. It is to be noted that the traditional plant set point temperature was 17°C.

The collector efficiency values calculated in this way were compared with the efficiency figures for the same operating conditions calculated using the theoretical curve shown in Figure III.3.6.

The collection efficiency calculated from the monitored data is typical of an air solar collector. However, the values do allow the instantaneous performance of the collector to be calculated. This is basically due to the fact that the monitored data are influenced by the behaviour of the building.

Consumption of natural gas burned by the back-up energy system in the building was also monitored. Seasonal consumption amounted to around 5,860 kWh, or an effective average saving of 40% compared with a traditional building.

## REMARKS

This building again demonstrates the energy-saving potential of this simple type of solar air system, operating in an open loop.

## ACKNOWLEDGEMENTS

Manufacturers of the solar collector: Coopsette div Infissi, Via A. Volta, 8 - S. Ilario D'Enza (RE), Italy

Architect: Giuseppe Costantini, Luino, Italy

Energy consultants: G. Solaini, B. Merello, G. Dall'O', Milano, Italy

Building owner: Alfieri, Luino, Italy

Chapter authors: G. Scudo, G. Dall'O', Politecnico di Milano, Italy

## BIBLIOGRAPHY

ASHRAE (1986). *Methods of testing to determine the thermal performance of solar collectors*. ANSI/ASHRAE Standard 93-1986.

Peck MK, Proctor D (1983). Design and performance of a roof integrated solar air heater. *Solar Energy*, **31**, 193

Solaini G , Dall'O' G, Carbonaro V (1996). *Multi-family solar houses in Luino, Italy*. 13th PLEA Conference, Louvain la Neue, Belgium.

# III.4 A Multi-Family Residential Building

Munich, Germany

*Munich*

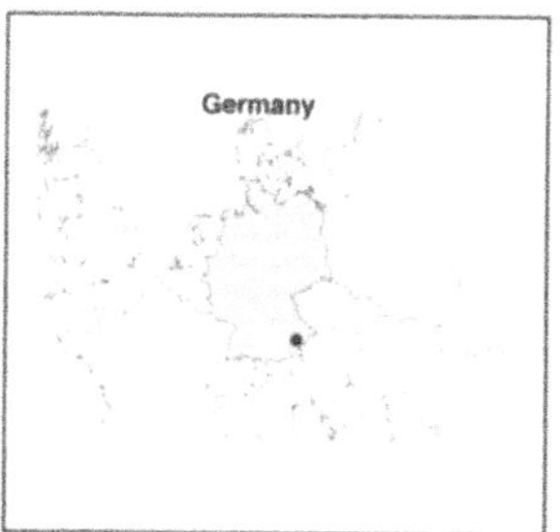


*System types 1 and 5*

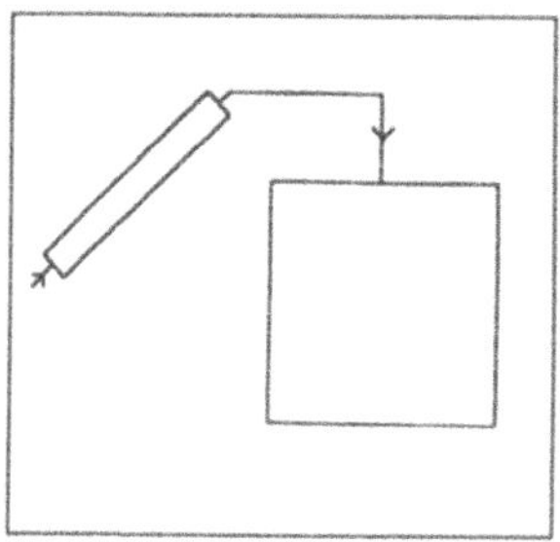

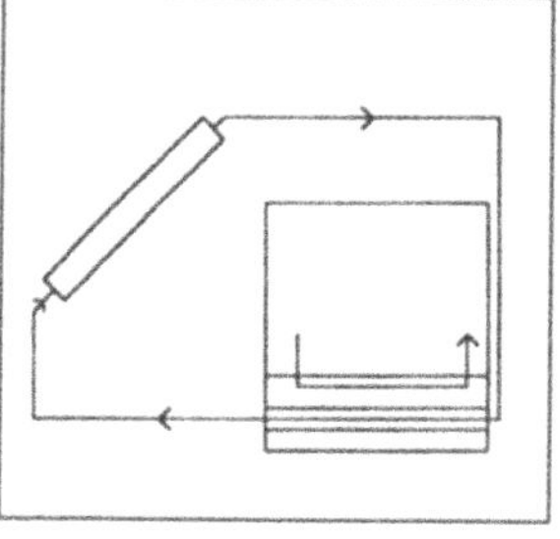

## PROJECT SUMMARY

This multi-family residential building in the inner city area of Munich is characterized by a compact architectural design and a high standard of thermal insulation. Three of the duplex flats, located on the fifth and sixth floors, have solar air systems, two with transparent insulation systems and one with a solar water system. To reduce the heating energy consumption, three of these flats are provided with solar air systems, two with transparent insulation systems and one with a solar-water system. In the first solar air system, the heated air is lead directly into the apartment. In the second, the air flows through a closed door, releasing the heat to the storage wall. The third system is a combination of both.

### Summary statistics

| | |
|---|---|
| System type: | type 5 |
| Total heating demand: | 54 kWh/m²a |
| Heated floor area: | 77.3 m² |
| Collector type: | air collector |
| Collector area: | 6.6 m² |
| Storage type: | concrete hypocaust |
| Storage capacity: | 1.0 kWh/K |
| | |
| System type: | type 1 |
| Total heating demand: | 37 kWh/m²a |
| Heated floor area: | 77.3 m² |
| Collector type: | air collector |
| Collector area: | 6.6 m² |
| Solar system output | |
| Basis: | monitored |
| Year solar system built: | 1996 |

## SITE DESCRIPTION

The multi-family residential building was built in the inner city area of Munich:

| | |
|---|---|
| latitude | 48°N |
| longitude | 12°E |
| altitude | 511 m above sea level. |

The annual mean ambient air temperature is 8°C, while it is 5°C for September to May. There are 4245 heating degree days on the base 20°C/15°C, annual. Horizontal global radiation, September to May, is 614 kWh/m²a (annual 1072 kWh/m²a).

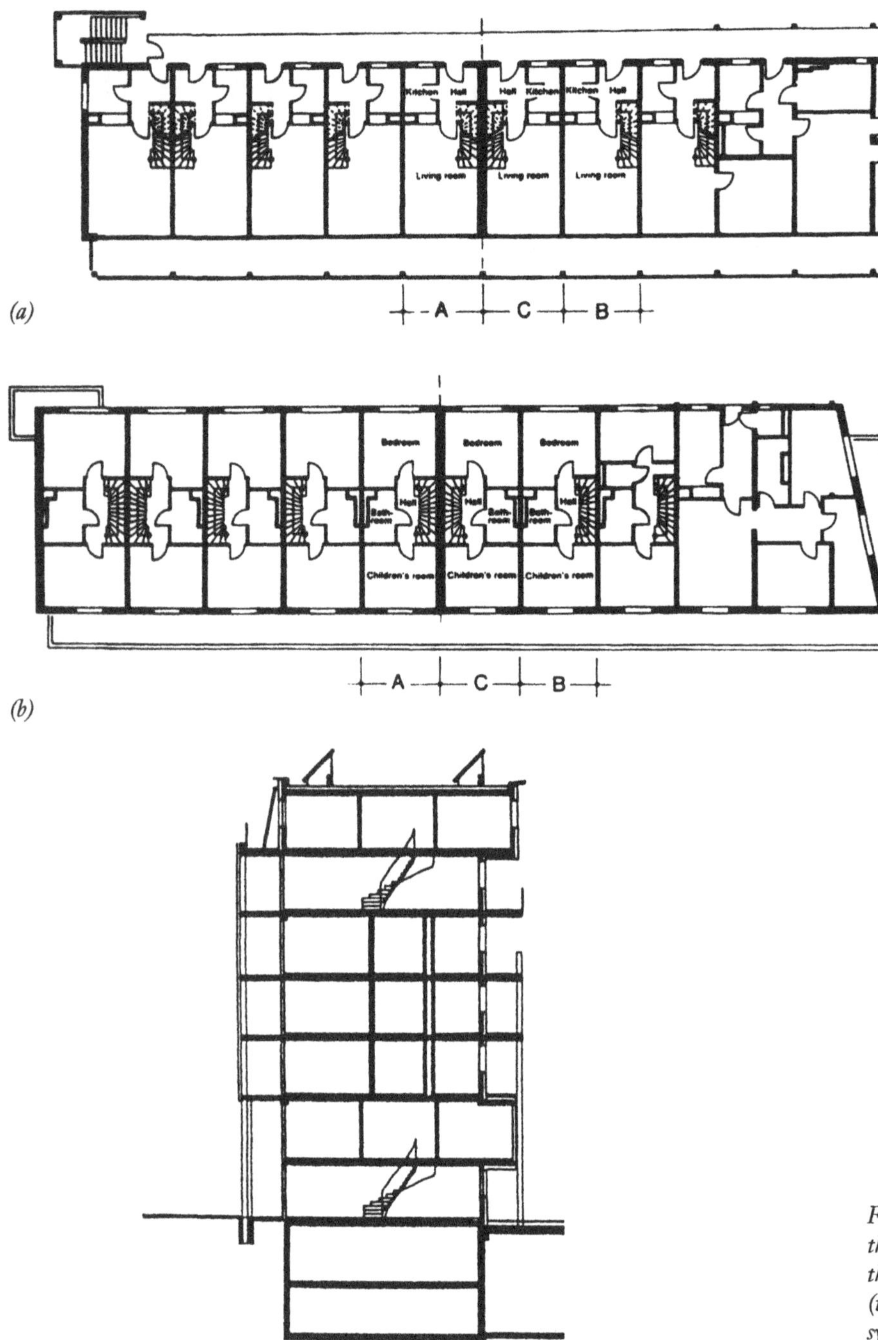


*Figure III.4.1. Ground plans of (a) the fifth floor and (b) the sixth floor of the multi-family residential building (the three apartments with solar air systems are indicated) and (c) a cross-section of the building*

## BUILDING PRESENTATION

The three apartments with solar air systems, A, B and C, are considered. The location of the apartments is shown in Figure III.4.1 (a, b). The heated floor area of each apartment is 77.3 $m^3$. The cross-section of the building presented in Fig. III.4.1 (c) shows the glazed balconies of the lower five storeys. The glazing protects the apartments from the heavy traffic noise emitted from the busy streets below on the building's east and west sides. Moreover, the glazing creates a buffer zone that reduces the heat losses. Figure III.4.2 shows the air collectors attached beside the windows on the front of the external wall of the sixth storey. The solar energy is stored in the fifth storey, because the bedrooms are located on the sixth floor

*Table III.4.1. U-values of the different parts of the building envelope*

| Building part | *U*-value (W/$m^2$K) |
|---|---|
| External wall | 0.27 |
| Basement ceiling | 0.28 |
| Roof | 0.20 |
| External window | 1.0 |
| Window to buffer zone | 1.5 |
| Window of buffer zone | 3.0 |

*Figure III.4.2. South view of the multi-family building*

The building is compact, with a surface to volume ratio of 0.36 $m^{-1}$. The building is much better insulated than required by the German regulations (WSVO) that were mandatory during the planning period. Table III.4.1 gives the *U*-values of the various parts of the building envelope.

The apartments are heated by radiators fed by a gas burner located in a central boiler room. Domestic hot water is heated by solar collectors and reheated – if necessary – by a high-efficiency burner. The apartment heated by solar system A has natural ventilation via the windows. The apartments with systems B and C are ventilated by fans during times of sufficient solar radiation. Here, the air is preheated in the air collector. During times without solar radiation the apartments have to be ventilated via the windows.

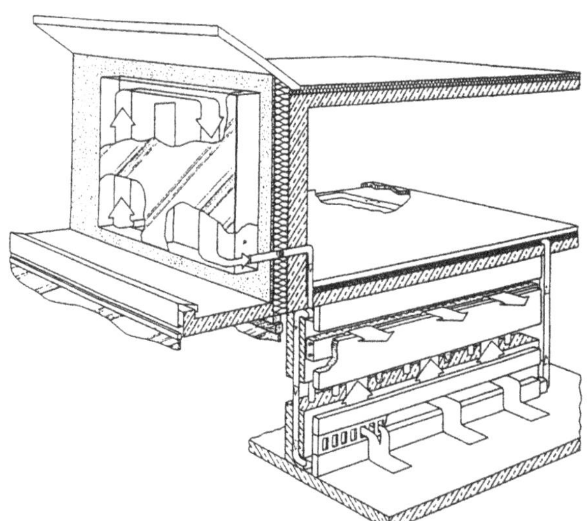

*Figure III.4.3. Isometric view of solar system A*

## SOLAR SYSTEM

The three apartments are provided with different solar air heating systems

### System A

In this system (Figure III.4.3), a solar air collector is in a closed loop with the storage. The air collectors are installed on the outside of the south-facing external wall of the bedroom on the sixth floor. The living room to which the solar energy is to be transferred is situated on the fifth floor. The internal partition wall between the flats provides the thermal storage.

In order to avoid the storage wall discharging when no heating is required in the living room, the storage wall is unloaded actively. For this, an insulated facing shell with slotted holes above and below stands 16 cm in front of the storage. Below the upper opening, a convector is installed. As soon as heating energy is required in the living room, the thermostat valve on the radiator and the lower flap of the facing shell are opened. The heated air between storage wall and facing shell ascends and exits through the upper opening into the living room. At the same time, cooler room air continues to flow through the lower opening. This free-convection cycle discharges the thermal storage.

### System B

In this solar air system (Figure III.4.4), the fresh air is preheated in the air collectors. As in system A, the air collectors are situated on the sixth floor on the front of the external wall of the bedroom. The fresh air is blown directly into the living room if the air temperature of the collector exceeds the indoor air temperature by approximately 5 K. To avoid uncomfortably high indoor air temperatures, the feedback control switches off the fan if the preset indoor air temperature, e.g. 23°C, is exceeded. Room air exits through a flap in the bathroom responding to pressure increase. The apart-

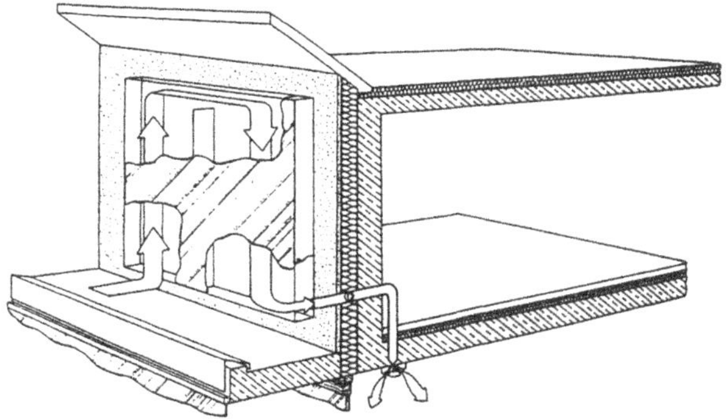

*Figure III.4.4. Isometric view of solar system B*

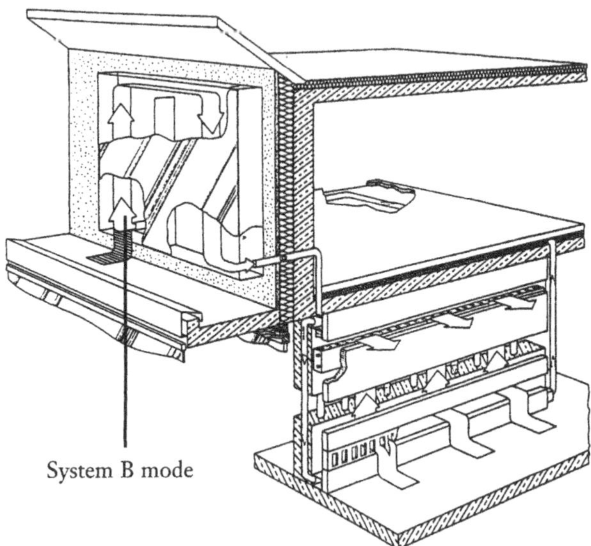


*Figure III.4.5. Isometric view of solar system C*

ment is equipped with radiators if the solar radiation supply is insufficient.

### System C

This solar air system (Figure III.4.5) is a combination of systems A and B. If solar energy is available and heat is required in the apartment, first the heated fresh air is blown into the living rooms. When no more heat is needed, the charging and discharging of the storage wall proceeds in the same way as in System A. The operating mode is shifted to storing operation by closing the fresh air flap at the collector and at the fresh air duct. During the charging of the storage, collector and storage wall form a closed system.

### The collectors

The systems A, B and C each have two air collectors, totalling 6.6 m$^2$ in area. Each collector unit has a width of 1.32 m, a height of 2.50 m, a depth of 0.16 m and is covered with 6 mm low-iron single glazing. The reverse side of the collector is insulated with an 90 mm-layer of mineral wool. To avoid having ducts outside the collector, the air ducts are integrated in the absorber. The absorber is profiled, each profile having a size of 50 mm/40 mm. The air volume rates are given in Table III.4.2.

### Storage

In systems A and C the solar energy absorbed in the air collector is stored in a concrete partition wall ($T \times H \times L$: 20 cm $\times$ 2.55 m $\times$ 3.55 m) between the flats on the fifth floor. In the partition wall are integrated 11 vertical sheet metal ducts 80 mm in diameter, with 0.28 m separation. In front of the storage wall, a square distribution manifold links the vertical ducts in the concrete wall. The air flows in a closed loop. The capacity is 1.0 kWh/K, 0.15 kWh/K per m$^2$ of collector area

*Table III.4.2. Air volume rates*

| | Air volume rate/collector area (m$^3$/m$^2$h) |
|---|---|
| System A: | 30 |
| System B: | 54 |
| System C: | 23 (in system A mode) |
| | 37 (in system B mode) |

### Distribution

Fans are installed in the exhaust and supply air ducts of the closed storage cycles A and C. System B has two supply air ducts, in each of which a fan is installed. In order to minimize noise, the fans are situated outside the living room in separate boxes beside the collectors. On the covering of the box, photovoltaic units are installed, supplying the fans with 24 V DC. There is no connection to the (mains) electrical supply of the building.

### Control

If the air temperature in the collector exceeds the storage temperature by more than 10 K, the fan is switched on. The power supplied to the fan is directly proportional to the solar intensity. Thus, with weak solar radiation, the fan runs more slowly, allowing the air to have more time to be warmed by the collector.

## PERFORMANCE

The building was completed and occupied at the end of 1996. In addition to the six apartments located on the fifth and sixth floors with solar systems, another five apartments were monitored and the results are presented in Table III.4.3.

The measurement results for systems A, B and C are compiled in Table III.4.4.

During the measurement period from 1 November 1996 to 31 May 1997 the solar radiation incident on the 6.6 m$^2$ collector surface totalled 3252 kWh. Out of this, 654 kWh (20%) could be loaded into the storage wall of System A.

Due to malfunction of some sensors, Systems B and C could not be measured continuously. Data gaps were calculated based on data measured so far. These computed results are 923 kWh for System B and 790 kWh for System C, corresponding to 28% (24%, respectively) of the incident radiation.

*Table III.4.3. Collector input and output*

Measurement period: Nov 1 1996 through May 31 1997

Solar system

| System | Input [kwh] | Output [kwh] | | |
|---|---|---|---|---|
| | total | total | per m$^2$ floor area | per m$^2$ collector area |
| A | 3252 | 654 | 8.5 | 99.1 |
| B | 3252 | 923 | 12.0 | 139.8 |
| C | 3252 | 790 | 10.2 | 119.7 |

*Table III.4.4. Collector energy inputs and outputs extrapolated for the 1996/97 heating season (based on measurements from 1 November 1996 to 31 May 1997)*

Extrapolated measurement period : Sept 1 1996 through May 31 1997

| | | Solar system | | |
|---|---|---|---|---|
| System | Input [kWh] | Output [kWh] | | |
| | total | total floor area | per m² | per m² collector area |
| A | 4233 | 840 | 10.9 | 127.3 |
| B | 4233 | 1197 | 15.5 | 181.3 |
| C | 4233 | 1019 | 13.2 | 154.4 |

To avoid overheating, the feedback control of System B switches off the fan if the preset indoor air temperature exceeds 23°C. It is to be noted that flat B is supplied with approximately 250 m³/h of air when the incident solar radiation reaches 1000 W/m². The resulting air change of 1.4 $h^{-1}$ is significantly higher than would be required for hygiene. Hence, the preheated fresh air also causes higher ventilation losses and the 923 kWh/a total in Table III.4.3 does not fully contribute to reduce the heating energy demand.

In System C, the fresh air is warmed first; then, once the indoor air temperature has reached 23°C, the control will shift to the closed circulation between the collector and storage wall.The portion of energy transferred directly to the room by way of warming fresh air depends on the preset indoor air temperature: the higher the set point, the lower the supply fraction from fresh air heating.

In Table III.4.4, the measured data of Table III.4.3 are extrapolated for the entire heating period. Calculations are based on radiation data obtained from a measuring station outside Munich.

The amounts of heating energy supplied to the three apartments by radiator heating are compiled in Table III.4.5. For the entire heating period, the energy consumption of each flat was extrapolated on the basis of heating degree days.

## RESULTS

- During the measurement cycle, indoor air temperatures were relatively high, e.g. flats A and C had an average indoor temperature of approximately 22.5°C and B averaged 21.8°C.
- The installed solar systems saved between 10 and 15 kWh/m²a of heating energy over the heating season.
- When fresh air heating can be fully used, the operating mode of direct air heating is more efficient than the closed loop via storage wall. Collector losses are lower since the outside air temperature level is lower than that of the air flowing back from storage.
- Preheating supply air by air collectors leads to a higher air change rate than necessary for hygiene. Thus, ventilation losses are also increased.
- The energy gained from preheating fresh air depends on the preset indoor air temperature. If the thermostat is set at a comparatively high temperature (21°C, for instance) preheating will only yield a small percentage of energy, since fresh air fans will be switched off at 23°C to avoid overheating.
- The d.c. fans which are powered by photovoltaic units are not connected to the electric supply of the building. Being solar powered they transport more air with increasing the solar intensity. Thus by weak solar radiation, the slower flow rate allows the air more time to be warmed in the collector.
- The apartment is not affected by flow noise, as the fans are installed directly at the collector output outside the flat.
- The entire system as built is too expensive and hence not yet cost-effective.

*Table III.4.5. Heating energy demand of the various apartments*

| Apartment | Measurement period: Nov. 1, 1996 through May 31, 1997 | | Extrapolated measurement period: Sept. 1, 1996 through May 31, 1997 | |
|---|---|---|---|---|
| | Energy demand | | Energy demand | |
| | kWh/a | kWh/m²a | kWh/a | Wh/m²a |
| A | 3511 | 46 | 4143 | 54 |
| B | 2398 | 31 | 2830 | 37 |
| C | 4009 | 52 | 4731 | 61 |

## ACKNOWLEDGEMENTS

Architects: Christian Raupach, Günther F. Schurk, Architekten BDA, Bauerstrasse 19, 80796 München, Germany

Development and scientific consulting: Fraunhofer-Institut für Bauphysik, Nobelstrasse 12, 70569 Stuttgart, Germany

Builder: GWG Gemeinnützige Wohnstätten- und Siedlungsgesellschaft mbH München, Sonnenstrasse 15, 80331 München, Germany

Funding: Bundesministerium für Bildung, Wissenschaft, Forschung und Technologie, PO Box 2006607, 53136 Bonn, Germany

Chapter authors: Johann Reiss and Hans Erhorn, Fraunhofer-Institut für Bauphysik, Nobelstrasse 12, 70569 Stuttgart, Germany

# III.5 Lützowstrasse Residential Building

Berlin, Germany

*Berlin*

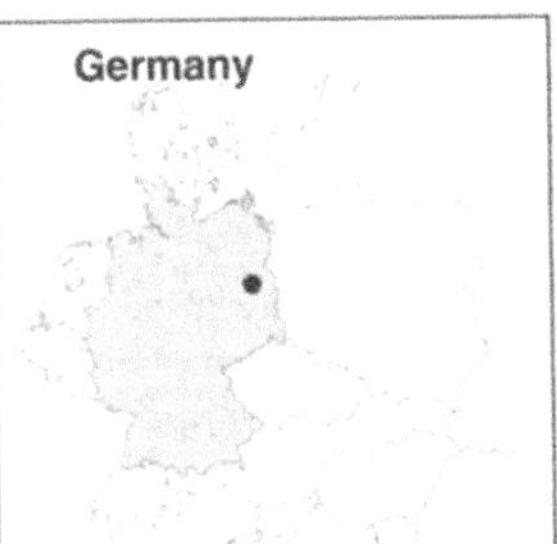


*System type 4*

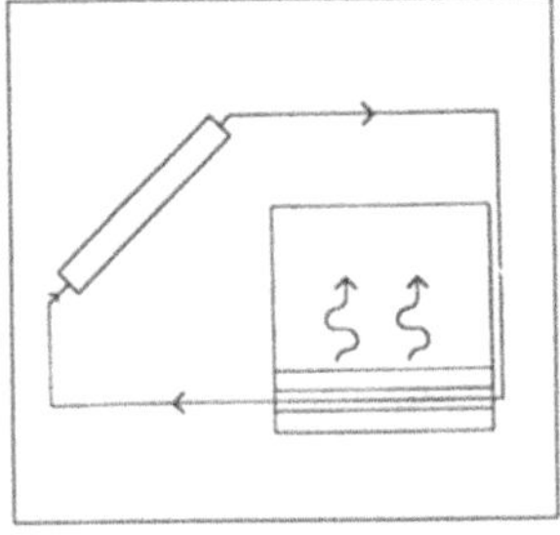

## PROJECT SUMMARY

Solar air collectors are integrated into the south facade of this apartment building in Berlin. The solar heat is distributed to the 31 apartments by floor hypocausts. Highly insulated walls and roofs minimize the heat losses.

### Summary statistics

| | |
|---|---|
| System type: | type 4 |
| Collector type: | facade-integrated solar air collector |
| Collector area: | 106 m$^2$ (total) |
| Storage type: | hypocaust |
| Annual contribution of the solar air system: | 12.6 kWh/m$_{floor}^2$ |
| Annual auxiliary heat consumption: | 39.6 kWh/m$_{floor}^2$ |
| Basis: | monitored |
| Floor area, volume (total): | 640 m$^2$,1696 m$^3$ |
| Year solar system built: | 1988 |

## SITE DESCRIPTION

The apartment building is in Berlin:

| | |
|---|---|
| latitude | 52°N |
| longitude | 13°E |
| altitude | 55 m above sea level |

The total vertical radiation for one year is 798 kWh/m$^2$. The horizontal radiation is 1024 kWh/m$^2$. The project has good solar access.

## BUILDING PRESENTATION

The building contains 31 dwellings, totalling 2474 m$^2$. Figure III.5.1 presents a plan of a typical floor and a section of the building. Figure III.5.2 shows the south facade of the building and the solar air collector and sun spaces.

## SOLAR SYSTEM

The system works as a closed charging loop with radiant discharge.

The solar air collectors are integrated into the south walls adjacent to the sun spaces. Each single-floor solar apartment has 9.3 m$^2$ of collector surface and each maisonette solar apartment has 18.6 m$^2$. Tubes embedded in the concrete floor of the solar apartments are connected to the collectors.

The construction of the solar-air-collector and hypocausts is shown in Figure III.5.3. Figure III.5.4 provides a schematic presentation of the whole system.

To simplify the illustration of the system, only two of the usual four tubes of the collector are shown in Figure III.5.4. The air coming from the ceiling reaches the collector at point (1) and leaves the collector at point (2) after being warmed up by the rib absorbers.

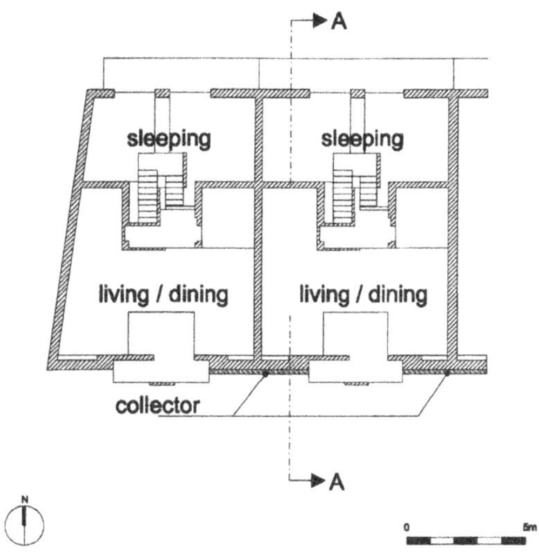


*(a) Typical floor plan*

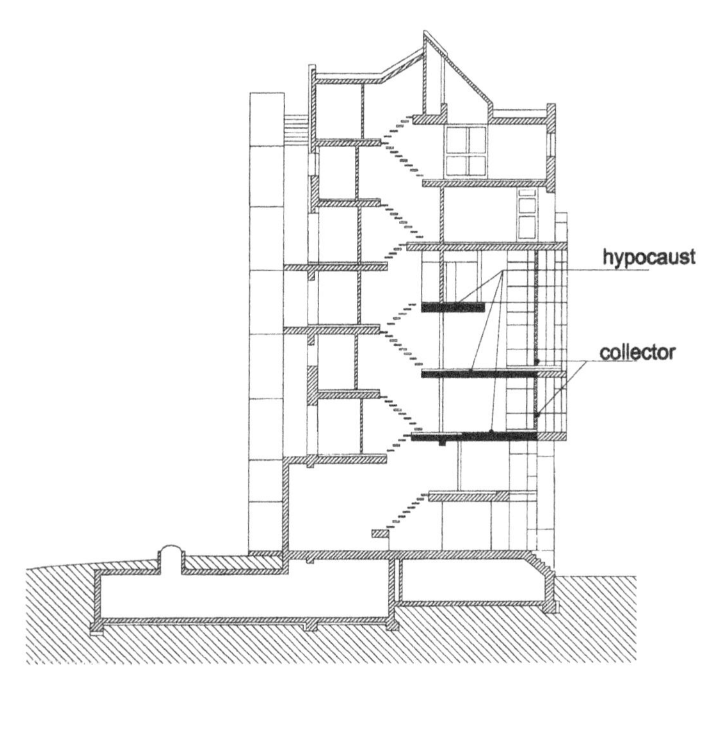


*(b) Section A-A*

*Figure III.5.1. (a) Plan of a typical floor and (b) a section of the building*

The heated air is pushed through the hollow tubes of the concrete ceiling and warms it (3). The concrete stores the heat for three to four hours and then the heat is delivered to the rooms in the evening, when it is needed. A fan (4) forces the air through the whole system.

Although the solar heating is very efficient, on sunless days conventional heating is still needed. The system is automatically controlled. As soon as the collector is 5 K warmer than the air inside the tubes, the fans start. The air flow through the collectors varies between 218 and 473 $m^3/h$. At maximum speed the ventilator power is 50 W.

## PERFORMANCE

The building was monitored during the heating period of 1988/89, ending in May 1989. The approximate 30

*Figure III.5.2. South facade of the building, street side*

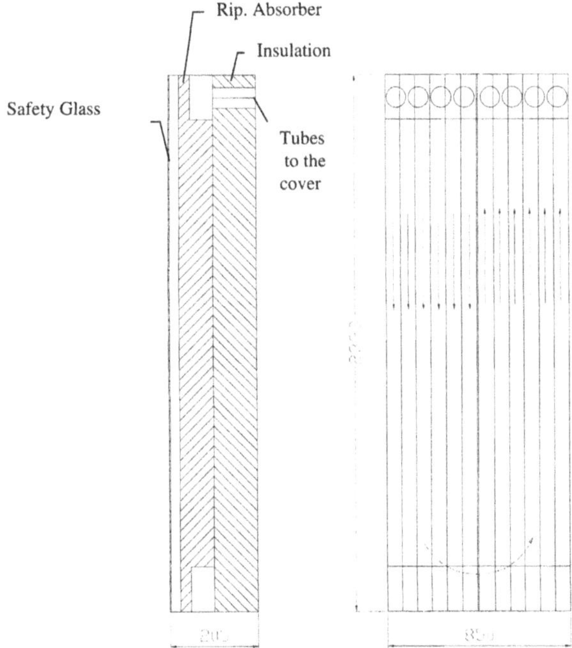


*Figure III.5.3. Construction of the Grammer solar air collector*

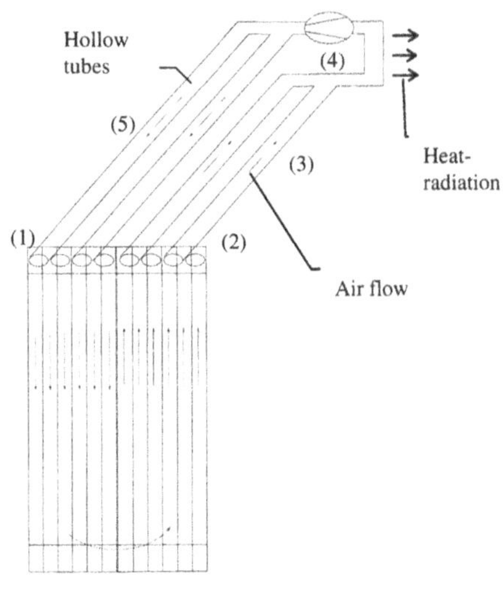


*Figure III.5.4. Schematic presentation of the solar heating system*

kWh/m² consumption of the solar apartments is 60% less than that of the reference apartment.

The atrium apartments, due to their location on the top of the building, have higher summer temperatures and lower winter temperatures than other apartments (see Figure III.5.5). The measured values clearly indicate that the different solar systems reduce auxiliary heating demand substantially, although the systems sometimes did not work as well as expected.

## REMARKS

From the two-year monitoring period the following observations are possible:

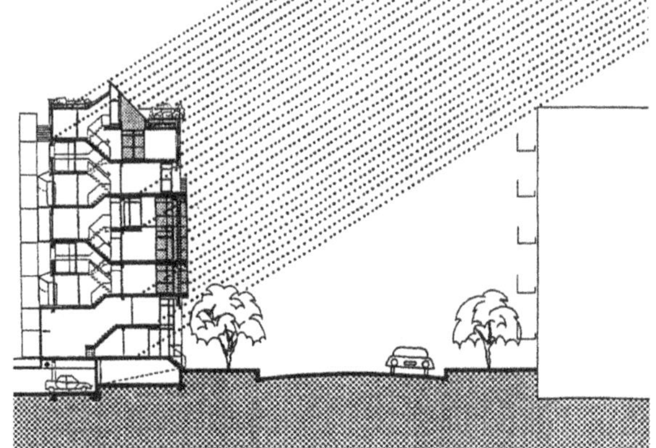

*Figure III.5.5. Schematic view of the building showing the influence of the house on the other side of the road*

- The solar apartments require up to 57% less auxiliary heating energy than the reference apartment.
- Between 20% and 33% of the radiation striking the solar air collector during the winter months was converted into heat. This heat was directly used for heating by fan-forcing it through the tubes of the hypocaust system.
- The electricity needed for the fans (6.8% of the thermal energy) can be reduced by optimizing the control system. Individual user manipulation should be avoided.
- Inadequate servicing of the fans greatly reduced the performance of thc collector.
- There was no overheating of the solar apartments during the summer.

## ACKNOWLEDGEMENTS

Manufacturers of the collector: W. Grammer KG, Wernher von Braun Strasse 6, D-92224 Amberg, Germany

Architects: IBUS (Institut für Bau-, Umwelt-, Solarforschung, Berlin, Germany)

Funding: BMBF (Bundesministerium für Bildung, Wissenschaft, Forschung und Technologie, Germany)

Analyses: Fraunhofer - Institut für Bauphysik, Nobelstraße 12, 70569 Stuttgart, Germany

Chapter author: Siegfried Schröpf, W. Grammer KG

## BIBLIOGRAPHY

Reiss J., Erhorn H., Stricker R (1991). *Passive und hybride Solarenergienutzung im Mehrfamilienwohnhausbau.* Messergebnisse und energetische Analyse des deutschen IEA-Task VIII-Gebäudes WB 64/1991.

W. Grammer KG (1987). *Projektunterlagen zum Bauvorhaben Berlin, Lützowstrasse.* W. Grammer KG, Wernher von Braun Strasse 6, 92224 Amberg, Germany

BINE (1990). *Energiespargebäude mit hybrider und passiver Sonnenenergienutzung.* BINE, Mechenstrasse 57, 53129 Bonn, Germany.

IBUS (1991). *Passive und hybride Solarenergienutzung im innerstädtischen Wohnungsbau. Demonstrationsprojekt Solarhaus Lützowstrasse 5 Berlin – Tiergarten.* Institut für Bau-, Umwelt- und Solarforschung GmbH, Caspar Theyss Strasse 14a, 14193 Berlin, Germany.

# III.6 Toftegård Multi-family Houses

Herlev, Denmark

*Herlev*

*System type 6*

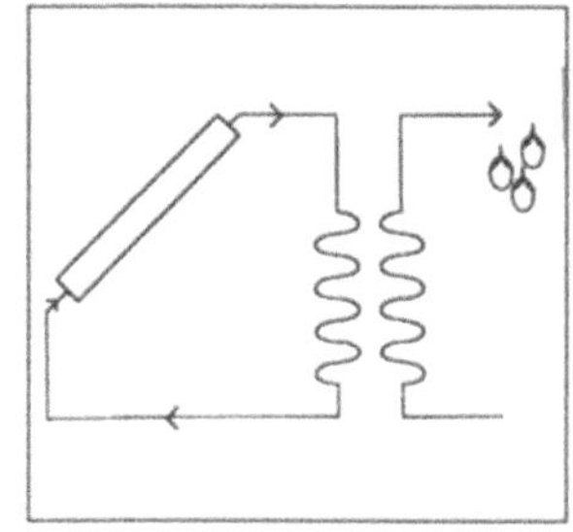

## PROJECT SUMMARY

Two roof space collector systems for preheating domestic hot water have been installed during the renovation of the Toftegård multi-family houses built in the early 1950s. As part of the renovation, instead of renovation of the old roofs, a new floor of apartments was erected on top of the old buildings. The renovation increased the living area by 25% without increasing the heating and water demand, thanks also to other energy-saving measures in the existing dwellings. No rent increase for the occupants of the older apartments has been necessary.

The roof-space solar air collectors cover 33% of the demand for domestic hot water for 184 apartments. The actual performance of the systems is as expected. The economies are the same as for solar heating systems with liquid collectors in Denmark. Toftegård received the environmental award of the Federation of Non-profit Housing in Denmark in 1994.

### Summary statistics

| | |
|---|---|
| System type: | type 6 |
| Collector type: | two roof-space collectors, 169 and 231 m$^2$ |
| Storage type, volume, capacity: | water, 6 and 7.5 m$^3$, 6.9 and 8.9 kWh/K |
| Annual contribution of the solar air system: | 175 kWh/m$^2_{collector}$<br>4.4 and 5.5 kWh/m$^2_{floor}$ |
| Annual auxiliary heating DHW and circulation: | 35 kWh/m$^2_{floor}$ |
| Basis: | monitored |
| Heated floor area: | 6680 m$^2$ and 7285 m$^2$ |
| Year solar system installed: | 1992 and 1993 |

## SITE DESCRIPTION

Toftegård is situated in Herlev, approximately 20 km east of Copenhagen, Denmark:

| | |
|---|---|
| latitude | 56°N |
| longitude | 12°E |
| altitude | 25 m above sea level. |

There is a temperate coastal climate.

## BUILDING PRESENTATION

The Toftegård area consists of 16 multi-family buildings from the early 1950s with 547 dwellings. The four-storey buildings are mainly constructed of yellow bricks, except for the new fourth floors, which are light-weight constructions of insulation within a wooden framework, plus an outer skin of grey fibre-reinforced concrete sheets. A roof-space collector system has been installed in the attic of two of the new floors. A section and a floor plan of the building with the roof space collector of 169 m$^2$ are shown in Figure III.6.1.

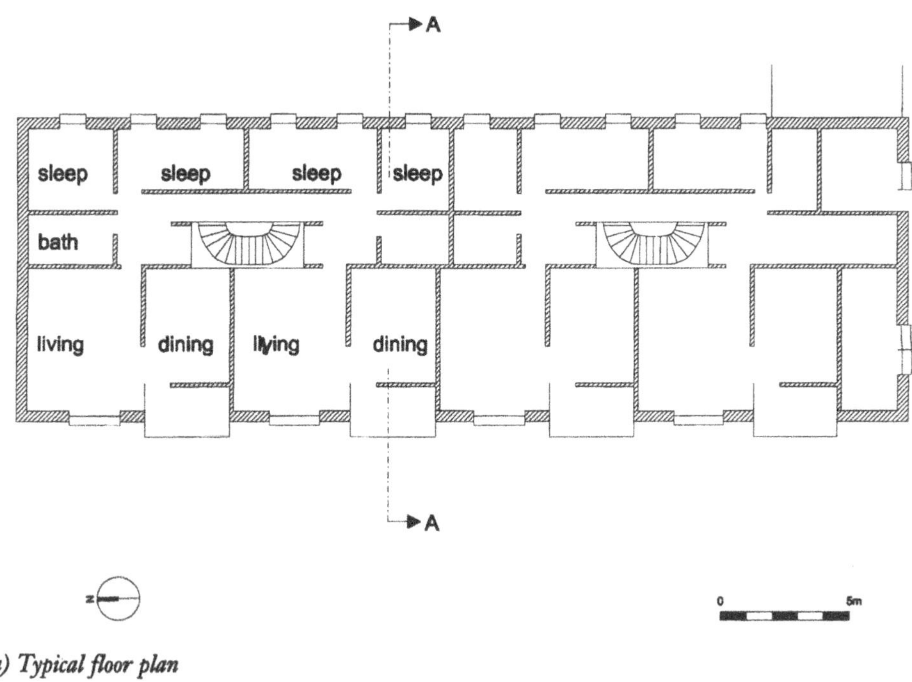


*(a) Typical floor plan*

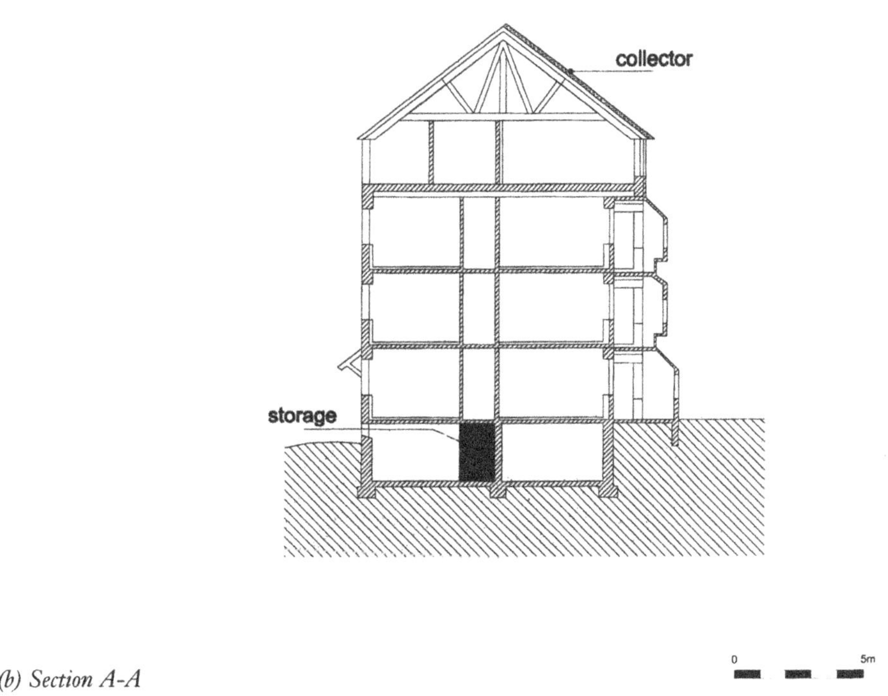


*(b) Section A-A*

*Figure III.6.1. (a) Plan and (b) section of the building with the small roof-space collector*

The new 112 dwellings on the fourth floor are low-energy dwellings with low-energy windows (double-glazed sealed units with low-emission coating), 200 mm of insulation in the walls and 300 mm on the ceiling. The space heating demand per m² of the new apartments is only 30% of the existing apartments.

The annual space heating demand in Toftegård is 4,900 MWh, while the annual domestic hot water demand including losses in the circulation circuits is 2,100 MWh/a.

## SOLAR SYSTEMS

The solar heating systems are roof-space collector systems, where the attics are utilized as solar collectors. The heat from the roof-space collectors is transferred

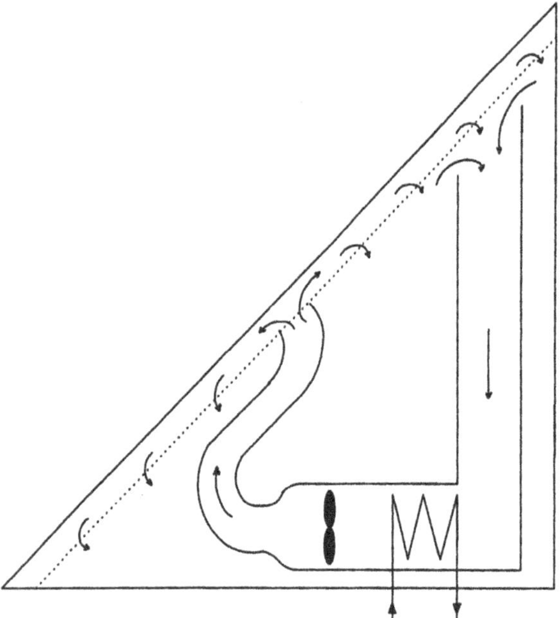

*Figure III.6.2. The principle of the roof-space collector systems in Toftegård*

to liquid-based domestic hot water systems by air-to-water heat exchangers located in the roof-space collectors.

## The system

The principle of the roof-space collector systems is shown in Figure III.6.2. The attic of the buildings is divided into two parts, a west-facing part being the roof space collector, while the east-facing part is a traditional attic. The traditional roof of the attic in the collector part has been replaced by a transparent cover allowing the solar radiation to penetrate into the attic. The solar radiation is absorbed in a black fibre cloth mounted behind the cover and air is heated when drawn through the fibre cloth, providing good heat transfer. The heated air is then forced through an air-to-water heat exchanger, where the heat is transferred to liquid and transported to a traditional hot-water storage system. After leaving the air-to-water heat exchanger, the air is returned to the space between the cover and the fibre cloth. In this way, the coldest part of the roof space is the space behind the cover, thereby reducing heat losses.

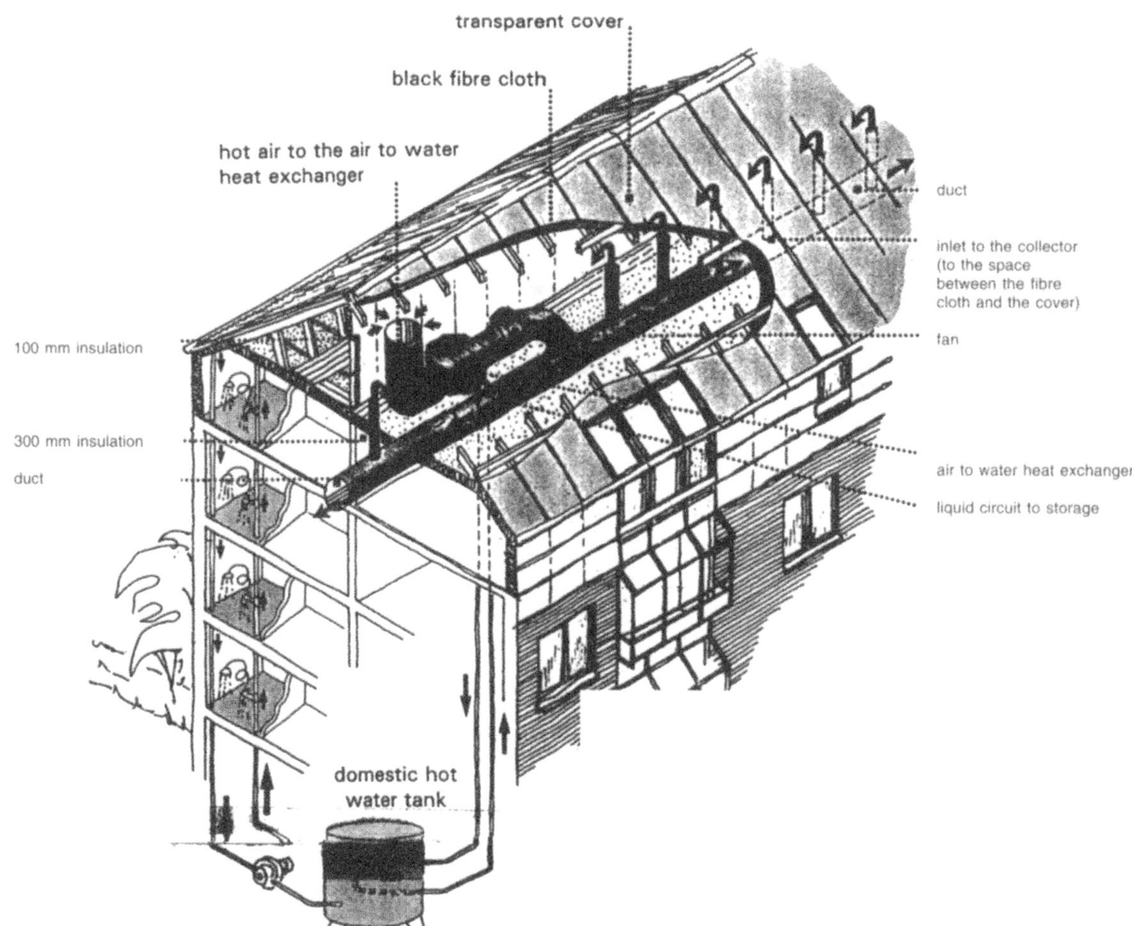


*Figure III.6.3. Isometric drawing of the heating system*

## The solar collectors

The collectors are located in the west-facing part of the attic of two of the buildings. The glazing areas of the collectors are 169 and 231 m². The collectors face west (84° from the south). The tilt of the cover is 37°.

The west-facing part of the attics is separated from the east-facing parts by an insulated wall with 100 mm mineral wool. The cover consists of triple-walled transparent UV-stabilized polycarbonate (of thickness 16 mm and length 5.3–6.7 m). The fibre cloth absorber is mounted approximately 100 mm behind the cover. The cloth is 2 mm thick and made of polypropylene.

The lower part of the collectors (11% and 14% of the collector area) – located below the attic between the roof windows of the dwellings (see Figures III.6.1 and III.6.3) – is less ventilated because of the smaller air gap behind the absorber. This is estimated to reduce the efficiency of these parts of the collectors by 20–40% compared to the rest of the collectors.

## Storage

Water tanks with built-in heat exchangers store the solar heat in the basement of the buildings (Figure III.6.3). The sizes of the tanks are 6 and 7.5 m³ with a diameter of 1.8 m and heights of 3 and 3.8 m. These are only preheating tanks. The domestic hot water is, if necessary, post-heated by district heating in the existing domestic hot water tanks.

## Distribution

The distribution consists of two circuits: an air circuit and a liquid circuit.

*Air circuit*

The principle of the air circuit is shown in Figure III.6.2, while Figure III.6.4 shows the actual air circuit in the smaller collector and Figure III.6.5 the larger collector. The main characteristics of the air circuit are shown in Table III.6.1.

The air is taken from the top of the roof space collectors and drawn through the air-to-water heat exchanger (a tube–fin-type heat exchanger) by a fan situated on the 'cold' side of the heat exchanger. The fan blows the air into the space between the cover and the fibre cloth through several smaller ducts as shown in Figures III.6.3 and III.6.4. In order to save electricity two-speed fans have been used. The two flow rates through the absorber are 15.5 and 31 m³/m²h. The flow rate through the collectors is controlled by the temperature difference between the bottom of the tank and the inlet temperature to the air-to-water heat exchanger.

The ventilation system of the larger collector has been optimized on the basis of the experience with the smaller collector. This has resulted in a lower energy demand for the fan of this system.

*Liquid circuit*

The liquid circuit is similar to traditional collector circuits in liquid-based solar heating systems with a pump and a submerged heat exchanger in the storage. The flow rates are respectively 1.8 and 3.3 m³/h, equivalent to 10.7 and 14.3 litres/m²h, for the smaller and the larger collector. The liquid is antifreeze-protected water with 10 wt% propylene glycol. The heat transfer coef-

*Table III.6.1. The main characteristics of the air circuit in the two roof-space collectors*

| | Smaller collector | Larger collector |
|---|---|---|
| Diameter of fan | 560 mm | 630 mm |
| Power of fan | 0.7 kW | 1.1 kW |
| Energy demand - fan + pump | 3,450 kWh/a | 3,000 kWh/a |
| Flow rates | 2,500 and 5,000 m³/h | 3,750 and 7,500 m³/h |
| Heat exchanger | 40 kW at 90–60/ 65–38°C | 72 kW at 74–46/ 49–29°C |

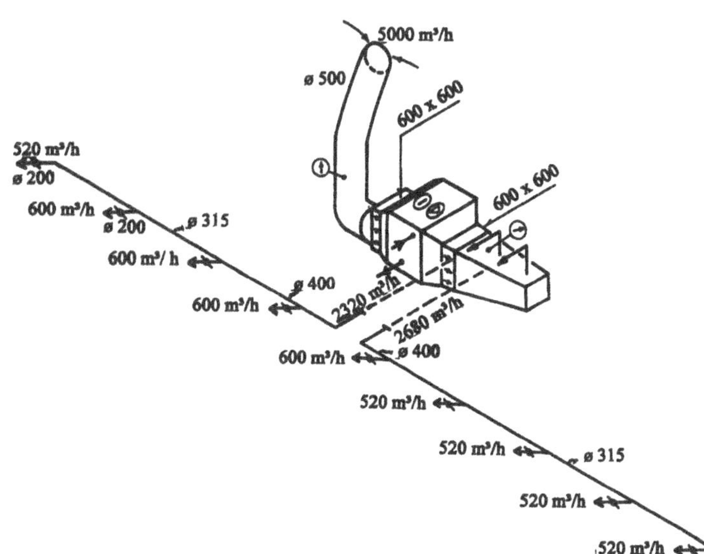


*Figure III.6.4. The air circuit in the smaller roof-space collector*

*Figure III.6.5. The ventilation unit in the larger roof-space collector*

ficient for the submerged heat exchangers has been calculated to be 4,200 and 5,000 W/K respectively for the systems with the smaller and larger collector.

### Control

The system is controlled by a differential thermostat function embedded in the computer-based control and monitoring system of the buildings. The systems respond according to the temperature difference between the bottom of the storage and the air inlet to the air to water heat exchanger, as follows:

| | |
|---|---|
| $\Delta T > 6K$: | The pump of the liquid circuit starts and the fan starts at half speed. |
| $\Delta T > 15K$: | The fan is switched to full speed. |
| $\Delta T < 10K$: | The fan is switched back to half speed. |
| $\Delta T < 3K$: | The pump and the fan stop. |

To prevent the roof space getting too hot, ambient air is blown through the roof space when the temperature exceeds 92°C until it falls below 87°C.

## PERFORMANCE

The systems have been monitored for two years, 1994/95, as part of an EU demonstration project. However, only the energy delivered to the storage from the collectors has been measured – not the energy delivered from the storage to the domestic hot water systems.

### Measured performance

The yearly delivered energy to the storage is 35,000 kWh and 47,000 kWh for the two systems. This is what the simulations predicted during planning.

The heat loss from the two tanks is estimated at 5,500 and 6,700 kWh/a. The performance of the two solar heating systems is thus 29,500 and 40,300 kWh/a, equivalent to 175 $kWh/m^2{}_{collector}$ a. This is a rather low performance compared to traditional liquid-based solar heating systems for preheating domestic hot water. The three main reasons for the low performance are:

- The collectors face west (84° from the south). This alone accounts for almost half of the reduction in performance.
- The part of the collectors between the roof windows of the dwellings on the new fourth floor has a lower efficiency.
- There are two heat exchangers in the collector circuit.
- The roof-space collectors are not entirely airtight. It has been measured that approximately 10% of the air circulated in the collectors is replaced by ambient air. This reduces the start efficiency of the collectors by 6–7% and increases the heat loss by 10%.

### Improvements

Calculations have shown that the performance of the systems could be improved by up to 25% if the fresh air to the buildings were also preheated by the roof-space collectors.

## REMARKS

An evaluation of the roof-space systems sponsored by the Danish Ministry of Energy has confirmed that the systems perform as expected. Inspections of the roof-space collector systems have revealed no degeneration of the main components of the collectors: cover, fibre cloth, fans, heat exchangers or the wooden construction after respectively three and four years of operation. This is impressive, given that the components are exposed to temperatures of up to 100°C in the summer and to a temperature swing of up to 80 K between day and night.

The systems costs rank in the lower range of traditional liquid-based solar heating systems in Denmark.

## ACKNOWLEDGEMENTS

Architect: KBI Consultants A/S, Copenhagen, Denmark
Energy consultants: Niels Radisch; KBI Consultants A/S, Copenhagen, Denmark; Søren Østergaard Jensen, DTI Energy, Taastrup, Denmark
Construction: Enemærke Petersen, Ringsted, Denmark
Building owner: Herlev 44, Herlev, Denmark
Chapter author: Søren Østergaard Jensen, DTI Energy, Taastrup, Denmark

## BIBLIOGRAPHY

Jensen SØ (1996). *Evaluation of the roof space systems in Toftegård, Herlev* (in Danish). Solar Energy Laboratory, Danish Technological Institute. Gregersensvej, PO Box 141, DK-2630 Taastrup, Denmark.

# III.7 Havrevangen Project

Hillerød, Denmark

*Hillerød*

*System type 6*

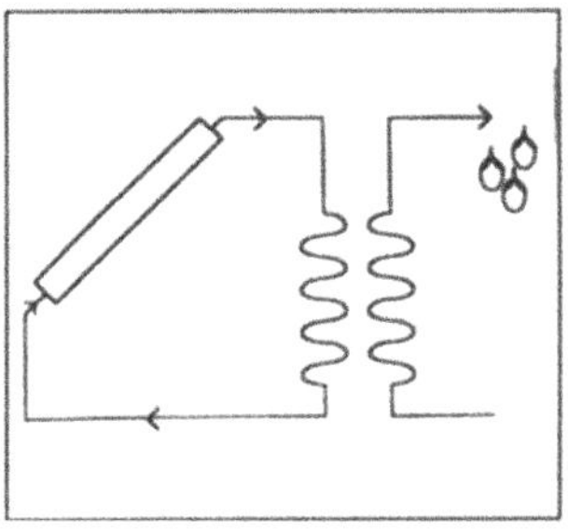

## PROJECT SUMMARY

A saving of 55% on the heating bill was achieved with these 50 solar low-energy dwellings. The project includes:

- air solar collectors and air-to-water heat exchangers;
- active solar room and domestic hot-water heating using:
  - high insulation levels, low *U*-value glazing;
  - heat recovery for ventilation air;
  - hydronic floor heating;
  - hybrid solar storage integrated in the floor construction;
  - low-temperature district heating;
  - DHW heating in the summer months by the air solar system.

### Summary statistics

| | |
|---|---|
| System type, variation: | type 6 |
| Collector type, area: | opaque, roof-integrated, 376 $m^2$ |
| Storage type, capacity: | water, concrete floors, 17.5 kWh/K, 3.5 kWh/K |
| Annual contribution of the solar air system, including DHW: | 21 $kWh/m^2_{floor}$ |
| Annual auxiliary heat consumption: | 80 $kWh/m^2$ |

| | |
|---|---|
| Basis: | monitored/calculated |
| Heated floor area: | 4300 $m^2$ |
| Year solar system built: | 1994 |

## SITE DESCRIPTION

The Havrevangen project is located on the outskirts of Hillerød, west of Copenhagen. Because of the topography and the lack of trees, the site is fairly exposed to wind. The row houses are fully exposed to the sun and placed a good distance apart in order to avoid mutual shading. The climate, which varies little throughout Denmark, is coastal with mild winters and relatively cool summers:

| | |
|---|---|
| latitude | 55°N |
| longitude | 12°E |
| altitude | 14 m above sea level |

## BUILDING PRESENTATION

The 50 houses are placed in five rows with 8–12 houses in each row, as shown in Figure III.7.1. The rows are 1½ and two storeys high.

The 50 dwellings comprise 18 two-room, 26 three-room and six four-room apartments. A variety of floor plans have been worked out for the apartments. One of these is shown in Figure III.7.2, together with a section of an apartment.

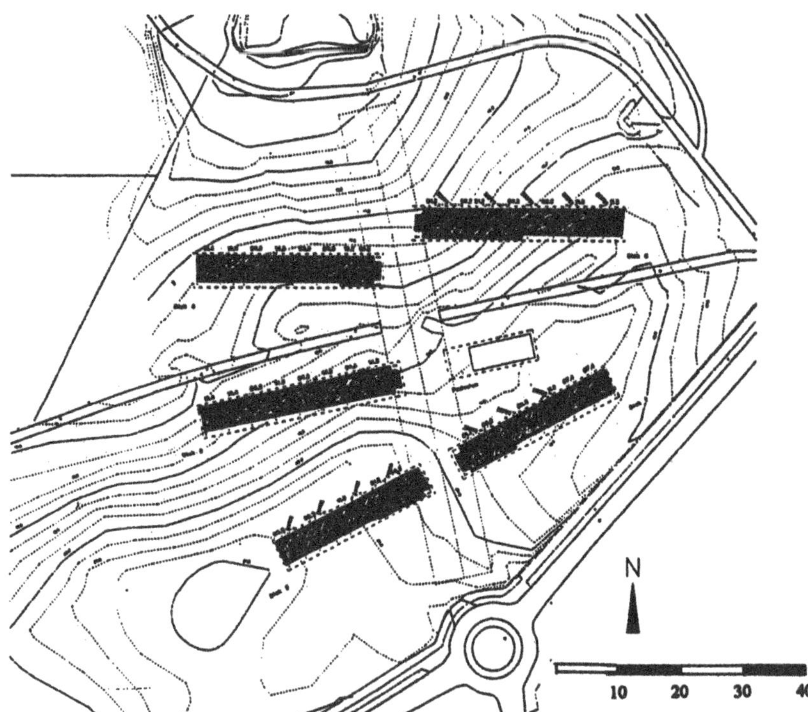


*Figure III.7.1. Site plan and view; scale 1:1000*

### Low-energy features of the buildings

The windows have low-emissivity glazing with a *U*-value of 1.4 W/m²K. The glazed area corresponds to 15% of the floor area in each dwelling and 70% of the windows are south-facing. The houses are insulated to a level well above the requirements in the Danish building regulations; *U*-values are: walls 0.17 W/m²K; roof 0.12 W/m²K; and floor 0.17 W/m²K. Special attention was given to weather-stripping to achieve a natural air exchange rate of 0.1 per hour. A mechanical ventilation system provides the necessary ventilation. The system includes an air-to-air heat-exchanger which recovers up to 80% of the heat in the ventilation air.

The building heat loss coefficients for all 50 dwellings are 0.53 W/m²K including ventilation losses and 0.38 W/m²K excluding ventilation losses.

## SOLAR SYSTEM

### The system

The active solar system incorporates 376 m² of roof-integrated air solar collectors, 7.5 m² per dwelling. The solar-warmed air from the collectors passes through a heat exchanger coupled to a storage tank in the boiler room and then to the space-heating and hot-water-heating circuits, see Figure III.7.3. During summertime the collectors provide DHW heating.

The system is designed to work in a number of different modes of which the three most significant are shown in Figure III.7.4. The stored solar heat can be used for space heating, DHW heating or both, depending on the temperature levels in the main storage. To be used for space heating the temperature at the top of the main storage tank must exceed 30°C; to be used for DHW heating it must exceed 60°C.

### The collector

The collector has an absorber assembled of elements which all have the same cross-section, as illustrated in Figure III.7.5. It is mounted in a standard Al-profile roof-light system (Vitral) covered by iron-free glazing and insulated on the back with 10 cm of mineral wool, see Figure III.7.6.

### Storage

During the heating season the collected solar heat can be directed to the floor heating system and the thermostats controlling the floor heating can be overruled to allow for hybrid solar storage in the concrete floors. Surplus heat is stored in the water tank and/or used for DHW heating. In each of the five boiler rooms a storage tank of 3 m³ is installed. The total storage capacity is 17.5 kWh/K in the water tanks and 3.5 kWh/K in the concrete floors.

### Distribution

Space heating is provided by a floor heating system to ensure comfort at the lowest possible consumption of heating energy. Auxiliary heating is supplied from the district heating network. The piping from the boiler room to the dwellings is placed inside the external insulation of the buildings to utilize losses in the heating season. A heat meter measures the total consumed

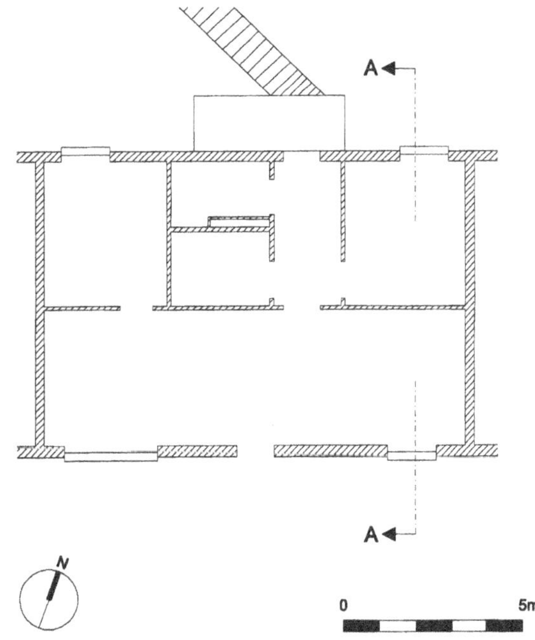


*(a) Typical floor plan*

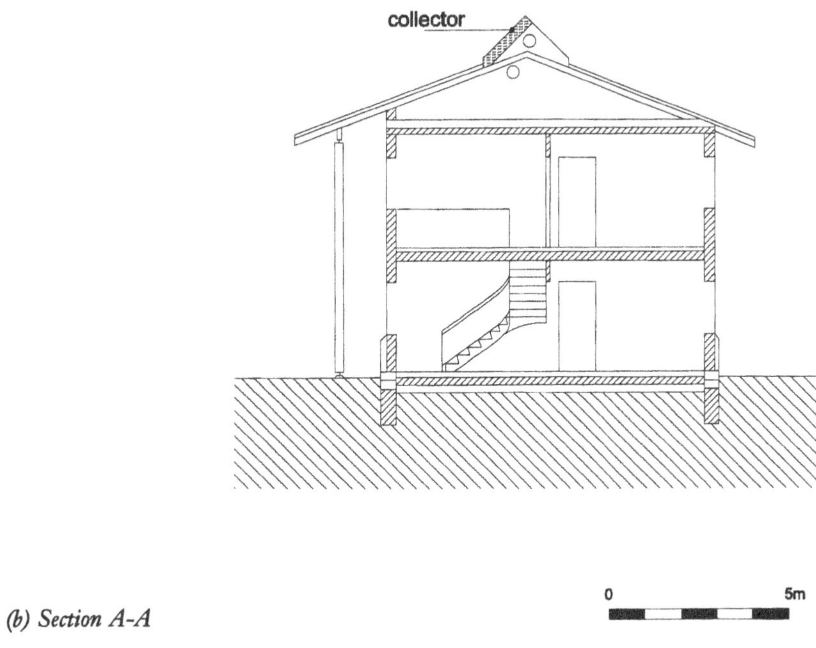


*(b) Section A-A*

*Figure III.7.2. (a) Floor plan of an example four-room apartment (area 104 m²) and (b) a section of a two-storey apartment.*

energy for heating and domestic hot water in each dwelling. Experience has shown that this stimulates user interest in saving energy. At the end of each row is a boiler room with pumps, fan, heat exchangers, storage tank and control system.

## Controls

An energy management system monitors and controls the active solar systems, the storage, the distribution network and the ambient weather conditions. The system allows delivery of data to a remote monitoring system.

## PERFORMANCE

### Overall performance

In Havrevangen the total purchased auxiliary energy for heating (degree-day adjusted) and DHW amounts

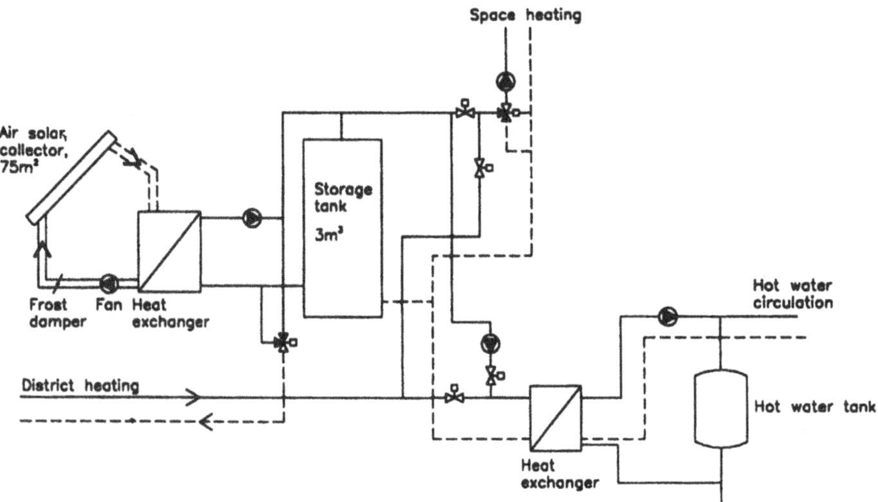


*Figure III.7.3. Schematic diagram of the solar and heating systems*

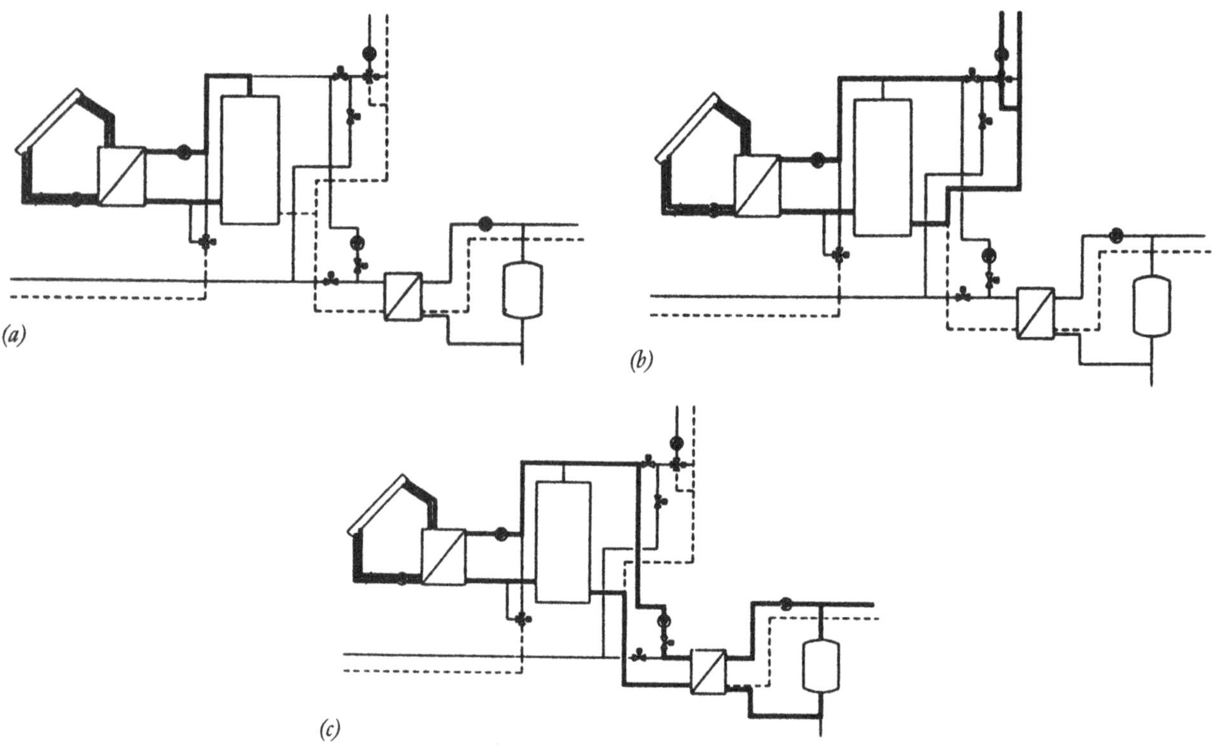


*Figure III.7.4. Three modes of the air solar system operation – (a) mode 1: solar collection to storage; (b) mode 2: direct space heating; (c) mode 3: Direct domestic hot water heating*

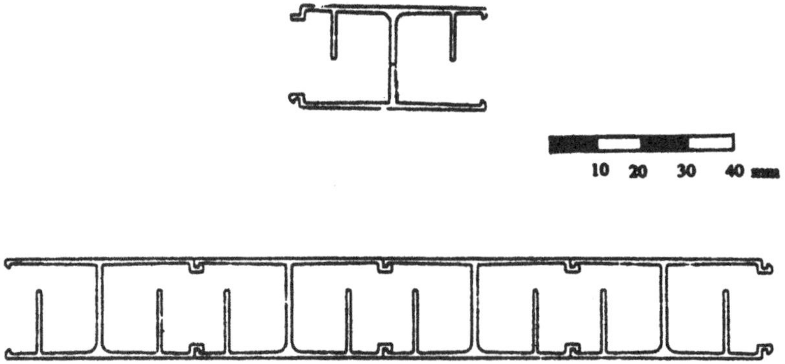


*Figure III.7.5. Cross-section of a single absorber element and an example of four elements assembled*

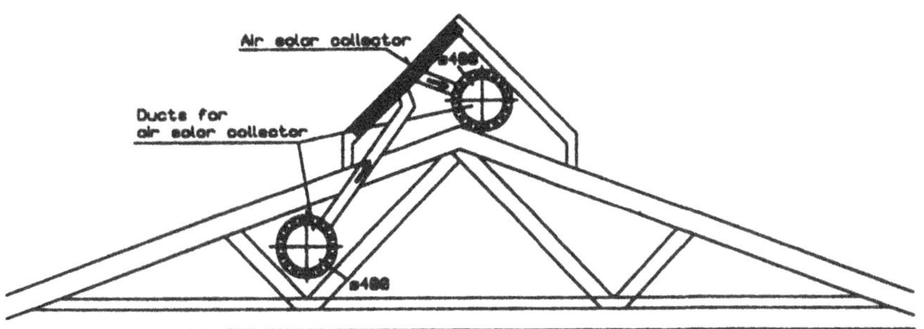


*Figure III.7.6. Section showing attic with the air solar collector and the ducts*

*Table III.7.1 Overall energy balance and obtained savings(degree-day corrected)*

| | kWh/m² | % |
|---|---|---|
| Energy conservation and passive solar | 79 | 44 |
| Active solar | 21 | 12 |
| District heating | 80 | 44 |
| Built according to standard (reference) | 180 | 100 |

to 80 kWh/m², which is 45% of a reference building project built according to the Danish building regulations (180 kWh/m² building area). This includes losses in local heating distribution networks and water preparation tanks, as well as in common house heating and DHW energy consumption use. The global horizontal solar radiation over the monitored year was 1009 kWh/m², which is very close to the solar radiation of an average year.

The contribution from overall energy saving measures and active solar system can be seen in Table III.7.1.

Figure III.7.7 shows the distribution of the actual heating load in Havrevangen. It should be noted that this figure includes the heating energy use in the five dwelling rows excluding the district-heating network losses and common house-heating energy consumption. From the figure it appears that the losses constitute much of the overall heating energy use. It should be noted, however, that these same losses (primarily in DHW preparation, circulation and storage) would correspond in a reference project to 11%.

The measured space-heating load totals 222 MWh corresponding to 53 kWh/m². The domestic hot water heating use totals 67 MWh or 1340 kWh per apartment.

*Figure III.7.7. Distribution of heating energy use*

## The solar system

The solar system was designed to deliver both space heating and domestic hot water. The heat for the space heating is delivered at low temperatures (< 40°C). The DHW is heated in a heat exchanger that needs temperatures above 55°C on its warm side. Thus the solar system will cover space heating as well as DHW heating, depending on its operating temperature.

The monthly active solar system contribution in row 2 is shown in Figure III.7.8. It can be seen that practically all the DHW load is covered in the summer months, July and August, and most of the total heating load in May, June and September.

The Danish 1995 summer was very warm and sunny. On several days the solar system was stopped because the temperatures in the storage tank had reached 95°C. This proved one of the advantages of the air solar system. The solar fan was simply stopped by the building energy management system (BEMS) with no risk of boiling off liquid from the solar circuit.

The solar system output was 240 kWh/m² for the monitored period, November 1994 to October 1995.

## The collector

The solar collector used in Havrevangen was developed especially for this project with funding from the Danish Energy Agency.

The aim was to develop an air solar collector that could be competitive with liquid collectors. The monitored performance of the solar system is comparable with the monitored performance of corresponding liquid solar systems for the same period. Thus the air solar collector that was developed has fully lived up to expectations. The efficiency curve found from the measurements is shown in Figure III.7.9.

## Hybrid storage

The concept includes a technique never tried in practice before, a hybrid solar system combined with air solar collectors. This system had to be very carefully designed and testing was a necessity. The project provided practical experience with the hybrid storage using a floor heating system.

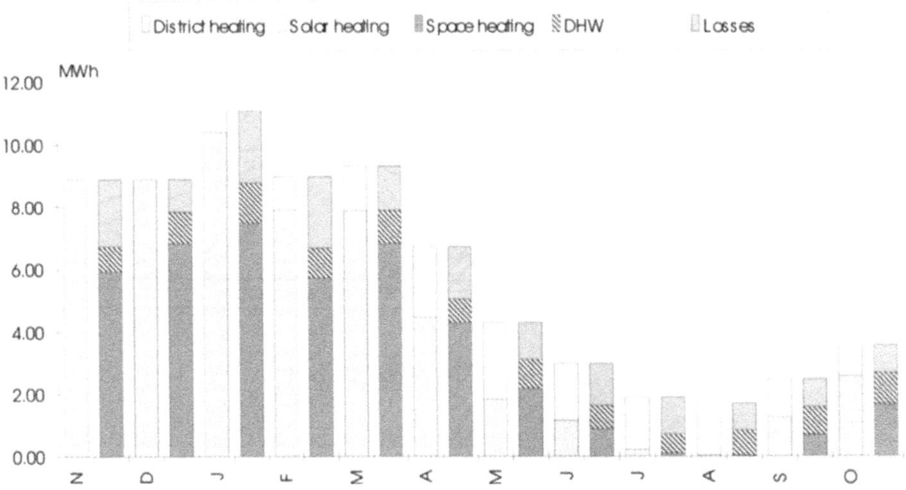


*Figure III.7.8. Monthly energy balances for row 2 in Havrevangen*

However, because of the individual billing system of the tenants in Havrevangen, the overruling of the room thermostats had to be given up. As installed, the BEMS system and the individual meters do not allow for this introduction of 'free' solar heat. This was an unexpected difficulty that must be overcome in future projects.

## Economy

The energy-saving features of Havrevangen cost US $700,000 (ECU 630,000) – according to the contract between the contractor and the builder. The Danish Energy Agency grants a standard 30% of the solar system price, which in this case amounted to US $97,700 (ECU 87,930). Thus the net-cost for the builder, excluding the support, was US $590,000 (ECU 531,000).

The price of the purchased heating energy from the district heating network consists of two parts: a fixed power charge and a variable charge for the energy consumed. Therefore, the calculation of a simple payback time cannot be expressed solely in terms of the amount of energy saved, but must also include the savings obtained by requiring a lower maximum power output from the district heating network. The simple payback time of the energy saving investments is 15.8 years. The cost per unit of energy saved is: US $0.08/kWh (ECU 0.072/kWh), which can be compared to the price of US $0.07/kWh (ECU 0.063/kWh) paid for the heating in Havrevangen.

Electricity for the fans in the solar system constitutes the major part of the operating costs. The fan electricity consumption corresponds to 10% of the heating energy output of the solar system, which is equal to what was estimated in the design phase. In view of the observed high efficiency of the solar system, it is likely that the system could be optimized in a way that allow the fans to work in the low-speed modes for a greater percentage of the time, thereby reducing the electricity consumption.

# REMARKS

## Acceptance

The tenants faithfully report their monthly energy and water readings on special forms and have the opportu-

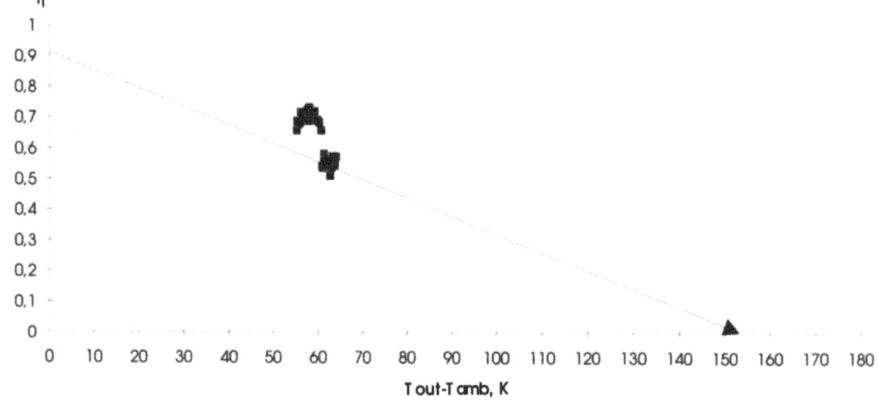


*Figure III.7.9. Efficiency curve for the air solar collector installed in Havrevangen*

nity to comment on the dwellings on these forms. Generally their reaction has been positive. The comfort of the floor heating system has been praised. Cold air from the ventilation system inlets in very cold weather has been criticized by some tenants. The overall feeling of fresh air and the avoidance of humidity condensation problems has been praised by others. Generally, it is the impression that the project has increased awareness about energy saving amongst the tenants.

### Barriers

The project revealed an important barrier for common hybrid solar-system storage control in combination with individual heat meters. Solar heat should not be metered, but given for free or at another rate to make it acceptable to the tenants.

During the first year of operation, the solar, ventilation and heating systems have shown high reliability. There have been no maintenance requirements in this period. A separate budget has been worked out for the long-term planned maintenance of the solar and low-energy features.

### Environmental benefits

The environmental impact is quite significant through the reduced $CO_2$, $SO_2$ and $NO_x$ emissions corresponding to the stated energy savings, as seen in Table III.7.2.

## ACKNOWLEDGEMENTS

Manufacturers of the solar collector: Vitral A/S, Havremarken 4, 3650 Ølstykke, Denmark

*Table III.7.2. Emission reductions*

| Emissions, kg/TJ | $CO_2$ | $SO_2$ | $NO_x$ |
|---|---|---|---|
| Emissions reference, kg: | 125,567 | 234 | 220 |
| Emissions actual, kg: | 64,556 | 120 | 115 |

Builder: Frederiksborg Boligselskab v. KAB, Copenhagen, Denmark

Architect: Lars Børjeson, Vilhelm Lauritzen A/S, Copenhagen, Denmark

Engineers: Cenergia Energy Consultants, Ballerup, Denmark

Contractor: KKS Entreprise A/S, Copenhagen, Denmark

Economic support: The EU THERMIE Programme and The Danish Energy Agency

Chapter author: Ove Mørck, Cenergia Energy Consultants

## BIBLIOGRAPHY

Det gode byggeri (1994). *Politikens Boligavis*, 16 October. Newspaper article.

Efficient Danish modular solar air collector. (1994). Article in *Sun at Work in Europe*.

Havrevangen H (1994). *Danish Ecological Buildings*. Example book issued by the Danish Ministry of Housing and Building, Arkitectur no. 7.

Mørck O (1994). *Solar Low Energy Buildings with Hybrid Storage*. ICUEB conference in Beijing

Lavenergibyggeri i Hillerød (1994). *Byggeri og økologi - status over dansk boligbyggeri*. Status report from the Danish Ministry of Housing and Building, April 1994.

Mørck O, Kofod P (1993). *Udvikling af luftsolfanger*. Cenergia Report, June 1993.

Pedersen PV (1993). *EC-supported low-energy housing projects in Denmark. European Directory of Energy Efficient Buildings*, 13-19. James + James Science Publishers, London.

# III.8 Rødovre Apartments

## Copenhagen, Denmark

*Copen-hagen*

*System type 3*

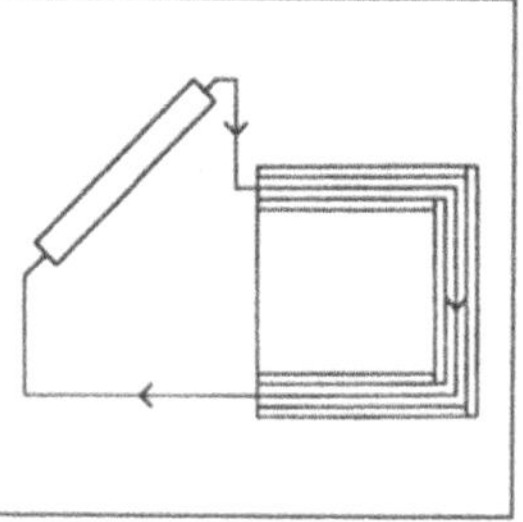

### PROJECT SUMMARY

A combined air/liquid solar collector system was retrofitted to a three-storey apartment building in Copenhagen, originally built in the 1950s. The building has been insulated outside with 100 mm mineral wool. Between the insulation and the original walls there is an air gap. In this air gap the hot air from the solar air collector is circulated to heat the building. Site-built combined air/liquid collectors cover the whole south-facing roof. The collectors were built on top of the existing roof, which serves as the back of the collectors.

The solar air system is a dual system, including both a closed collector loop to an air to water heat exchanger and a double envelope.

#### Summary statistics

| | |
|---|---|
| System type: | type 3 |
| Collector type and area: | air–liquid, 390 m$^2$ |
| Storage type, volume: | water tanks, 1500, 2,800 litre |
| Annual contribution from the solar system: | 16 kWh/m$^2_{floor}$ |
| Annual auxiliary heat consumption: | 117 kWh/m$^2_{floor}$ |
| Basis: | monitored |
| Heated floor area: | 2377 m$^2$ |
| Year solar system installed: | 1992 |

### SITE DESCRIPTION

The Solar House is located in Rødovre, approximately 10 km south-west of the centre of Copenhagen:

| | |
|---|---|
| latitude | 56°N |
| longitude | 13°E |
| altitude | 20 m above sea level |

The climate is temperate coastal.

### BUILDING PRESENTATION

The building, one of three apartment blocks, has 45 apartments. The buildings had to be renovated in the early 1990s. In the two other blocks, the walls were insulated with 100 mm mineral wool and the buildings were re-roofed. The solar building was also insulated with 100 mm mineral wool, but externally. Between the insulation and the existing walls an air gap was made, in which the hot air from the collectors flows to heat the building. The solar building is a three-storey apartment building facing nearly east–west, so it has a south-facing roof, on top of which the entire 390 m$^2$ of site-built air/liquid collectors have been mounted. The tilt of the roof is about 20°. A plan of a typical floor and a cross-section of the building are shown in Figure III.8.1.

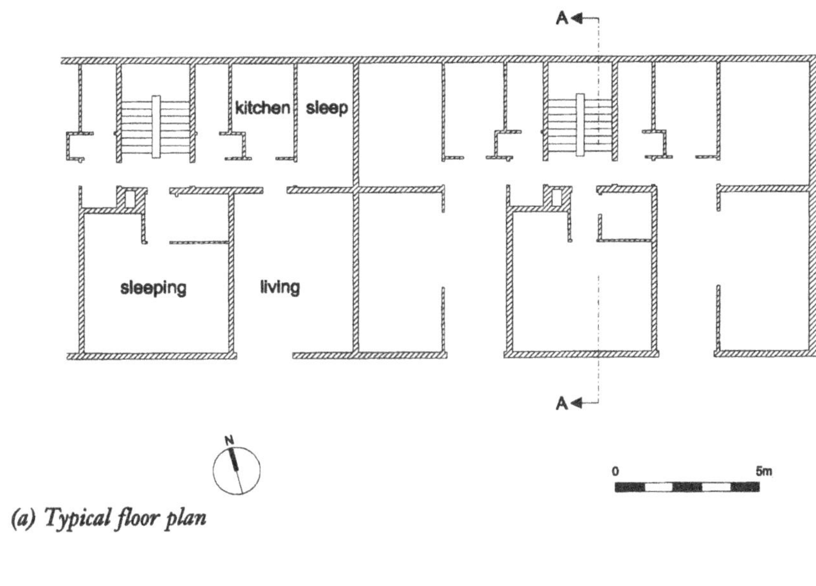


*(a) Typical floor plan*

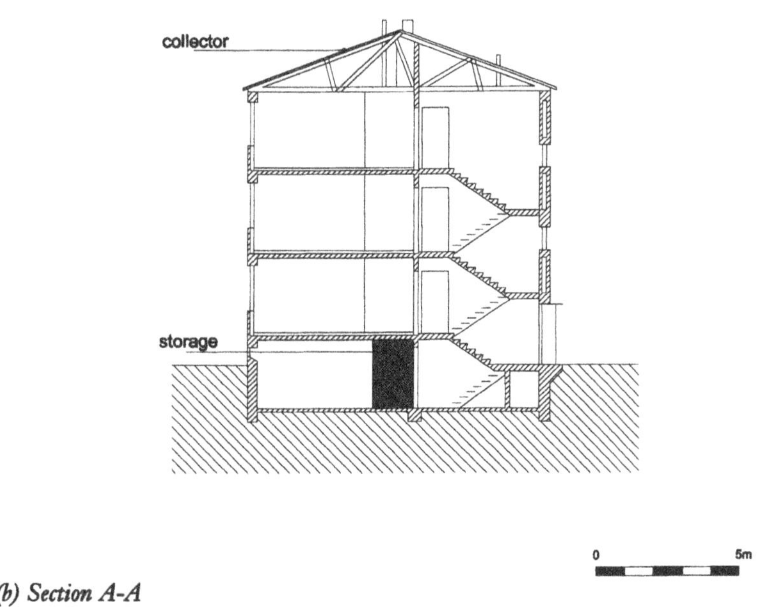


*(b) Section A-A*

*Figure III.8.1. (a) Plan of a typical floor and (b) cross section of the solar building, seen from the east*

## SOLAR SYSTEM

### The system

The combined air and liquid collectors were built in 1.24 m wide sections, each 5.2 m long. The air absorber has an area of about 360 m$^2$ and the liquid absorber of about 80 m$^2$. The air heats the house and the liquid heats the domestic hot water. The air is blown from the base of the roof, down the walls, and back to the collectors in a set of ducts (Figures III.8.2 and III.8.3).

The space and the hot-water heating are backed up by a district heating system.

### The collector

The collector is covered with double-walled polycarbonate sheets, 1.22 m wide, running from the ridge to the base of the roof, so that only sideways connections were necessary. The sideways connections were made with aluminium profiles. The operation of the absorbers in the combined air and liquid solar collector is shown in Figure III.8.4.

*Figure III.8.2. The air circulation system.*

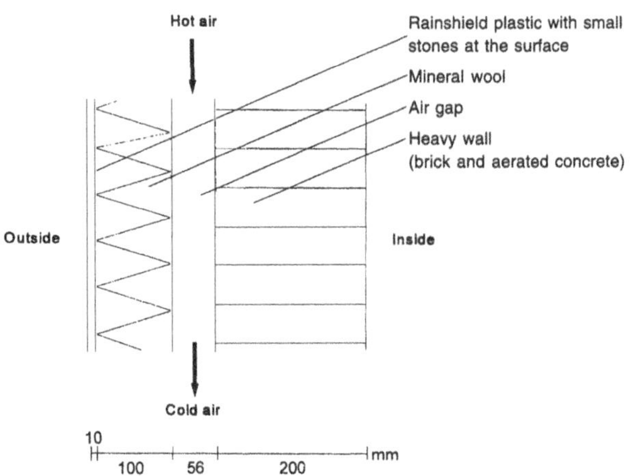


*Figure III.8.3. Cross-section of the facade, showing the principles involved*

The air collector has a black felt surface, 2 mm thick, acting as a porous absorber. The air flows through the felt twice. Because of the large surface area of the fibres in the felt, the heat transfer to the air is excellent.

The liquid collector has eight liquid absorbers (sunstrips), with 50 mm openings between the strips, over a distance of 2 m at the top of the collector box above the porous absorber. When the air flow is stopped in the summer, natural convection transfers heat from the porous absorber to the liquid absorber. In this way the collector can work as a water-heating collector without using the fan and still benefit from the air part of the collector.

## Storage

Some of the energy from the air collector system is stored in the heavy walls and some energy offsets an increased heat loss from the hot air in the air gap, while the remainder heats the building.

The liquid system absorber is connected to a 1,500 litre solar domestic hot-water tank for preheating a 2,800 litre main tank. The energy from the liquid-collector system is transferred to the solar tank via an external heat exchanger. The system has a total tank volume of 52 litres per m$^2$(liquid solar absorber).

## Distribution

As shown in Figure III.8.5, in the air system the distribution is handled by five fans with a total power of about 9.4 W/m$^2$(air collector). The hot air is extracted from the collectors and blown into the air gap in the walls with a total flow of about 5400 m$^3$/h.

A pump distributes water mixed with 40% polypropylene glycol in the liquid system with a flow of approximately 70 litres/min or about 0.85 litres/m$^2$min. On the secondary side of the heat exchanger a smaller pump circulates water from the solar domestic hot-water tank through the heat exchanger and back to the tank. The total power of the two pumps is about 660 W.

The hot-water tanks deliver hot water to three buildings, the solar building and the two others. The total hot-water demand and heat loss from the circulation pipe are about 264,000 kWh per year.

## Controls

The air and liquid collector systems are controlled by differential thermostats. The fans are started when the temperature in the collector is 3–4 K higher than the

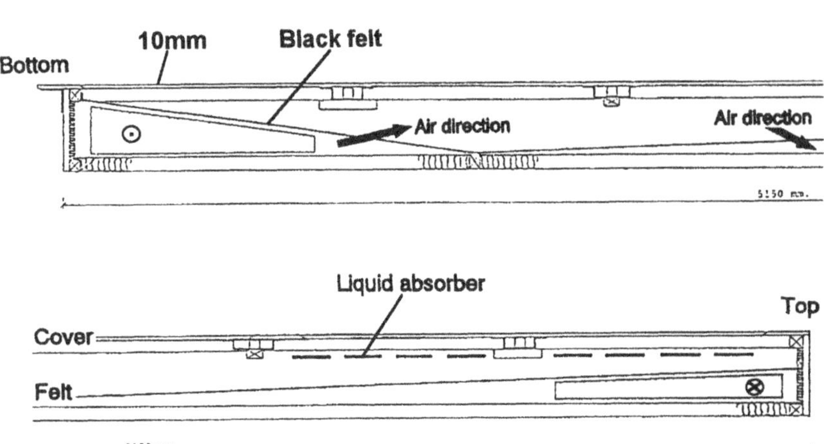


*Figure III.8.4. The operation of the combined air and liquid solar collector*

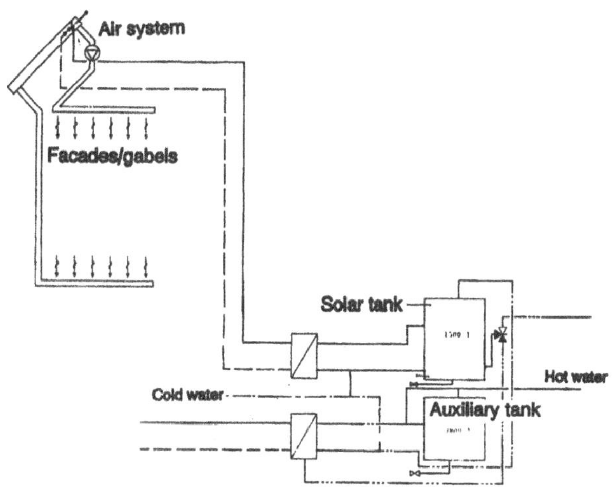


*Figure III.8.5. Schematic diagram of the air and liquid systems*

temperature in the air gap. To ensure indoor thermal comfort, the fans are stopped when the wall temperature exceeds 20°C to 30°C, depending linearly on the outdoor temperature in the range +20°C to –12°C.

The radiator system is regulated by individual radiator thermostats. In the liquid system the high-temperature sensor is located at the top of the liquid collector and the low-temperature sensor is placed in the bottom of the solar tank. When the temperature difference is 2–3 K the pumps are started.

## PERFORMANCE

### The air system

The temperatures in and out of the collectors were measured by thermocouples. The airflows in and out of the collectors were also measured. On the basis of these measurements, the performance of the air system has been calculated (Figure III.8.6).

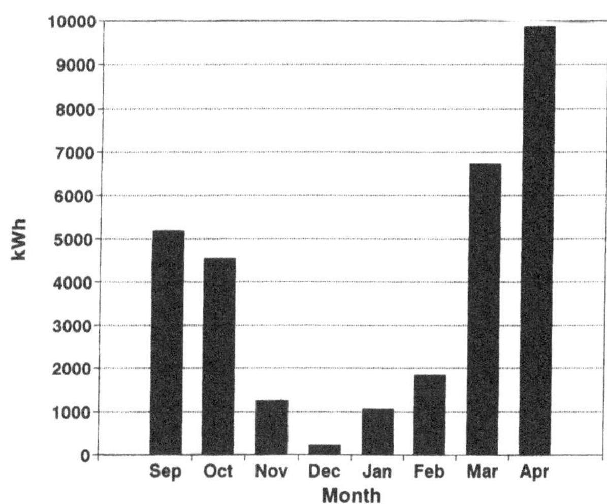


*Figure III.8.6. The monthly performance of the air system from September 1992 to April 1993*

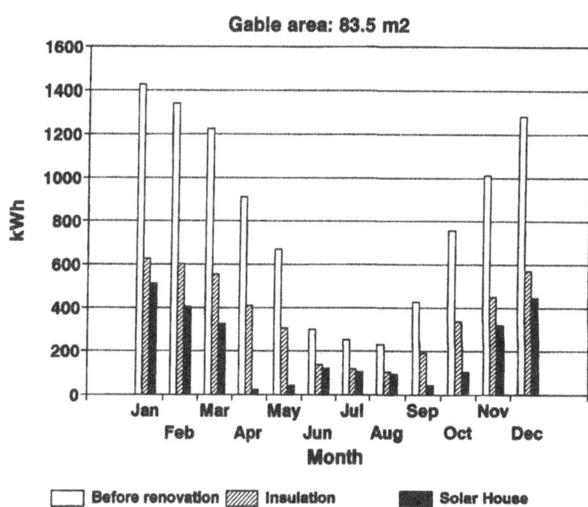


*Figure III.8.7. Heat loss from the building gable, before renovation, after insulation version and in the solar building*

The performance of the air system is low in the winter, when the heat demand is largest. The performance is high at the beginning and end of the heating season, i.e. September and October, and March and April.

The loss of energy through the end of the solar house has been compared to computer simulations of the loss of energy through a similar wall with traditional insulation and the gable as it was before the energy renovation. Figure III.8.7 shows that additional insulation reduces the wall heat loss to half of what it was before renovation. A further, but smaller, reduction results with the air collector.

From the calculations, it was estimated that approximately 50% of the energy produced from the air collector is utilized. This amounts to approximately 14,000 kWh per year or 38.9 kWh per year/m$^2$(air collector).

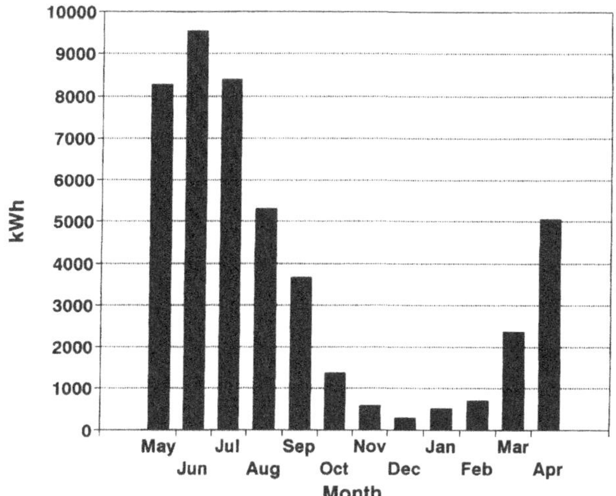


*Figure III.8.8. Monthly performance of the liquid collector from May 1992 to April 1993*

*Table III.8.1. Power consumption*

| | |
|---|---|
| Saving in heating of domestic hot water | 46,000 kWh |
| Saving in space heating (from the air system) | 14,000 kWh |
| Savings due to insulation | 35,000 kWh |
| Total savings | 95,000 kWh |
| *Less* electricity consumption by pumps and fans | 4,700 kWh |

### The liquid system

The contribution from the liquid collector to the heating of the domestic hot water was recorded by heat-flow meters. Figure III.8.8 shows the energy flowing from the collector to the tank for each month.

The auxiliary heating demand to heat the domestic hot water and cover the heat loss from the circulation pipe was 218,000 kWh in the period from May 1992 to April 1993. The collector contributed 46,000 kWh (560 kWh/m$^2$) or about 17 % of this.

The district heating for space heating after the renovation was about 170,000 kWh per year, compared to 219,000 kWh/ per year before the renovation, a 22% saving.

The reductions in power consumption, due both to insulation and to the solar system, are shown in Table III.8.1, together with the power consumed by the pumps and fans. With a price of US $0.08/kWh (ECU 0.07/kWh), the savings is about US $7360 (ECU 6624). The total cost of the heating system was US $178,000 (ECU 160,000). This yields a payback time of 24 years.

## REMARKS

The performance of the air system is lower than expected, largely due to leaks in the collector seals. The liquid system acts very satisfactorily, with a yearly performance of 560 kWh/m$^2$(liquid collector).

This project shows that it is important to find an inexpensive way of building this type of system and to increase the output of the air system. Tighter construction is easier for new buildings than in the retrofitting of older buildings. An energy saving of 20% was achieved, including the insulation of the walls.

Measurement of the airflows in and out of the collector has shown that the outlet flow from the air collector is 20% to 33 % higher than the inlet flow. Air from the outside is thus drawn into the air system through leaks in the air collector. Sealing the air collectors better would greatly improve performance, particularly in the coldest months. An increase of the tilt of the roof or the collectors would also improve the performance of the solar collector system.

Computer simulations indicate that the performance could be increased by 100% under the best conditions.

## ACKNOWLEDGMENTS

Architect: Jensen + Jørgensen + Wohlfeldt, Denmark

Energy consultant: Danakon a/s, Taastrup, Denmark; Thermal Insulation Laboratory (now IBE), Lyngby, Denmark

Construction: Enemærke & Petersen A/S, Svend Breil and Steni Danmark ApS, Denmark

Building owner: Cooperative Housing Society Lejerbo, Denmark

Chapter author: Finn Kristiansen, Department of Buildings and Energy (IBE), Technical University of Denmark, Lyngby, Denmark

## BIBLIOGRAPHY

Svendsen B, Vejen NK (1993). *Projekt Solhus (Project Solar House)* (in Danish). Danakon a/s, consulting engineers and Thermal Insulation Laboratory, Technical University of Denmark, Building 118, DK-2800 Lyngby, Denmark. Report number 258, November 1993.

# III.9 A Solar Air Apartment Block

Gothenburg, Sweden

*Gothen-burg*

*System type*

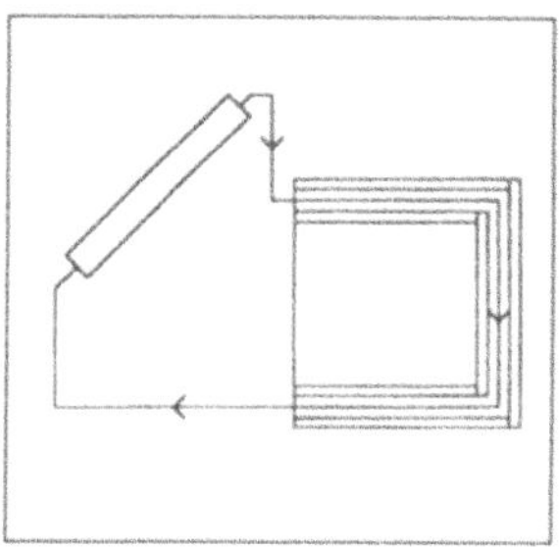

## PROJECT SUMMARY

A double envelope solar air system was installed in a 1950s three-storey apartment block when it was renovated in 1986. The system saves 40% compared to identical conventionally renovated buildings, thanks to the following:

- The south-facing side of the roof has been converted to a 350 m$^2$ solar air collector.
- An extra insulated facade with an integrated air-space of 900 m$^2$ has been added to all surrounding walls.
- A DHW system including a 12 m$^3$ storage tank has been added in order to provide solar heated water to four blocks (96 apartments) during the summer.

### Summary statistics

| | |
|---|---|
| System type: | type 3 |
| Collector type, area: | Opaque, roof-integrated, 350 m$^2$ |
| Storage type, volume: | Double envelope walls, 180 m$^3$ |
| Annual contribution from the solar system: | 248 kWh/m$^2$a |
| Annual auxiliary heat consumption: | 66 kWh/m$^2$ $_{floor}$ |
| Basis: | monitored |
| Heated floor area: | 1445 m$^2$ |
| Year solar system installed: | 1986 |

## SITE DESCRIPTION

The apartment block is located in Järnbrott, a suburb 5 km south of the city of Gothenburg on the west coast of Sweden. The building is situated in a row of four identical houses on a smooth south-facing slope:

| | |
|---|---|
| latitude | 58°N |
| longitude | 12°E |
| altitude | 38 m above sea level. |

## BUILDING PRESENTATION

This three-storey apartment block, with 24 public housing units, was built in 1952. The reconstruction and the solar installations were completed in December 1986. A community greenhouse has been built along the south facade in order to 'explain' the principles of solar heating to the occupants.

A site plan, showing the housing area and the solar house, is given as Figure III.9.1, while Figure III.9.2 shows the building before and after renovation. Figure III.9.3 shows the plan of a typical floor, together with a section of the building.

### Energy-saving features

The following energy-saving installations were made:

- An extra 8 cm of mineral wool was added to all external walls.

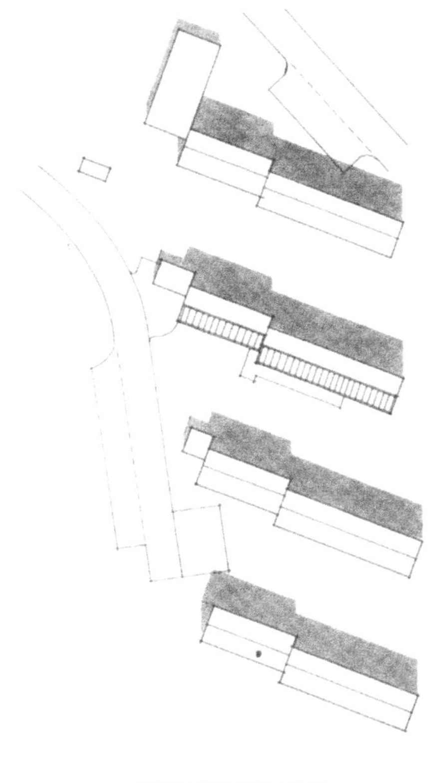

*Figure III.9.1. Site plan*

(a)

(b)

*Figure III.9.2. The apartment block (a) before and (b) after the reconstruction was completed in 1986*

- An air-space of 5 cm was created between the old wall and the additional insulation.
- A new 6 cm brick facade was added to all walls.
- The old two-pane windows were replaced by new three-pane windows.
- The auxiliary heating system was rebuilt and a new control unit was installed.
- The natural ventilation system was cleaned.
- A solar air system was installed for space heating during winter and DHW heating during summer.

## SOLAR SYSTEM

The system is a closed loop (Figure III.9.4). Energy is saved by decreasing the transmission losses through the building envelope and sometimes by transferring solar heat to the interior of the building. The system is divided into separate sections, each supplied by its own solar collector, air space and fan.

### The solar air collector

The old roof tiles on the south-facing side of the roof were replaced with black corrugated steel sheet, which was covered with a double transparent polycarbonate glazing. Under the sun-warmed corrugated steel sheet, and above the insulation, air is heated and circulated to the air gap between the old and new exterior walls. A 1 cm still-air gap between the transparent cover glazing and absorber reduces the absorber's heat loss to the ambient air. Details are given in Table III.9.1 and a cross-section is shown in Figure III.9.5.

### The double-envelope walls (storage, distribution)

On all the external walls, an air gap has been created between the old facade and the new additional wall insulation. The solar-heated air is circulated through the air gap, transferring heat to the old concrete wall. The heat is stored in the old walls and slowly conducted to the interior of the house. After having passed through the double envelope-walls, the air is transported back to the solar air collector to regain heat. Even with weak solar radiation, the temperature of the circulated air is above ambient temperatures and therefore reduces the heat losses from the walls. Details are given in Table III.9.2.

*Table III.9.1. Details of the solar collector*

| | |
|---|---|
| Collector area | 350 m$^2$ (0.242 m$^2$ collector area/ living area) |
| Roof tilt | 27° |
| Absorber/duct system | Corrugated steel |
| Absorber surface | Standard black coating (PVF2) |
| Glazing | 6 mm UV-treated, double polycarbonate |
| Insulation | Mineral wool under the absorber |

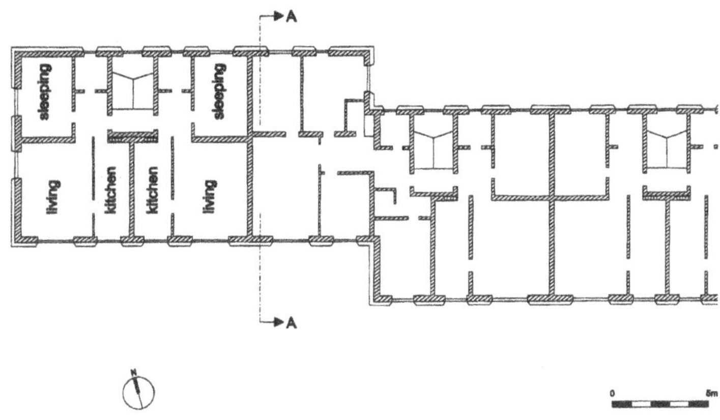


*(a) Typical floor plan*

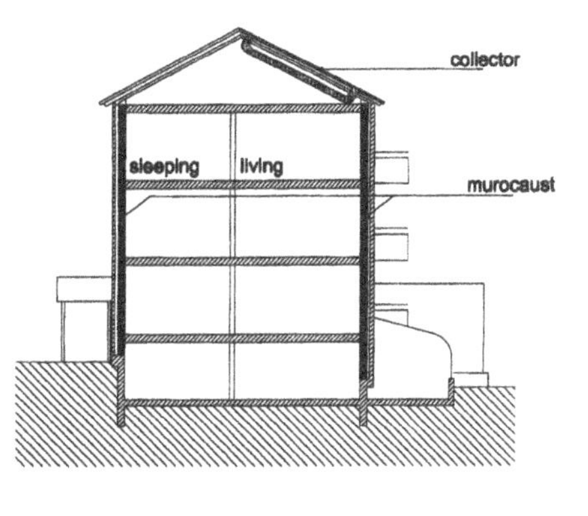


*(b) Section A-A*

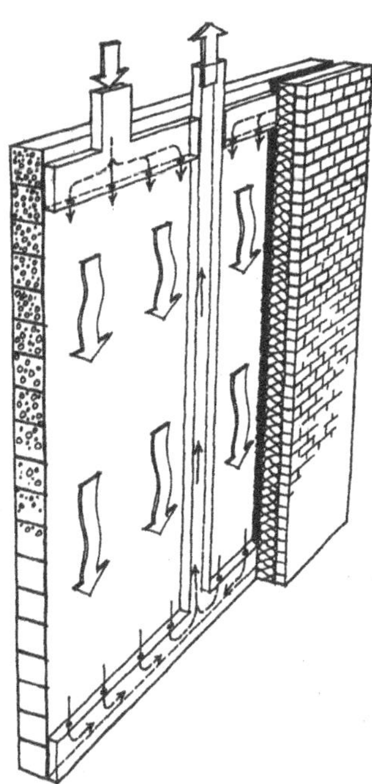

*Figure III.9.4. Section of the solar system*

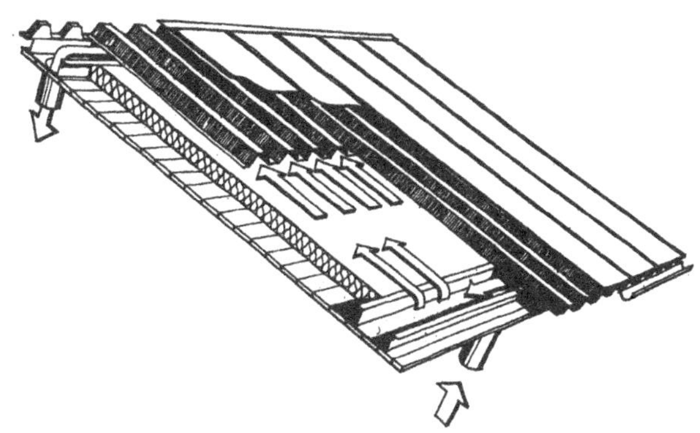

*Figure III.9.5. Cross-section of the site-built solar collector*

*Table III.9.2. Details of the double-envelope walls*

| | |
|---|---|
| Double-envelope wall area | 900 m$^2$ |
| Additional wall | Air space 50 mm<br>8 cm mineral wool.<br>Air space 50 mm |

## The DHW system

During the summer months the solar heat is diverted to air-to-water heat exchangers (one in each section of the system), to pre-heat domestic hot water. The solar-heated water is stored in a 12 m$^3$ storage tank in the basement (Figure III.9.6).

### Controls and fans

The solar-heated air is distributed to the air gap in the double-envelope external walls through a duct system installed in the attic. The air is moved by a number of small fans (Figure III.9.7; Table III.9.3) during the daytime, when the air temperature in the solar collector exceeds the temperature in the air space surrounding the house. The air-flow system has to be manually adjusted between winter and summer modes in April/May and September/October.

### Auxiliary heating system

The building is connected to the local district heating network. The experimental building has been furnished with a separate radiator circuit in order to make the adjustments necessary to adapt the solar system. Radiators are placed under the windows in each room.

## PERFORMANCE

### Monitoring results

The solar house and three identical reference buildings were renovated from 1989 to 1991. Apart from the solar system, the solar house and the reference buildings

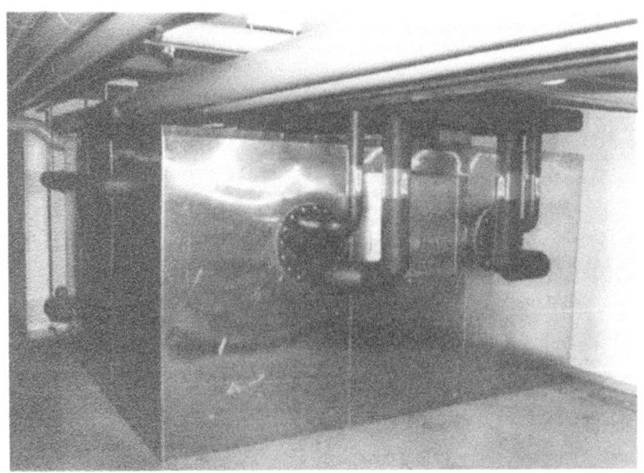

*Figure III.9.6. The storage tank*

*Table III.9.3. The fans and sensors*

| | |
|---|---|
| Number of fans | 13 |
| Type, power of fans | centrifugal, each 200W |
| Capacity of each fan | 800 m$^3$/h at 150 Pa |
| Total capacity | 10 400 m$^3$/h at 150 Pa velocity drop |
| Duct cross section | 250 mm |
| Control system | One differential thermostat |
| Sensors | One in the top of the collector, one in the air space. The fans start when the temperature in the solar collector exceeds the temperature in the double-envelope wall. |
| Other controls | One fire detector protects each section and a central unit turns off all the fans if there is a fire. |

were all rebuilt in the same way, with the same amount of extra insulation, new three-paned windows etc. A separate radiator circuit has been installed in the solar house.

During the heating season in 1990, the annual need for heat to the radiator circuit in the experimental building was shown to be 59.2 kWh/m$^2$(net dwelling area). Extrapolated to an average year with regard to outdoor temperature this corresponds to 66 kWh/m$^2$. This amounts to a 40% lower heating demand than for the neighbouring reference buildings (Figure III.9.8).

Beside reducing the need for purchased heating by 56,400 kWh/a, the solar system has contributed 30,000 kWh/a to the heating of domestic hot water or 40–50% of the total energy need for DHW during the summer. Equivalent results for both heating and domestic hot water were obtained for 1989 and 1991. During the heating season in 1990, the fans of the solar air system ran for more than 700 hours, using 1,930 kWh electricity (approximately 3% of the energy gain).

*Figure III.9.7. The small centrifugal fans in the attic*

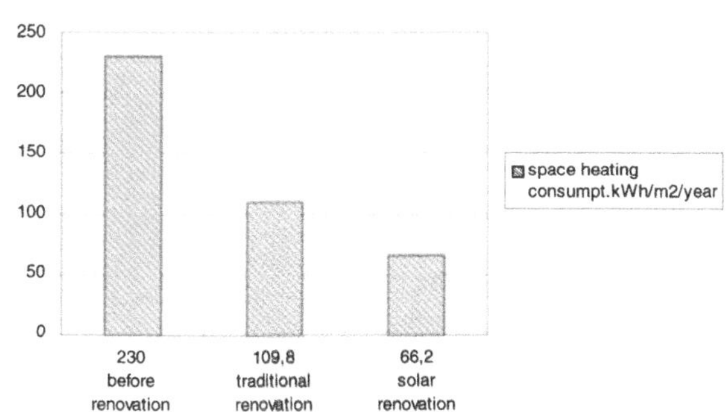


*Figure III.9.8. Space heating before and after conventional and solar renovation*

## REMARKS

The project demonstrates that it is possible to achieve a 40% reduction of energy need for space heating and DHW heating using a solar air double-envelope system. If the system is well integrated into the building structure and uses traditional building materials and standard components, the costs of the energy savings are reasonable.

Since a double-envelope solar air system with no seasonal storage will not cover the energy need during wintertime – at least not in Sweden – the system must be combined with an auxiliary heating system. Attention must be given to adjusting the auxiliary heating system to the solar system in order to achieve a maximum output from the solar system. Poor adjustment will result in a waste of energy.

The owner is very pleased with the project and has had no problem in understanding and maintaining the system. The system is run and maintained by the housing company's employees. Since the start-up in 1986 the system has been functioning continuously. Some fans have been replaced during routine maintenance.

No significant signs of degradation of the materials have been detected to date. However, it is known that the polycarbonate glazing of the solar collector will become slightly discoloured after 10–15 years. Since this is a closed system, no dirt in the duct system has been noticed.

## ACKNOWLEDGMENTS

Architect: Christer Nordström, Arkitketkontor AB, Askim, Göteborg, Sweden

Chapter author and energy consultant: Christer Nordström

Building owner: Bostads AB Poseidon, Göteborg, Sweden

Monitoring and evaluation: Jan Olof Dalenbäck and Jan Gustén, Chalmers University of Technology, Göteborg, Sweden

## BIBLIOGRAPHY (IN ENGLISH)

Ecocycles – *The Basis of Sustainable Urban Environment* (A report from the The Environmental Advisory Council-SOU 1992:43), Stockholm 1992. Allmänna Förlaget, S-106 47 Stockholm, Sweden. ISBN 91-38-13048-3

*European Conference on Architecture (1987) - proceedings, Munich 1987.* HS Stephens ans Associates, Agriculture House, 55 Goldington Road, Bedford MK40 3LS, England. ISBN 0-9510271-2-3. Munich, Germany.

*ISES Solar World Congress (1987).* Hamburg, Germany.

Malbert B (editor). *Ecology-based planning & construction in Sweden.* The Swedish Council for Building Research, Stockholm, Sweden.

# III.10 Ouellette Manor Senior Citizens' Building

Windsor, Ontario, Canada

*Windsor, Ontario*

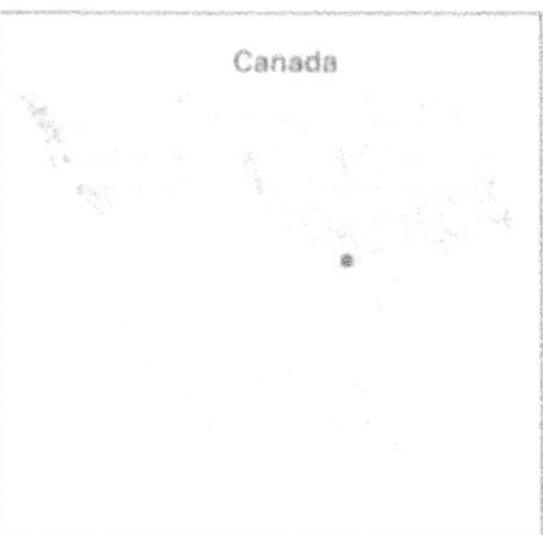

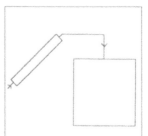

*System type 1*

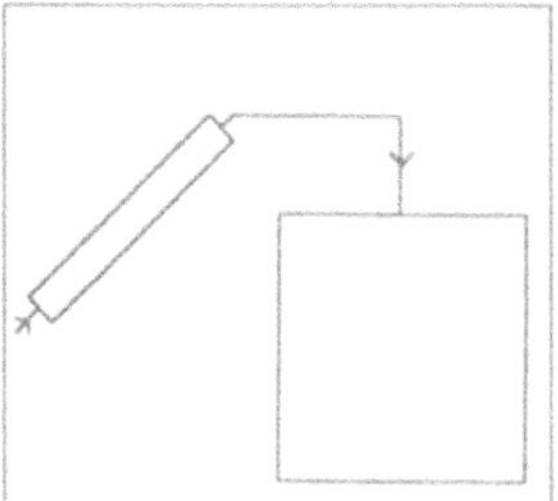

## PROJECT SUMMARY

The world's tallest solar air collector was installed on this building as part of a renovation project. The brick exterior of the 24-storey building was deteriorating from moisture penetration and freezing. Recladding was necessary to prevent further damage. An unglazed brown perforated-plate solar collector was chosen to cover the facade. This solution prevents further moisture damage, has an attractive appearance and provides free heating of fresh air required in the building.

The unglazed solar air heater (61 m high × 5.5 wide) looks like conventional metal cladding but it is mounted 250 mm from the main wall and is perforated to allow outside air to travel through the metal, where it picks up solar heat. The heated air is then directed to the fresh air intake of the ventilation system and supplies 28% of the ventilation heating requirements. The solar heating system has a six-year payback.

### Summary statistics

| | |
|---|---|
| System type: | type 1 |
| Collector type, area: | unglazed brown, 335 m$^2$ |
| Storage type: | no storage |
| Year solar system installed: | 1994 |

## SITE DESCRIPTION

There are no other tall buildings nearby and thus the south-facing wall receives full sunlight. Windsor is located in Canada across the river from Detroit, Michigan:

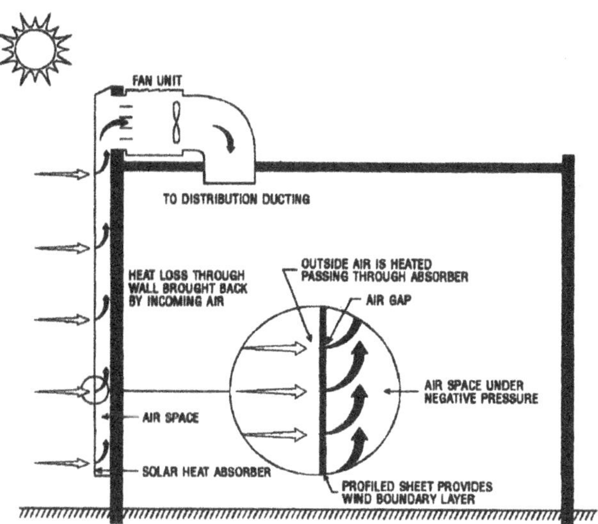


*Fig. III.10.1. Cross-section of the solar ventilation heating system*

latitude 42°N
longitude 83°W

Heating is required 8–9 months a year.

## BUILDING PRESENTATION

This 24-storey building was built in the 1960s. The ventilation fan is located on the roof and provides 23,000 m$^3$/h of fresh air on a continuous basis. The air is used to ventilate and pressurize the corridors of each floor. (Corridors are pressurized in North America to prevent cooking odours and also smoke, in case of fire, from entering the corridors and travelling to other apartment units.) Six existing gas-fired heaters provide auxiliary heat when solar heat is not sufficient.

A common problem with high-rise buildings built in the 1960s and early 1970s is damage caused by the movement of moisture through walls. High-rise technology was known, but building physics was not well understood. Room humidity penetrates exterior walls from inside and rain penetrates from outside. South walls in particular can be exposed to severe freezing and thawing cycles in cold weather, causing expansion and brick-crumbling. The walls of this building were clad with rain screens in 1993 and the south wall was completed in early 1994 with a solar-wall rain screen.

## SOLAR SYSTEM

### The system

It was decided to install solar wall panels on the south side as the cladding material at a small additional cost. In winter, the solar panels heat the ventilation air. In summer, the solar panels shade the south wall from heat gain, reducing the cooling requirements. A cross section of the system is given in Figure III.10.1.

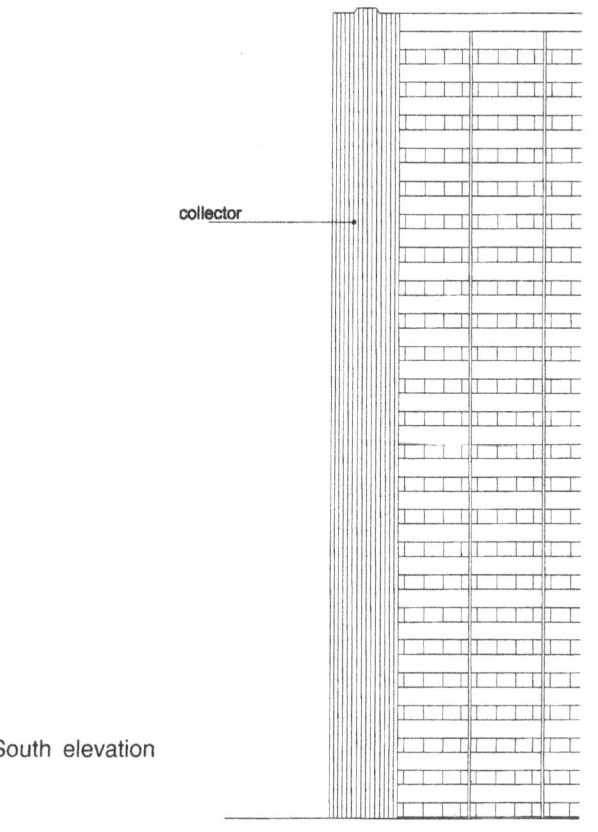


*Figure III.10.2. The south elevation, showing the unglazed solar air heater at the corner of the wall*

### The collector

The collector is 61 m high by 5.5 m wide, giving a total area of 335 m$^2$ (Figure III.10.2). The solar air panels are made from corrugated dark-brown aluminium resembling conventional metal wall cladding. They can be supplied in many dark colours and are not restricted to black. The solar heat absorption of the brown colour is 0.9, compared with 0.95 for a black coating. When installed, the solar wall heater looks like any other wall of the building; this was important to the building owners. Outside air is drawn through the holes (1.5 mm diameter) distributed over the entire metal surface. The aluminium panels are 0.8 mm thick and are mounted approximately 250 mm from the main wall using a clip system and horizontal members to support the cladding. The air flow rate through the solar collectors is 69 m$^3$/m$^2$h.

### Distribution

The solar-heated air is collected at the top and ducted along the roof to the intake of the existing gas heaters (Figures III.10.3 and III.10.4). No additional fan power is necessary since the existing fans are used. The heated air is then distributed to the corridors of each floor of the 24-storey building. The high building creates a stack effect in the solar panel. Hot air rises on its own

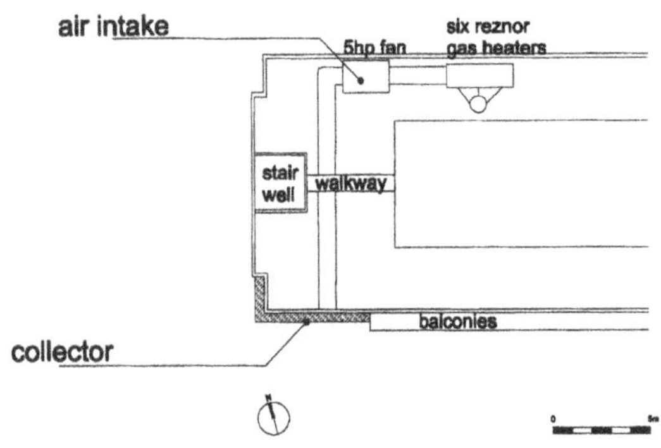


*Figure III.10.3. Plan view of the roof, showing the duct from the solar collectors to the existing fan and gas heaters*

*Figure III.10.4. The top of the solar collector connected to the duct on the roof*

without the need of a fan which gives higher air flows or reduced fan power consumption than when air is drawn through the bypass damper.

### Controls

When the ambient temperature is below 20°C, a damper in the collector plenum opens to allow the solar heated air to enter the fan intake. When the ambient temperature exceeds 20°C, the solar damper closes and a bypass damper opens to draw in outside air from the roof.

### Storage

There is no storage included in the system, although the thermal mass in the wall does provide some heat to the air cavity in the evening. Another heat source is the heat loss through the main wall, which enters the air cavity. The heat loss provides a few degrees of heating to the incoming air at night.

## PERFORMANCE

A computer simulation carried out by the supplier of the solar panel predicted an annual energy savings of 195,700 kWh for a dollar savings of $3525 (ECU 3173). The cost of the solar wall system over the conventional wall cladding was an additional $21,700 (ECU 19,530) according to the maintenance manager. This results in a payback of approximately six years. The cost of natural gas in Canada is very low; for this project under 1 cents per kWh or 1.5 cents when burner inefficiency is included (ECU 0.0015 and 0.0018 respectively). The energy savings are not being monitored, but another apartment building which has been monitored showed results very close to the predicted values. The solar collector generates an average of 584 kWh/m$^2$ of heat each year. The incremental cost is $65/m$^2$ (ECU 58.5/m$^2$).

## REMARKS

The solar heating system is operating as designed. Any apartment building that requires new cladding, as a rain screen, to improve insulation levels, or for any other reason, should be considered for the solar-wall cladding system. Most apartments in Eastern Europe with planned renovations could benefit from the energy savings, improved air quality and removal of dampness with solar air heating. The heated air can either be ducted through a central ventilation system, as in the Windsor project, or be supplied with individual variable speed wall fans for each flat or dwelling unit.

Other applications could be where mechanical ventilation air, heat or dampness removal is required. If more south-wall area is available, the solar-heated air could be used to heat both the corridors and common areas as well as the apartments adjacent to the solar wall panels. Roof-mounted panels are possible where snow is not a problem or if the roof is steep.

The Ouellette Manor installation has received extensive press coverage in the USA and Canada because it is the tallest panel in the world. Both the building superintendent and manager of the Housing Authority have been interviewed on television. The Ontario Ministry of Housing which oversees the Housing Authorities already has implemented similar solar wall installations in other communities and now routinely contacts the supplier whenever a recladding project is being proposed.

## ACKNOWLEDGMENTS

Manufacturer of the solar panels: Conserval Engineering Inc., 200 Wildcat Road, Toronto, Ontario M3J 2N5, Canada; Tel. +1 416 661-7057; Fax +1 416 661-7146
Solar panel: High performance SOLARWALL®
Building owner: Windsor Housing Authority, Windsor, Ontario Canada
Location: 920 Ouellette Avenue, Windsor, Ontario
Chapter author: J. C. Hollick, Conserval Engineering Inc., Toronto, Canada

# III.11 Weinmeisterhornweg Row Houses

Berlin, Germany

*Berlin*

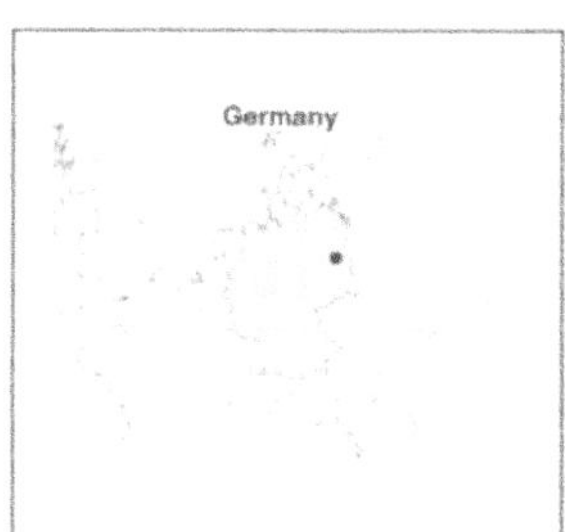


*System type 5*

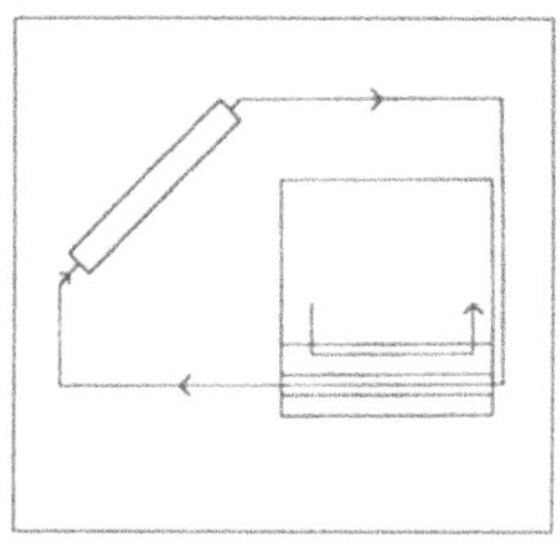

## PROJECT SUMMARY

The 60 residences in this project require significantly less heating than stipulated by regulations, thanks to a compact form, good thermal insulation of the building envelope and a deliberate use of solar energy by south-facing windows and sun spaces. Furthermore, hybrid solar systems were used in 36 residential units to store additional solar heat in the building.

### Summary statistics

| | |
|---|---|
| System type: | type 5 |
| Collector type: | air collector |
| Collector area: | 6.6 $m^2$ |
| Storage type: | concrete hypocaust |
| Storage capacity: | 1.1 kWh/K |
| Solar system output: | 90 kWh/$m^2$a |
| Total heating demand: | 46.2 kWh/$m^2$a |
| Heated floor area: | 71.3 $m^2$ |

## SITE DESCRIPTION

The row houses are located in Berlin-Spandau (50 m above sea level):

| | |
|---|---|
| latitude | 52°N |
| longitude | 13°E |

The ambient mean air temperature (September to May) is 7.0°C and the annual ambient is 9.5°C. There are 3750 annual heating degree days (base 20°C/15°C) and the horizontal global radiation is 566 kWh/$m^2$a (September to May) and 978 kWh/$m^2$a annually.

## BUILDING PRESENTATION

One of the seven rows of houses is described in detail. It is a 2½ storey building which includes three blocks, each having two apartments, one above another. The envelope surface to volume ratio is 0.51 $m^{-1}$ for the total row building, 0.57 $m^{-1}$ for the corner block and 0.40 $m^{-1}$ for the middle block. The building is orientated about 35° west of south. Figure III.11.1 shows one floor plan and a cross-section of the building.

Building statistics are given in Table III.11.1, details of construction in Table III.11.2 and heating demand in Table III.11.3.

The building is heated with radiators. Space heating and domestic hot water are produced by four gas-fired condensing boilers for all 60 residential units. There

*Table III.11.1. Building statistics*

| | |
|---|---|
| Number of apartments | 6 |
| Heated floor area of the total building | 465 $m^2$ |
| Heated floor area of the upper apartment, middle block | 71.3 $m^2$ |

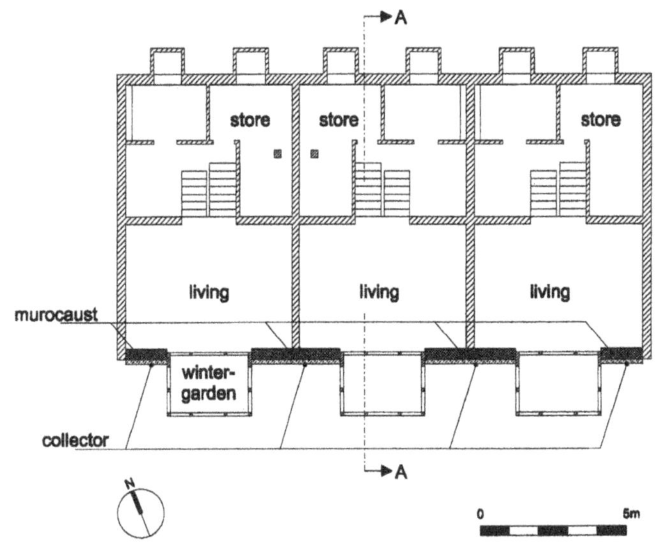


*(a) Ground floor plan*

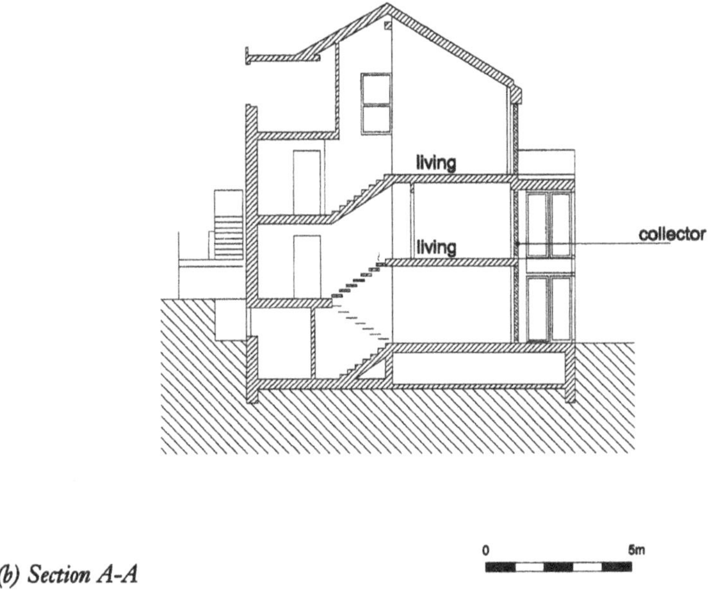


*(b) Section A-A*

*Figure III.11.1. (a) Ground floor plan of the building's basement storey and (b) cross-section of the building*

are no solar collectors installed to heat the domestic hot water. The apartments are ventilated naturally.

## SOLAR SYSTEM

Apart from the passive use of solar energy through the south-facing windows and sun spaces, a hybrid system presents another way of using solar energy.

### The system

The hybrid system consists of air collectors and thermal storage as a closed system. The air collectors are part of the south facade. The total area of the south facade of the upper apartment of the middle block is 18.6 m$^2$ – 6.8 m$^2$ window area (37%), 5.3 m$^2$ uncovered wall area (28%) and 6.6 m$^2$ wall area covered with collectors (35%). The storage wall is situated behind the collectors. The principle of the system's operating mode is shown in Figure III.11.2.

The air heated in the solar collectors is fan-forced through a duct register in the external wall. During the heating period, the fan only operates if air in the collector is about 5 K warmer than the concrete storage wall. To ensure that the heat from the concrete wall is only released to the living room when heating is actu-

*Table III.11.2. Details of construction*

| | | |
|---|---|---|
| Walls: | | |
| South | $U$ = 0.29 W/m²K | Concrete 24 cm<br>Insulation 12 cm<br>Exterior plaster 1 cm |
| North, east and west | $U$ = 0.35 W/m²K | Interior plaster 2 cm<br>Lime sand bricks 24 cm<br>Insulation 8 cm<br>Ventilated exterior lining |
| Roof: | $U$ = 0.30 W/m²K | Earth<br>Insulation 12 cm<br>Timber 2.4 cm |
| Floor | $U$ = 0.37 W/m²K | Screed 6 cm<br>Insulation 10 cm<br>Concrete 15 cm |
| Windows: | | |
| Building | $U$ = 1.5 W/m²K, $g$ = 0.62 | |
| Sun space | $U$ = 2.6 W/m²K, $g$= 0.79 | |

*Table III.11.3. Heating energy demand (calculated values)*

| | |
|---|---|
| Overall building | 68 kWh/m²a |
| Upper apartment, middle block | 50 kWh/m²a |

ally required, there is an insulated facing shell on the room side of the concrete. The warm cavity behind the facing shell can be ventilated by openings at ceiling and floor level. The lower opening can be closed by a motor-driven air grating. A radiator is installed in the upper part of the cavity.

The discharging of the wall is controlled such that, if heat is required in the living room, the radiator valve and the lower air grating open at the same time. The air heated by the radiator flows upwards and into the living room. Through the lower opened air grating, cooler air enters and is warmed by the storage wall as it rises. Figure III.11.3 shows the storage wall with facing shell.

## The collector

Each of the six flats of the building is provided with two air Grammer collectors, each 1.32 m wide, 2.50 m high and 0.14 m deep. The collector covering is a 6 mm pane of safety glass. The collector area per apartment is 6.6 m² and the air volume rate/collector area is 76 m³/m²h.

## Storage

The external walls behind the two collectors have a depth of 24 cm and are used as storage. In the middle of the concrete wall, which has a width of 1.50 m and a height of 2.60 m, eight vertical sheet metal ducts with a diameter of 80 mm are situated. They are connected above and below by a horizontal, unbroken main line. These main lines are joined to the collector by two ducts. Thus a closed cycle is formed between storage and collector. As can be seen in Figure III.11.2, the air flows through the collector from below to the top and through the storage wall from the top to the bottom. The volume flow rate per collector amounts to 250 m³/h. The air speed in storage ducts is 1.7 m/s, the capacity 1.1 kWh/K and the capacity/collector area 0.17 kWh/m²K.

The storage capacity could not be increased as there was not enough space available on the exterior south-facing wall. Figure III.11.4 shows a picture of the duct register and the storage wall taken during the construction phase.

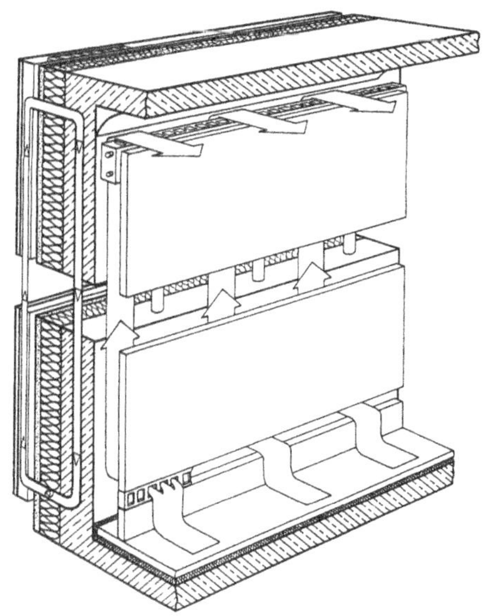

*Figure III.11.2. Isometric view of the solar system*

*Figure III.11.3. Picture of the storage wall with facing shell.*

*Figure III.11.4. The duct register and storage during construction phase* (source: *Hillmann and Rudorf, 1997)*

## Distribution

In the two ducts connecting the lower main line in the storage wall with the air collector, a radial double fan is installed. The electric energy demand of the fan amounts to 125 W, according to the manufacturer's information. To minimize noise, a reduced air flow rate mode was set. The measured electric power demand with this reduced operating mode comes to 65 W with an air flow rate of 250 m³/h. Details are given in Table III.11.4.

## Control

If the air temperature in the collector exceeds the storage temperature by 5 K, the fan switches on automatically. The temperature sensors are integrated into the storage wall and the collector. Discharging is according to the heating energy demand of the room. If the indoor air temperature drops below the required value, the lower flap in the facing shell is opened by a control motor. This allows the cool indoor air to flow through the lower opening behind the facing shell and through the upper opening back into the room again. Thus the air is heated while being discharged at the storage wall.

*Table III.11.4. Details of fan operation*

| | |
|---|---|
| Type of fan | Dual blowers, ENG 2-5D |
| Pressure (max.) | 240 Pa |
| Fan power/absorber area | 20 W/m² |
| Fan power/air rate | 0.26 W/m³ |

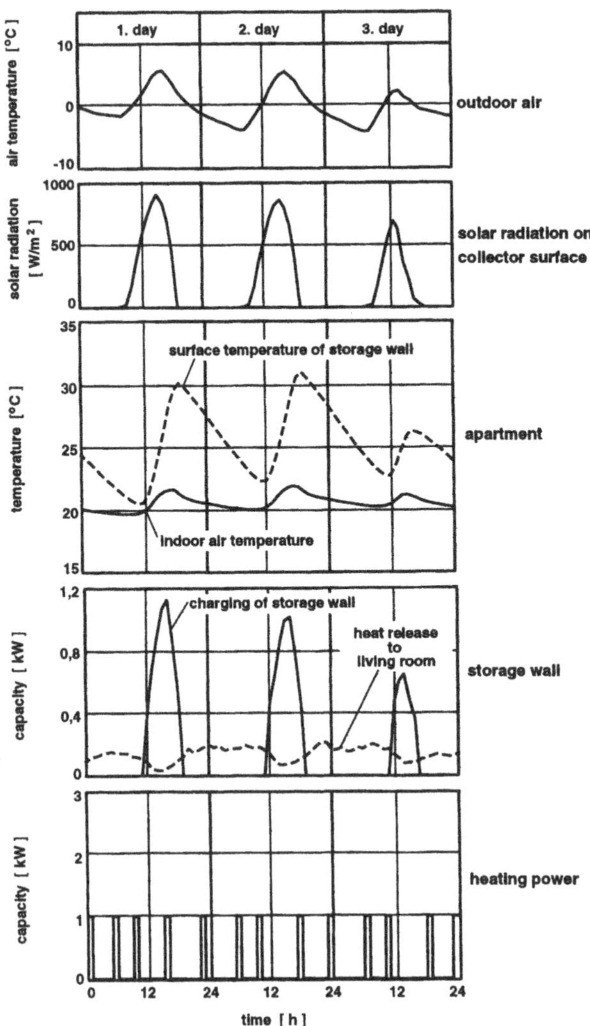


*Figure III.11.5. The time-related behaviour of air and surface temperatures, the operating state of the hybrid system and the required heating power with indication of the climatic conditions for a typical cold three-day period with high radiation intensity (8–10 April 1996) measured in the upper apartment of the middle block*

*Table III.11.5. Input and output values of the solar system energy and of electric power consumption by the fans*

| Energy inputs/outputs | Total energy (kWh) | Energy (kWh) per m²(collector area) |
|---|---|---|
| Solar energy incident on the collector surface | 3353 | 508 |
| Electric power consumed by fans | 150 | 23 |
| Solar system output (including electric power consumed by fans) | 597 | 90 |

## PERFORMANCE

The upper apartment of the middle block was monitored from December 1994 to May 1996. During the heating period 1995/96 (September–May) the solar energy that was incident on the collector surface amounted to 3559 kWh. 537 kWh of heat was supplied from the system, including the 150 kWh consumed by the fans. The degree of utilization of the collector comes to 13%. Table III.11.5 presents the total energy values and the values related to 1 m² of collector area.

The energy gained by the air collector and fed into the storage wall and the energy release to the room for heating purposes is shown in Figure III.11.5 as an example for a cold three-day period in April 1996.

The outdoor air temperature presented in the top diagram of Figure III.11.5 lies between −5°C and +5°C. The maximum solar radiation of about 900 W/m² can be rated as relatively high. As the building is orientated about 35° west of south, the maximum temperature does not occur exactly at noon, but later in the afternoon. The indoor air temperature of the living room adjacent to the charged storage wall lies between 20°C and 22°C. That the maximum temperature occurs at the same time as the maximum solar radiation can be explained by the solar gains collected by the windows. The maximum values of the solar wall's surface temperature exceed 30°C. During the period considered, its minimum values did not drop below the indoor air temperature. The charging of the storage wall starts at about 11.00 a.m. and continues until 6.00 p.m. If the maximum energy amount fed into the storage wall occurs at about 4.00 p.m., the storage wall surface shows maximum temperatures approximately three hours later. The energy that was transferred at the two storage walls during the three-day measurement period amounted to 13 kWh. The storage wall is discharged optimally such that the energy release is minimal during times of high radiation intensity and high solar gains through the windows, whereas, during times without radiation intensity, the energy release reaches a maximum power of approximately 200 W. The bottom diagram shows the heating power supplied to the apartment via radiators. The total heating energy of 13 kWh equals the solar energy amount supplied to the storage wall.

*Table III.11.6. Survey of all energy balance portions measured during the heating period 1995-96, related to the floor area*

| Energy balance portions (kWh/m²a) | Upper apartment of the middle block |
|---|---|
| Heating energy | 46.2 |
| Gains: | |
| passive | 26.6 |
| solar system gains | 8.4 |
| internal gains | 15.9 |
| Losses: | |
| transmission | 45.1 |
| ventilation heat losses | 51.9 |

Table III.11.6 shows the portions of the energy balance related to the floor area and measured during the heating period 1995/96.

## REMARKS

The evaluation of the measurements offers the following insights:

- The solar system reduced the heating energy consumption by 8 kWh/m²a to 46 kWh/m²a during the heating period 1995/96.
- The active discharge of the building's storage wall proved to be efficient. The energy is released to the apartment with a delay relative to the solar gains through the windows, thus reducing the risk of overheating.
- The hybrid systems are the least cost-effective of all thermal solar systems. They are approximately comparable to photovoltaic systems.
- The power consumption of the fans, which constitutes 25% of the transported energy, is too high. Despite the energy savings obtained, the total energy costs per year for heating and electric energy for the fans and pumps cannot be reduced.

To improve the effectiveness of future solar hybrid systems, the following is recommended:

- The efficiency of fans has to be improved substantially in order to ensure that the solar gains, which are to be used to reduce the heating energy, are not cancelled out by the cost of the electricity for the fans. The electrical power input has to be cut down from 75 W to less than 20 W.
- The fans should be operated with direct current and in relation to the solar radiation, so that the speed is reduced during times of low solar radiation, thus allowing a higher temperature level to be reached.
- The efficiency of the solar collectors should be improved. The determined operating ratio of the collectors of 13% should be improved to achieve the same efficiency as hot-water collectors.
- Air ducting systems cause noise due to air flow and fans. This should be given more attention when hybrid systems are designed, because a high noise level may cause the occupants to switch off the system. Fans should never be rigidly anchored to a component of the building's structure to prevent noise transmission.
- For reasons of economy, the use of hybrid systems should be considered only if the building is highly insulated, thus requiring only minimum heating energy demand. In the present case, further reduction of the heating energy by 8–10 kWh/m²a as a result of adding insulation may be very expensive. Combined with extreme insulation systems, an improved hybrid system can be competitive with other technical systems.

## ACKNOWLEDGEMENTS

Architects: IBUS Architekten im Ingenieurbüro für Bau- u. Stadtplanung GbR, Caspar-Theyss-Strasse. 14A, 14193 Berlin, Germany

Development and scientific consulting: Fraunhofer-Institut für Bauphysik, Nobelstrasse 12, 70569 Stuttgart, Germany

Builder: GSW Gemeinnützige Siedlungs- und Wohnungsbaugesellschaft Berlin mbH, Kochstrasse 22/ 23, 10969 Berlin, Germany

Research Commission by: Senatsverwaltung für Bauen, Wohnen und Verkehr, IV Ö1, Württembergische Strasse 6, 10707 Berlin, Germany

Chapter authors: Johann Reiss and Hans Erhorn, Fraunhofer-Institut für Bauphysik, Nobelstrasse 12, 70569 Stuttgart, Germany

## BIBLIOGRAPHY

Hillmann G, Rudorf W (1997). *Forschungsvorhaben Hybridwand. Weinmeisterhornweg 170–178, Berlin-Spandau.* Endbericht der IBUS GmbH, Berlin.

Lehmann B, Pinnig J, Berger F (1997). *Energieeinsparungen im sozialen Wohnungsbau am Beispiel des Projektes Weinmeisterhornweg 170/178.* Abschlussbericht der Wohnstadt GmbH, Berlin.

Reiss J, Erhorn H (1997). *Solare Hybridsysteme in einer Reihenhaus-Wohnanlage am Weinmeisterhornweg in Berlin.* Report WB 88/97, Fraunhofer-Institut für Bauphysik, Stuttgart.

# IV Schools

# IV.1 Introduction

## BUILDING CHARACTERISTICS

Schools and kindergartens are ideal buildings in which to demonstrate solar energy use. Normally, schools are two or three storeys high, with net floor areas between 3000 and 20000 $m^2$ (20 $m^2$ per student) and kindergartens are one or two storeys high with floor areas of 300 $m^2$ to 1500 $m^2$. Modern schools often include large atria or glazed spaces for breaks and activities. To maximize daylighting, classrooms often face south, especially in colder climates. Hence, there are many possibilities for solar air heating applications.

## HEATING AND VENTILATION REQUIREMENTS

A high level of comfort concerning air quality, room temperatures and lighting has to be guaranteed during the short periods of occupancy. The following are important factors:

- High ventilation rates to ensure good air quality may account for 30 to 60% of total energy demand.
- Night setback causes high heating load peaks during the mornings.
- Kindergarten floors should be heated to between 19°C and 26°C because children play on the floor.
- Entrance vestibules can be designed to minimize infiltration losses due to student traffic.

Internal gains are high, typically 25–40 W/$m^2$ from students and 10–30 W/$m^2$ from lights. In well insulated buildings, these gains may cover much of the heating load during occupancy. A dual heating system is sensible: a basic low-capacity system and a fast system for morning peaks or extremely cold weather. Shading devices are essential for rooms facing in directions from south-east to west.

## SOLAR AIR HEATING APPLICATIONS

The major applications of solar air heating systems in schools include:

- solar air collectors to heat mechanically supplied fresh air;
- sun spaces and atria for supplying air heating and/or room heating by venting warmed air into the colder parts of the building while keeping glazed areas comfortable;
- a closed-loop system to warm floors and walls, providing a comfortable indoor climate; double envelopes may be appropriate for retrofitting the envelopes of badly insulated schools;
- domestic hot water heating from the above-mentioned systems (via an air-to-water heat exchanger) during periods when no heating is required.

The schools and kindergartens in Germany and Austria provide insights from long-term measurements and detailed computer simulations.

## ACKNOWLEDEMENTS

Author:
C. Muss, Kanzlei Dr. Bruck

# IV.2 Secondary Modern School

## Koblach, Austria

*Koblach*

*System type 1*

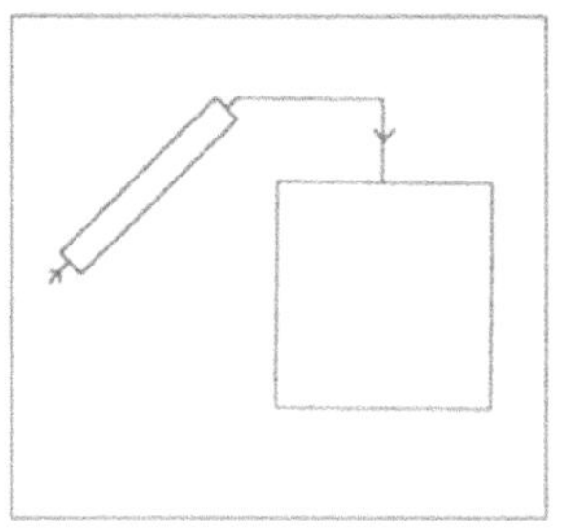

### PROJECT SUMMARY

This school has eleven classrooms, as well as rooms for special lessons, a library, a three-court sports hall and several non-educational rooms. A large area (178 m$^2$) of single-glazed solar air collectors has been sited between the windows on the south facade. A combination of reheating of ventilation air by means of the solar air collectors, control of ventilation rates and air quality, and heat recovery from return air considerably reduces the auxiliary heating demand.

If no heating of ventilation air is needed, the air collectors are used for DHW heating. Cooling of the building during periods of hot weather is achieved by night ventilation.

#### Summary statistics

| | |
|---|---|
| Solar system | type 1 |
| Collector type: | glazed, permeable and facade-integrated |
| Collector area: | 178 m$^2$ |
| Storage type: | none |
| Annual contribution of solar air system: | 45 MWh/a |
| Basis: | monitored |
| Gross heated floor area, volume: | 6,250 m$^2$, 32,530 m$^3$ |
| Annual auxiliary heat consumption: | 253 MWh/a |
| Total space heating load: | 66 kWh/m$_{gfa}^2$a (gfa = gross floor area) |
| Ventilation heating load: | 27 kWh/m$_{gfa}^2$a |
| Year solar system built: | 1994 |

### SITE DESCRIPTION

The building, located in the valley of the Rhine at Koblach in western Austria, is surrounded by single-family houses. The south facade is free from shading by trees, buildings etc. An aspect worth mentioning is the light-coloured concrete terrace on the south side of the building that provides additional solar energy gains for the south (albedo approximately 60%):

| | |
|---|---|
| latitude | 40 N |
| longitude | 10 E |
| altitude | 445 m above sea level. |

From October until April, the mean ambient air temperature is 4°C (annual mean 9°C), the mean global radiation on the horizontal plane is 390 kWh/m$^2$ (annual mean 1077 kWh/m$^2$) and the mean heating degree days (basis 20°C/12°C) are 3335.

### BUILDING PRESENTATION

Figure IV.2.1 shows the plan, a section and an elevation of the south facade with the integrated air collectors. The plans of the ground floor and of the second floor

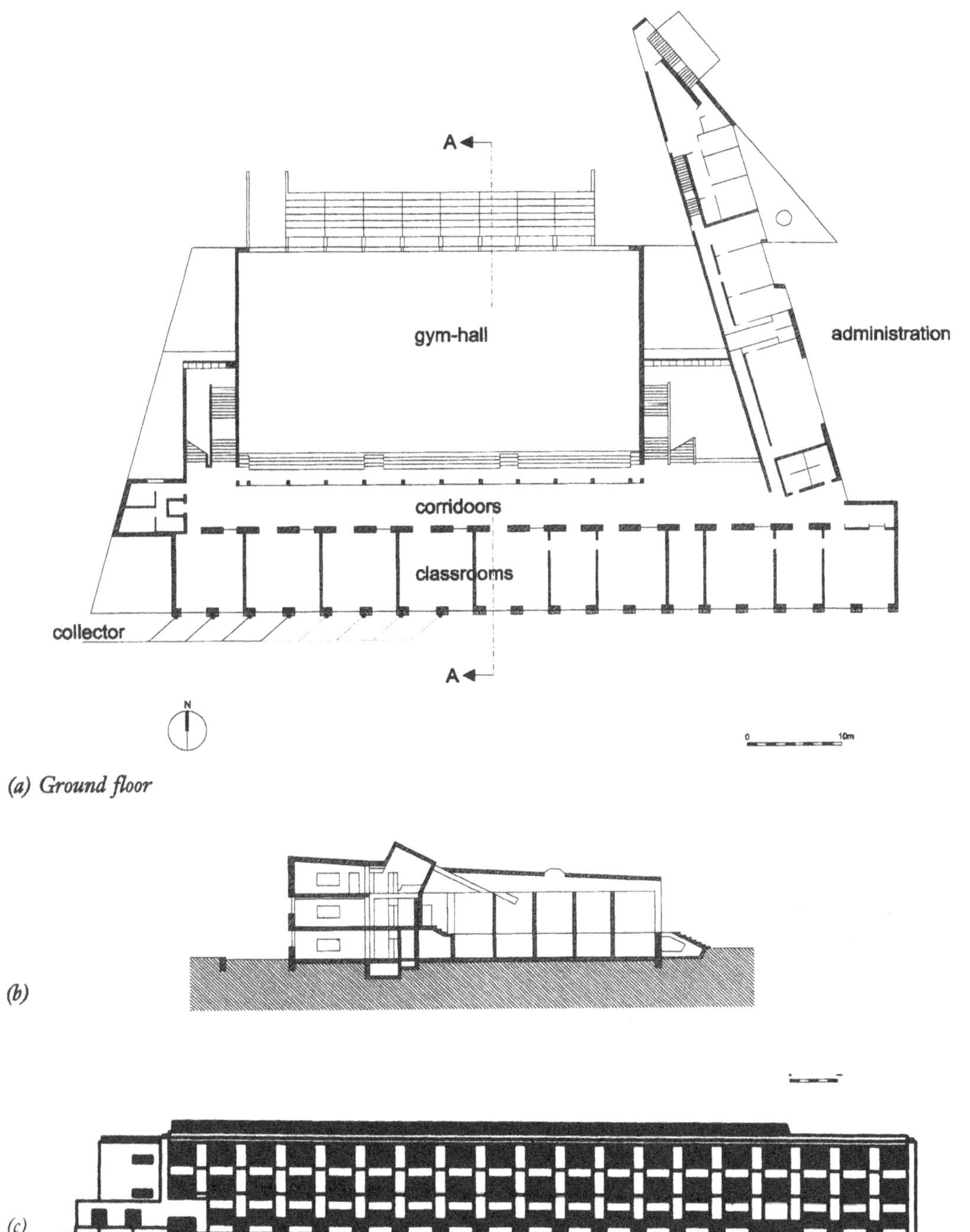


*Figure IV.2.1. (a) Plan of the ground floor, (b) a section and (c) an elevation of the south facade showing the solar air collectors placed between the windows*

correspond to the plan of the first floor (except for some differences in the usage of the rooms and grass roof areas on some northern parts of the building above the first storey).

Building statistics are given in Table IV.2.1 and details of the building envelope in Table IV.2.2.

*Table IV.2.1. Building statistics*

| | |
|---|---|
| Gross heated floor area (gfa), gross heated volume | 6,250 m$^2$, 32,250 m$^3$ |
| Total building heating load coefficient | 7790 W/K |
| Building heating ventilation load coefficient | 4750 W/K |
| Auxiliary energy for hot water production | 10 kWh/m$_{gfa}{}^2$a |
| Electricity demand for appliances and lighting | 16 kWh/m$_{gfa}{}^2$a |

*Table IV.2.2. The building envelope* k *(W/m²K)*

| | |
|---|---|
| Glazing (windows, glass roof of foyer<br>U = 1.3W/m$^2$k, g = 0.62 | 0 |
| Wall (plaster, 30 cm hollow brick, 8 cm insulation, plaster) | 0.3 |
| Concrete grass roof, 12 cm insulation: | 0.3 |
| Concrete floor on the ground, 10 cm inside insulation: | 0.4 |

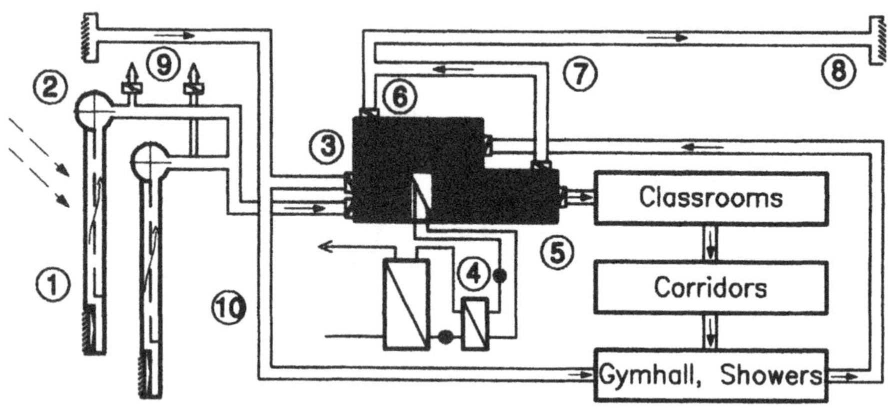


*Figure IV.2.2. The solar system: (1) solar air collectors; (2) inlet for outside supply air; (3) mixing of solar-heated air and outside supply air; (4) solar-supplied service water heating; (5) auxiliary heating of supply air; (6) heat recovery; (7) bypass during hot-water heating without supply air requirements; (8) exhaust air outlet; (9) outlet dampers for thermosyphon collector venting in summer; (10) extra supply air for sports hall (when there are evening activities) and for night-cooling*

*Table IV.2.3. Control of the ventilation system*

| Characterization | Control/conditions |
|---|---|
| Ventilation rates | Air quality controlled; max. air change in the classrooms (9000 m³/h) |
| Supply of air heating via collectors | Classroom temperatures < 20°C |
| Supply of air from outside | Classroom temperatures > 22°C |
| Mixing of solar heated air with outside air | 20°C < Classroom temperatures < 22°C |
| Service water-heating mode (if there is no ventilation air-heating demand) | Collector ventilation flow rates of 5300 m³/h, with return air directed to the exhaust air outlet |
| Auxiliary heating of supply air (during the morning and on cold winter days) | If the heating load for space heating cannot be met completely by the radiator heating system |

## SOLAR SYSTEM

The solar system of Koblach school provides solar-heated air either for ventilation air heating or for hot-water production via an air-to-water heat exchanger. A schematic view of the system is shown in Figure IV.2.2. The supply air is drawn through the collectors (1) or comes directly from outside (2) according to the control table (Table IV.2.3).

### Ventilation air-heating mode

The solar-heated air is distributed to the classrooms and then used to ventilate the corridors and the sports hall. Afterwards, the return air passes through the heat exchanger for heat recovery (6). No problems of air quality resulting from the multiple usage of heated air have ever occurred. Outside the hours of occupancy, if the temperature in the classrooms is less than 19.5°C and if the temperature of the solar heated supply air exceeds 29.5°C, solar-heated air is also used for space heating.

### Solar hot-water production

In summer and if no ventilation air heating is required, the system is used for solar hot-water production via an air-to-water heat exchanger unit (4).

When the system is not in use, the solar air collectors are vented through flaps that open as a result of thermosyphonic pressure gradients and close as a result of an underpressure in the collector when the supply air fan turns on.

### Collector construction

The site-built collectors consist of a black-coated aluminium sheet with small holes ($d$ = 1.2 mm, 1% of the

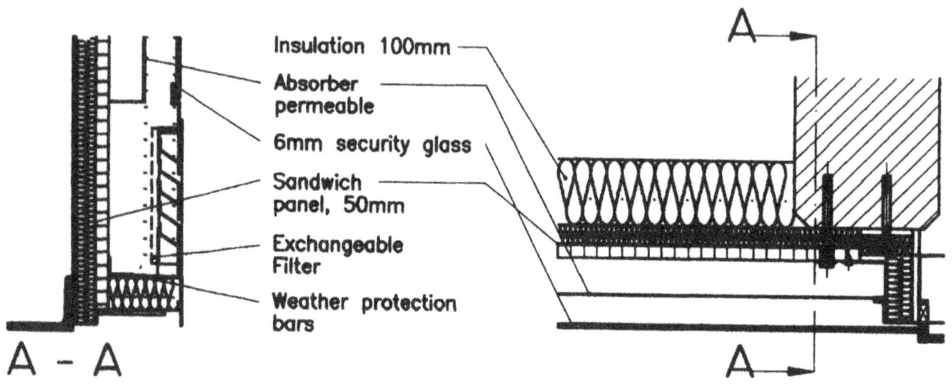


*Fig. IV.2.3. Collector construction*

*Table IV.2.4. Specifications of the collector components*

| | |
|---|---|
| Collector area | 178 $m^2$ |
| Collector elements | 17 elements, height 8 m, width 1.25 m<br>1 element, height 6 m, width 1.25 m (see also Figure IV.2.1(c)) |
| Collector air flow rate | Up to 50 $m^3/hm_{coll}^2$ |
| Mean air velocity in collector | 2.4 m/s |
| Solar contribution | 191 $kWh/m_{coll}^2 a$ for ventilation air heating,<br>62 $kWh/m_{coll}^2 a$ for service water heating |
| Power of ventilation fans (for the whole ventilation system) | 2.4 kW |
| Annual energy consumption by fans | 3000 kWh |
| Controls | Differential controllers |

absorber area) that functions as a permeable solar absorber (Figure IV.2.3). Supply air enters from the bottom of the collector via weather protection bars and an exchangeable filter.

The glazing of the solar air collectors was chosen for architectural reasons (glazed south facade). Compared with the unglazed permeable collectors, lower air flow rates and thus higher collector outlet temperatures are possible. This is an advantage for the solar hot-water production. The components and their specifications are given in Table IV.2.4.

## PERFORMANCE

Both the ventilation system and the weather conditions were monitored from May 95 to April 96. In addition, the consumption of gas for auxiliary heating and hot-water production and the electricity demand were recorded monthly. The school was simulated with TRNSYS and the results were validated against the monitoring data. Figure IV.2.4 shows the measured and simulated system energy flows.

The following are results from the simulations and monitoring:

- With 34 MWh/a (191 $kWh/m^2_{coll}a$) the solar air collectors cover 46% of the annual heating demand for ventilation air heating. The total collector energy gains from October 95 until April 96 were measured as 55 MWh (309 $kWh/m_{coll}^2a$), but 24 MWh was collected during service hot-water production and contributed just 2.5 MWh via the air-to-water heat exchanger (see below).
- The solar contribution to service hot-water production was 11 MWh/a (62 $kWh/m^2_{coll}a$). This rather low value I due to the poor efficiency of the air-to-water heat exchanger and a low demand, especially in summer when the solar water tank stays hot and provides high inlet temperatures to the heat exchanger, reducing the effectiveness of the heat exchanger.
- High passive solar energy gains in the classrooms, the corridor and the sports hall, together with the internal energy gains, reduce the auxiliary heating demand and limit the active solar energy gains (see *Remarks*).
- The energy amount saved by the heat recovery is rather small, because the mean supply air temperatures are higher with the air collector system and, therefore, the heat transfer from exhaust air to supply air is lower. Note that ventilation and infiltration losses due to opening the windows are included in Figure IV.2.4. These losses occur in the administration rooms (which have no mechanical ventilation) as well as in the classrooms and in the corridor on warm days.

A rather cold and cloudy spring in 1996 resulted in 28 MWh more auxiliary heating than was needed in spring 1995. Mean weather data for Koblach showed that the simulated auxiliary heating demand was 10% less than in 1996, while the solar energy gains (restricted by shading and room overheating) were at the 1996 level.

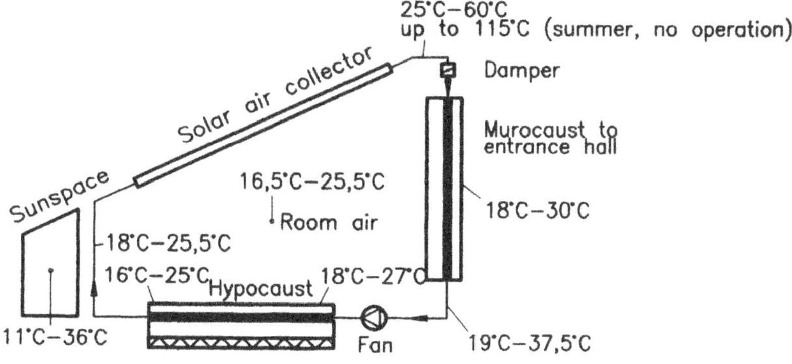


*Figure IV.2.4. Measured and simulated system energy flows*

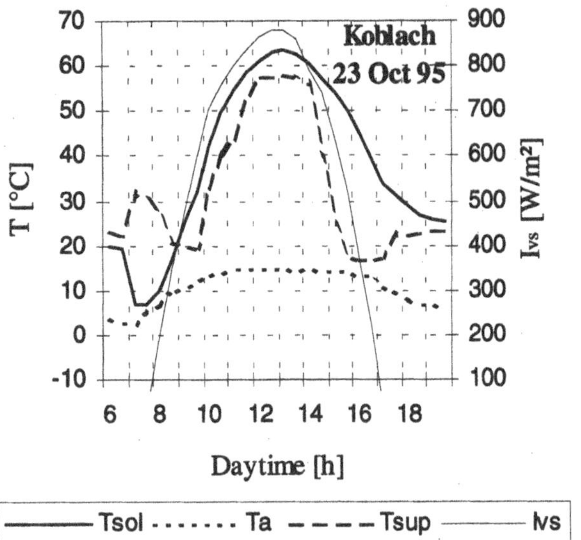


*Figure IV.2.5. System performance on 23 October 1995*

Figures IV.2.5 and IV.2.6 show the typical performance of the solar system on a warm sunny day in autumn and a cold sunny day in winter. $T_{sol}$ is the temperature of the solar-heated air, $T_a$ the ambient air temperature, $T_{sup}$ the supply air temperature (after auxiliary ventilation heating or after hot-water production and transfer to the exhaust air outlet) and $I_{vs}$ the radiation on the collector plane.

On the warm sunny day (Figure IV.2.5) the ventilation started at 6:30 and the supply was air heated by the auxiliary heater so as to support the radiators during the heating up in the morning. From 9:00 to 10:00 outside air was added to the solar-heated air from the collector, so that the supply air temperature was kept at about 20°C. Afterwards, the solar hot-water production started. The air flow rate was reduced to 5300 m³/h, the collector outlet temperatures increased to 63°C and the air transferred directly to the outlet after passing through the air-to-water heat exchanger.

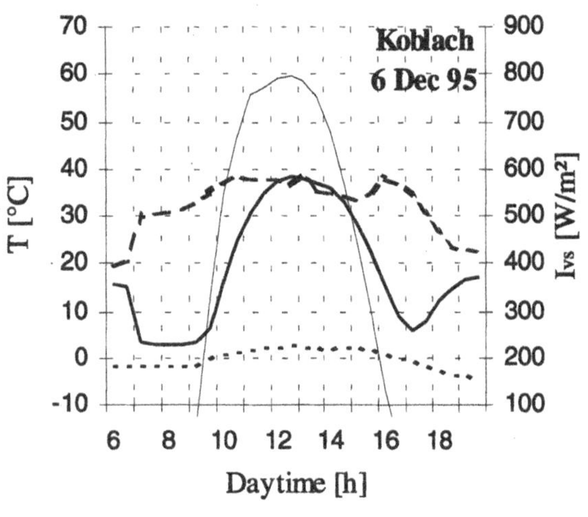


*Figure IV.2.6. System performance on 6 December 1995*

On the cold winter days, such as shown in Figure IV.2.6, the solar-heated supply air from the collectors can be completely directed into the classrooms. Supply air temperatures stay between 30°C and 40°C and hence the system can also be used for space heating.

The instantaneous efficiency of the solar air collector was determined to be 65% for ventilation air heating and 35% for hot-water production at mean working conditions (i.e. ventilation air heating at $I_{vs} \approx 300$ W/m², $T_a \approx 6$°C; hot water production at $I_{vs} \approx 800$ W/m², $T_a \approx 14$°C; see Figure IV.2.7). This corresponds to temperature rises in the collectors of 13 K (ventilation air heating mode) or 36 K (hot-water production mode), respectively. If the temperatures of the solar air collectors are lower than the temperatures inside the building, the system benefits from building transmission losses and preheating of the ventilation air occurs anyway.

The architectural positioning of the collectors (on the extended south facade and the light coloured terrace) offers a particular advantage in that the collector inlet temperatures in front of the south facade are up to 14 K higher than the ambient air temperatures. On a clear sunny day the temperature difference between collector outlet temperature and ambient air temperature may reach 50 K (see, for example, Figure IV.2.5). In summer and during hours of no air heating from the solar system, the thermosyphon collector venting operates well; the collector temperatures rise to 85°C.

The useful energy output of the active solar air heating system is limited by the passive and internal energy gains. Figure IV.2.7 shows the mean solar system power (including hours of no operation) dependent on ambient air temperature and global radiation on the collector plane. At ambient air temperatures between 10°C and 18°C and/or with incident radiation between 300 and 500 W/m², solar energy is available but often cannot be used either for ventilation air heating or for solar hot-water production. This is because the classrooms have large south-oriented windows and high internal energy gains and do not need

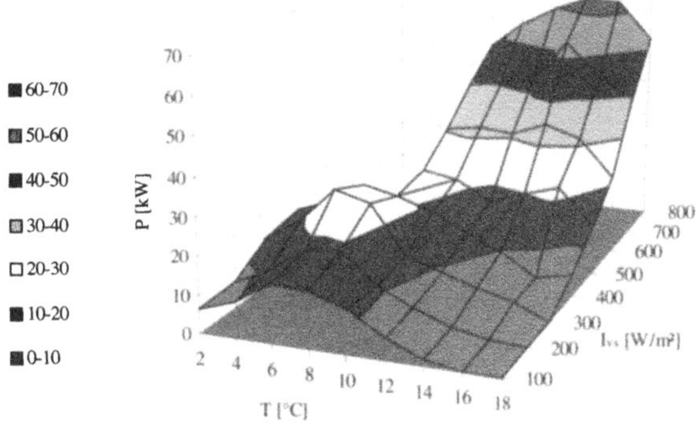


*Figure IV.2.7. Mean power* P *of the solar air heating system (including hours with no operation) as a function of the ambient air temperature* $T_a$ *and the radiation on the collector plane* $I_{vs}$

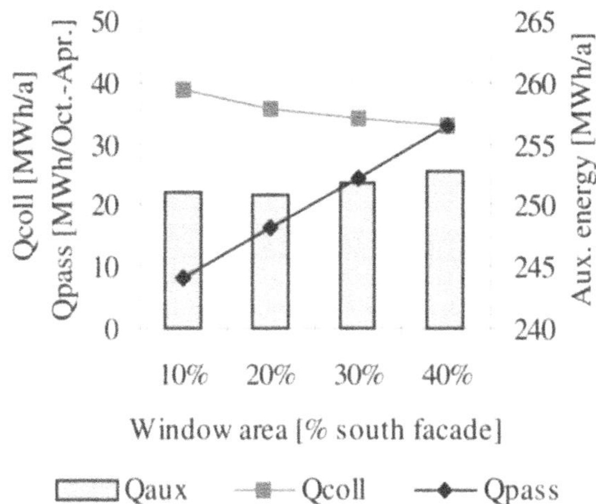


*Figure IV.2.8. Variation of the south-oriented window area*

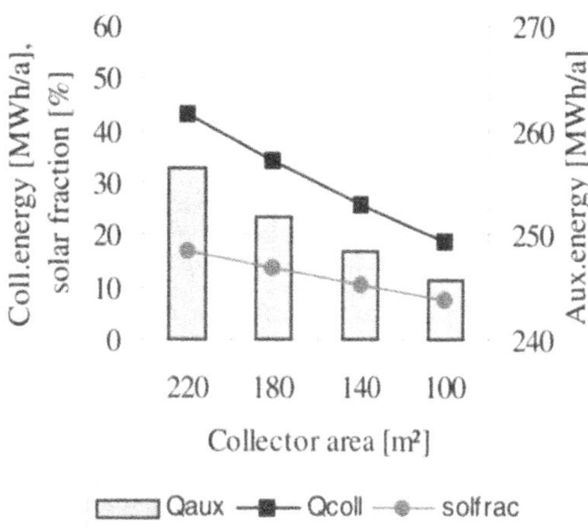


*Figure IV.2.9. Variation of the collector area and ventilation requirements*

ventilation air heating. However, at the same time, the collector outlet temperature is too low for solar hot-water production. If a bypass of the classrooms (high internal and solar passive energy gains) had been installed as part of the ventilation system, the utilization of the solar system would have been substantially greater (see simulation results).

## Simulations

In order to determine the influences of the different system parameters, computer simulations were done using mean weather data for Koblach. The TRNSYS input was calibrated against monitoring data. $Q_{aux}$ is the auxiliary heating demand, $Q_{coll}$ the solar collector energy gains, $Q_{pass}$ the incoming passive solar energy gains in the classrooms and the solar fraction (solfrac) is $Q_{coll}/Q_{aux}$.

*Design of the solar ventilation system*

In variation (a) solar-heated air bypasses the classrooms during periods of high passive solar gains and is directed to the corridors. In this case the useful output of the solar system would increase by 23% or 44 kWh/$m_{coll}^2$. The auxiliary heating consumption would decrease by 8 MWh per year.

In variation (b) solar-heated air is directed to the sports hall (gym). Here the useful output of the collector would increase by 29% or 56 kWh/$m_{coll}^2$ and the auxiliary heating savings would be 7.5 MWh per year. This value is less than variation (a) because the gym requires less heat and excess solar collector heat is simply vented to the ambient air .

*South-oriented window area*

Varying the south-facing window area from 30% of the facade had little effect on the auxiliary heating demand, but did affect the usable contribution of the solar air system, see Figure IV.2.8. The calculated 'optimum' window area of about 20% of the south facade (the system having solar active and passive energy gains in the south-oriented rooms) yielded better comfort.

*Collector area*

Figure IV.2.9 shows the influence of the size of the collector area. The specific air volume flow rate through the collector has always been kept constant at 50 $m^3$/$m^2$h. Accordingly, the ventilation requirements have been varied with the collector area. This leads to a higher auxiliary energy demand with a bigger collector area because of the higher ventilation load. However, the solar fraction covered by the solar air system rises with a bigger collector area and ventilation requirements.

*Upper limit of active solar energy gains*

A maximum possible solar output from the air collectors for ventilation air heating was calculated by setting no upper limit of the classroom temperature for the fan to turn off (ventilation occurs from Monday until Friday from 7:00 to 17:00 and on Saturday from 7:00 until 13:00). In this case, solar energy gains for ventilation air heating increase by 53 MWh/a (298 kWh/$m_{coll}^2$a), but classroom temperatures rise to 26°C, even with shading and opening the windows. From October 1995 until April 1996 the solar collectors generated 55 MWh, but 24 MWh was collected during the hot-water production mode and actually contribute only 2.5 MWh via the air-to-water heat exchanger.

With a higher hot-water demand, especially during the summer months, the useful annual solar energy gains for hot-water production could increase to 21 MWh/a or 120 kWh/$m_{coll}^2$a.

## REMARKS

The integration of site-built solar air collectors into the facade and the ventilation system proved to be cost-effective. Solar gains contributed substantially to heating the supply air. The thermosyphon collector venting during hours of non-operation proved to be an effective way of avoiding high collector temperatures and condensation problems. Monitoring and simulation results suggest the following:

- If solar-heated air is delivered to rooms with high internal and passive solar energy, ventilation ducts that can bypass these rooms should be installed so that the heated air can be directed to other parts of the building. With an optimum usage of active solar energy, a contribution of about 300 kWh/$m_{coll}^2$a for solar ventilation air heating has been calculated and confirmed by measurement data for a system like that in the Koblach School.
- Because of a low demand for hot water, the amount of solar energy used in Koblach school for hot-water production is minimal. In such a case, a solution with solar air collectors just for ventilation air heating would be preferable. With a higher demand for hot water, glazed solar air collectors may be used for ventilation air heating and hot-water production. An optimum solar contribution of about 120 kWh/$m_{coll}^2$a for hot-water production has been calculated for this system variation.

## ACKNOWLEDGEMENTS

Architect: Bösch & Bösch, Innsbruck, Austria
Initial solar simulations: Kanzlei Dr. Bruck, Vienna, Austria
Solar engineering: Christof Drexel Solarlufttechnik, Bregenz, Austria
Measurements: Energiesparverein Vorarlberg, Dornbirn; Christof Drexel Solarlufttechnik, Bregenz, Austria
Chapter author, simulations: Christoph Muss, Kanzlei Dr. Bruck, Vienna, Austria

# IV.3 Lochau Kindergarten

Austria

*Lochau*

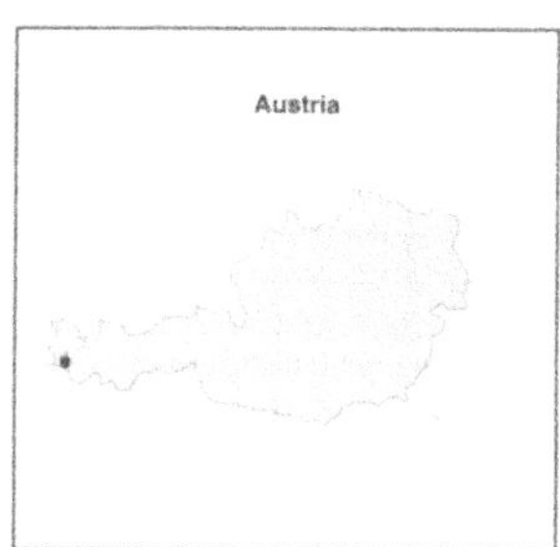


*System type 4*

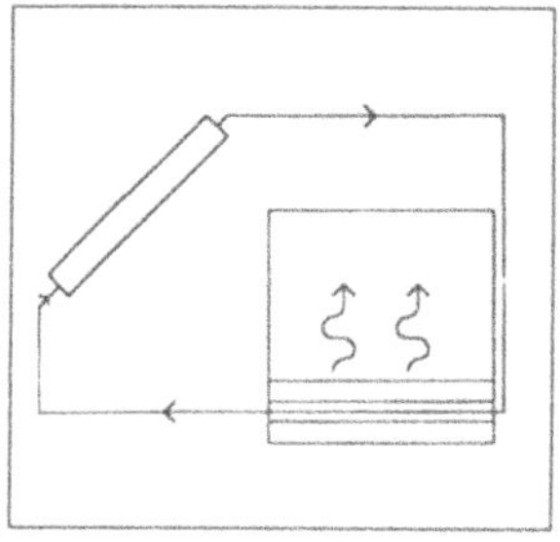

## PROJECT SUMMARY

This solar air system works with three separated closed loops between roof-integrated collectors and murocausts and hypocausts. High thermal mass provides a sufficient time delay between passive solar gains from a sun space and the passive discharge from the hypocausts. Except during December and January, the mean temperatures of the large inside surfaces warmed by the solar system stay 0.5 to 4 K above the mean room air temperatures. Uncomfortably high floor temperatures are avoided by conducting the solar-heated air first to the murocausts and then to the hypocausts. Thus the system provides a comfortable indoor climate and reduces the space-heating loads. The collector inclination of 16° leads to rather low collector outlet temperatures (due to low radiation on the collector plane). With the exception of minor overheating problems after a succession sunny and warm winter days, the system performs satisfactorily.

### Summary statistics

| | |
|---|---|
| System type, variation: | type 4 |
| Collector type and area: | Grammer, 90 m$^2$ |
| Storage type, area: | concrete hypocausts, 99.6 m$^3$ |
| Storage capacity: | 75 kWh/K |
| Total space heating load coefficient: | 510 W/K |
| Ventilation load coefficient: | 290 W/K |
| Electricity demand, appliances and lighting: | 15 kWh/m$^2$a |
| Annual solar system output: | 13.6 MWh/a |
| Annual auxiliary heat consumption: | 23 MWh/a |
| Basis | monitored |
| Gross heated floor area, volume: | 427 m$^2$, 1560 m$^3$ |
| Auxiliary heating demand: | 54 kWh/m$^2$a |
| Total space heating load: | 85 kWh/m$^2$a |
| Ventilation heating load: | 41 kWh/m$^2$a |
| Year solar system built | 1992 |

## SITE DESCRIPTION

The site in Lochau, about 500 m from Lake Constance, is surrounded by houses and is thus protected against the wind. A 19 m high building to the south of the kindergarten causes shading of parts of the collector in December and January:

| | |
|---|---|
| latitude | 47°N |
| longitude | 9°E |
| altitude | 415 m above sea level. |

Lochau has an ambient air temperature of 4°C (October until April – annual mean 9.1°C). The horizontal global radiation from October until April totals 390 kWh/m$^2$ (annual total 1080 kWh/m$^2$).

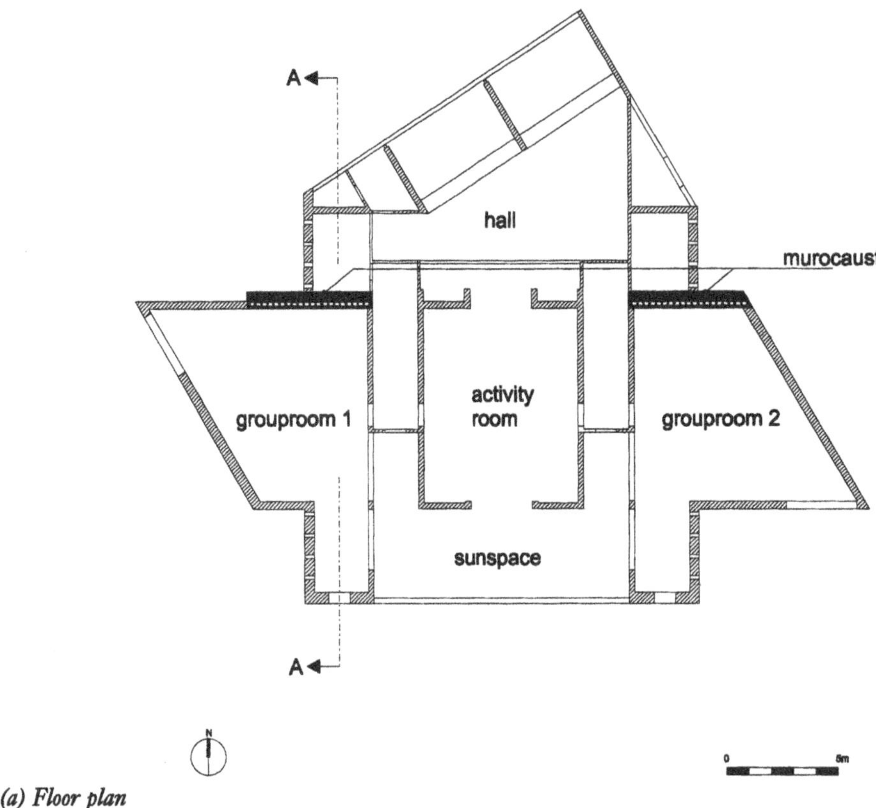


*(a) Floor plan*

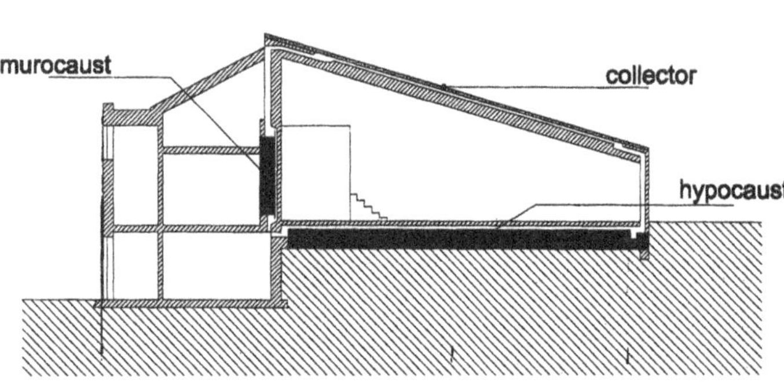

*(b) Section A-A*

*Figure IV.3.1. (a) Floor plan and (b) north–south section of the building*

## BUILDING PRESENTATION

Figure IV.3.1 shows the floor plan and a north–south section of the building. Table IV.3.1 lists the components of the roof, Table IV.3.2 gives the building statistics and Table IV.3.3 the envelope specifications.

*Table IV.3.1. The components of the roof*

| Roof with collector: |
|---|
| 0.6 cm security glass |
| 12 cm air collector |
| 0.5 cm bituminous roof covering |
| 2.5 cm wooden formwork |
| 12 cm isolation |
| 0.1 cm foil |
| 20 cm concrete roof |

## SOLAR SYSTEM

A schematic view of the solar air heating system of the activity room and the measured temperature ranges are shown in Figure IV.3.2. In the group room the warmer surfaces of the murocausts and hypocausts face into the same room (see Figure IV.3.1), whereas in the activity room system, the warmer murocaust surface faces into the entrance hall on the first storey.

*Table IV.3.4. Specifications of the system components*

| Specification | Group rooms | Activity room |
|---|---|---|
| Collector area | 30 m$^2$ | 30 m$^2$ |
| Collector length in flow direction/collector width | 10 m/3 m | 5 m/6 m |
| Height of the collector ducts | 50 mm | 28 mm |
| Air volume flow rate through the collectors* | 1560 m$^3$/h | 1690 m$^3$/h |
| Specific air volume flow rate through the collector | 52 m$^3$/hm$_{coll.}^2$ | 56 m$^3$/hm$_{coll}^2$ |
| Mean air velocity in the collector | 2.9 m/s | 2.8 m/s |
| Fan power | 320 W | 320 W |
| Hypocaust area | 56 m$^2$ | 58 m$^2$ |
| Murocaust area | 16.5 m$^2$ | 22 m$^2$ towards entrance hall |
| Number of ducts in the hypocaust and the murocausts | 30 | 30 |
| Diameter of the ducts in the hypocaust and the murocausts | 8 cm | 8 cm |
| Distance between the ducts in the hypocausts and in the murocausts | 10 cm (hypocaust), 8 cm (murocaust) | 15 cm |
| Heat capacity of the hypocausts | 17.5 kWh/K | 18.2 kWh/K |
| Heat capacity of the murocausts | 7.1 kWh/K | 8.2 kWh/K |

* All fan and ventilation data for full ventilation rate.

*Figure IV.3.2. Schematic view of the solar air heating system with measured data for the surface temperatures, the temperatures in the solar system loops and the room air temperatures*

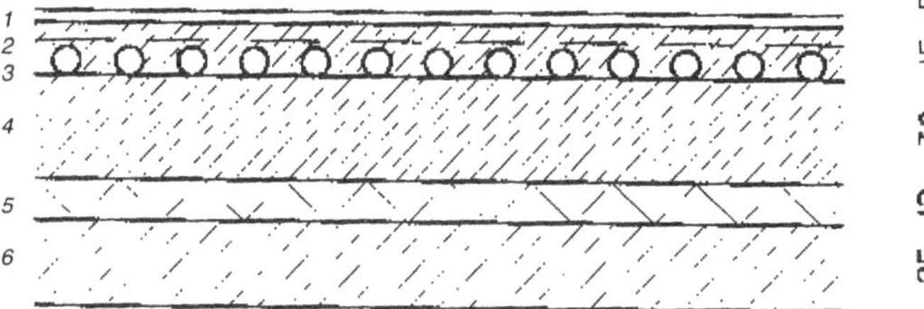


*Figure IV.3.3. Section through the hypocaust floor (the separated construction of the hypocausts and the foundation could have been avoided, see remarks): 1 floor covering (linoleum); 2 cement floor; 3 concrete with a 5 mm iron grate and 8 cm aluminium ducts; 4 concrete B160; 5 insulation; 6 concrete foundation*

This has been done because of higher passive solar energy gains and a lower heating load in the activity room.

Each of the three closed air collector loops is equipped with one fan that provides two different ventilation rates and is switched according to the following control:

- *Half ventilation rate*: collector outlet temperature $T_{coll}$ minus hypocaust outlet temperature $T_{ho}$ greater than 4 K and $T_{coll}$ > 25°C.
- *Full ventilation rate*: $T_{coll}$ > 36°C; reset to half ventilation rate if $T_{coll}$ < 28°C.
- *Turn off*: $T_{coll} - T_{ho}$ < 1 K or if the room air temperature exceeds an upper temperature limit (usually 24°C during the heating season).

The specifications of the system components are listed in Table IV.3.4. Figure IV.3.3 shows a section through the hypocausts and Fig. IV.3.4 shows the integration of the solar air collectors into the roof.

## PERFORMANCE

Both the system and the weather conditions were monitored from February 1995 to April 1996. The building and its solar air heating system were analysed with TRNSYS. The simulation inputs were calibrated against measurements. Figure IV.3.5 shows the energy balance of the system based on the measurements complemented by the simulation data. A graph of the measured mean weekly temperatures in group room 1 is presented in Figure IV.3.6 (hypocaust surface temperature = mean of inlet and outlet temperatures).

From the simulations and monitoring the following observations were possible:

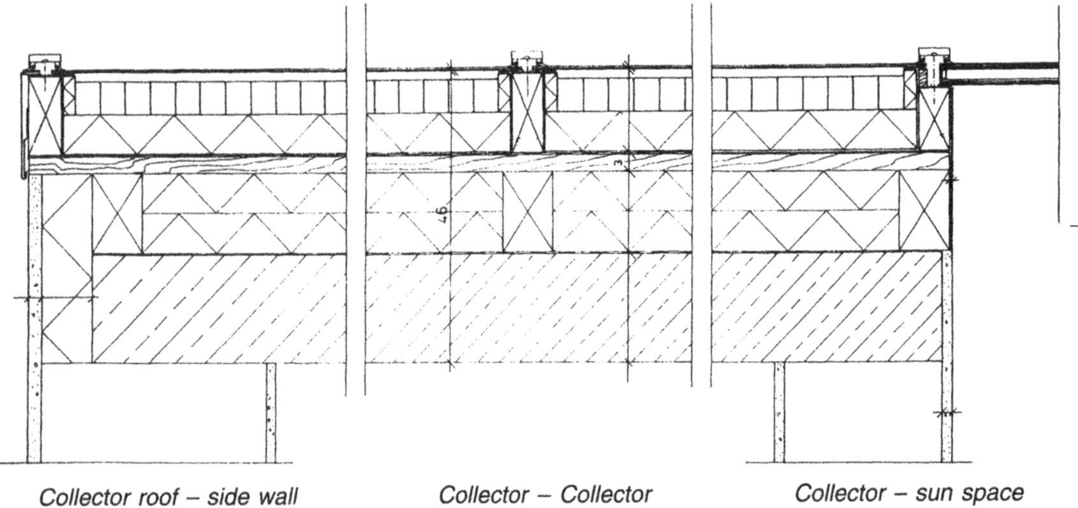


*Figure IV.3.4. Details of the collector integration into the roof and connections to the sun space*

- In spite of high passive solar energy gains, a substantial amount of energy comes from the active solar air system. A comfortable indoor climate results from the surface temperatures, which are significantly above those which would occur without the active system. Also, the high thermal masses help to avoid overheating from the active and passive solar energy gains.
- In general, the system functions best during cold and sunny weather conditions but the passive hypocaust discharge also provides solar heating during overcast periods following sunny days. The system is able to work with rather low incident radiation values (Figure IV.3.7).
- During September and October solar energy from the collectors is used for space heating and for the charging of the thermal masses. However, as a result of the warm ambient air temperatures in the autumn, the system is not activated until the middle of October. For that reason, the surface temperatures and the collector energy gains are significantly lower than in spring (Figure IV.3.8).
- In November, solar heating occurs from the radiant discharge of hypocausts and murocausts charged by the air collectors.
- Losses from the rooms to the floor occur first in December. The active solar energy gains are of minor importance in December and January.
- In February and March sunny periods are used efficiently for space heating and recharging of the hypocausts and murocausts.
- High collector energy gains result in rather high room air temperatures in April. This could have been avoided by lowering the turn-off temperature for the fan.

**Lochau, Oct 95 - Apr 96**

Collectors: 12.8 MWh  Auxiliary: 25.3 MWh

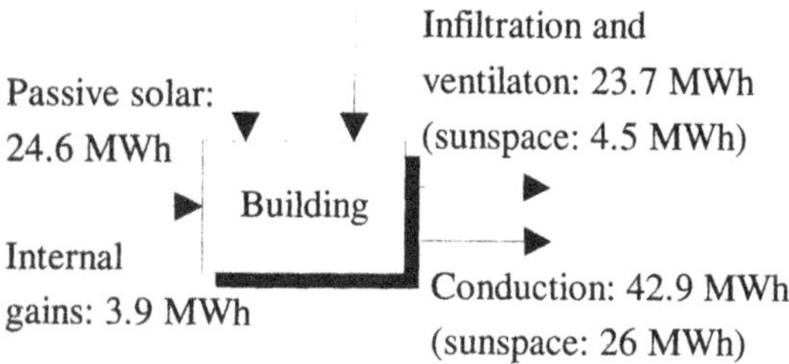


Energy flows for the sunspace included in the data without brackets

*Figure IV.3.5. System energy balance; collector energy and auxiliary heating measured and simulated; other data simulated*

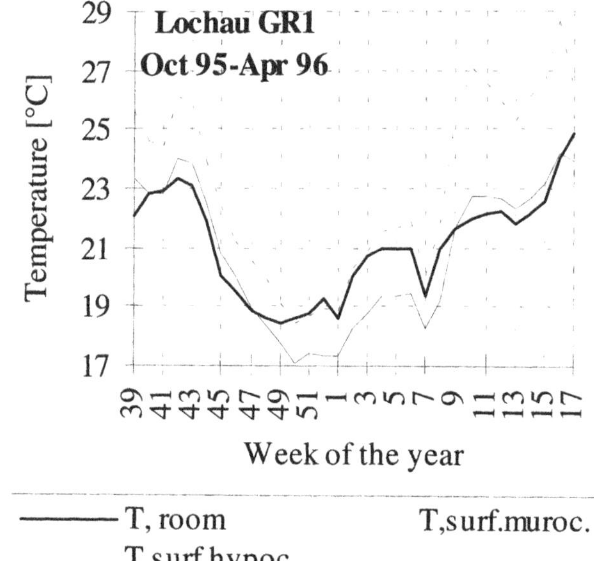


*Figure IV.3.6. Mean weekly temperatures of the room air and of the surfaces of the murocausts and hypocausts in Group room 1*

From October 95 until April 1996, the global radiation on the horizontal plane (420 kWh/m²) has been 8%

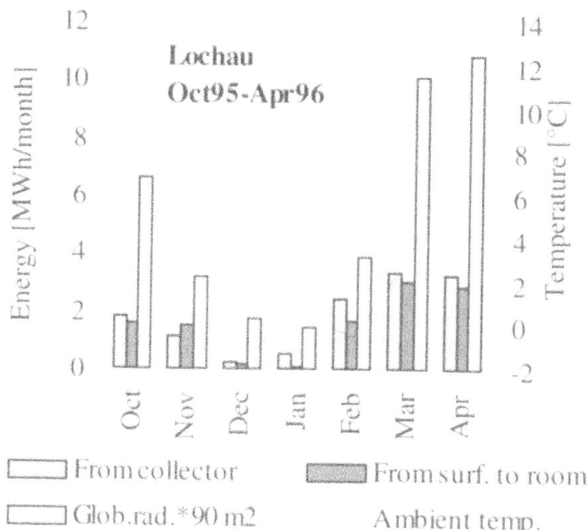


*Figure IV.3.7. Decrease of the room air temperatures and of the surface temperatures during a week with low solar radiation after a rather sunny period (5-11, 6-11: weekend).*

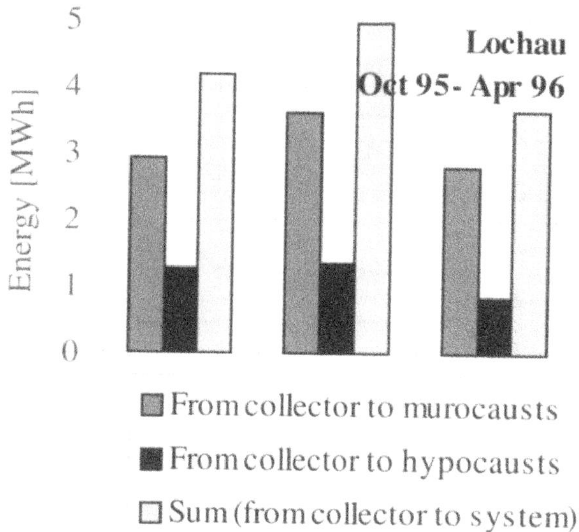


*Figure IV.3.9. Distribution of solar energy gains from the collectors to the murocausts and to the hypocausts*

higher than the mean value for Lochau (390 kWh/m²), although the ambient air temperature has been almost the same (4°C). Nevertheless, it was a rather untypical year with a warm autumn and a long and cold spring, which caused a 20% higher annual heating demand. For mean weather data at Lochau the solar energy gains from the collectors were simulated to 11.6 MWh from October until April (13.6 MWh from September until May). The auxiliary energy demand for mean weather data has been determined to 23 MWh/a. (Appropriate handling of the solar system control has been assumed: upper room-temperature limit for the solar system fan to be turned off: 22.5°C in September, April and May, and 24°C from October until March.)

The distribution of collector energy gains to murocausts and hypocausts (Figure IV.3.9) shows that

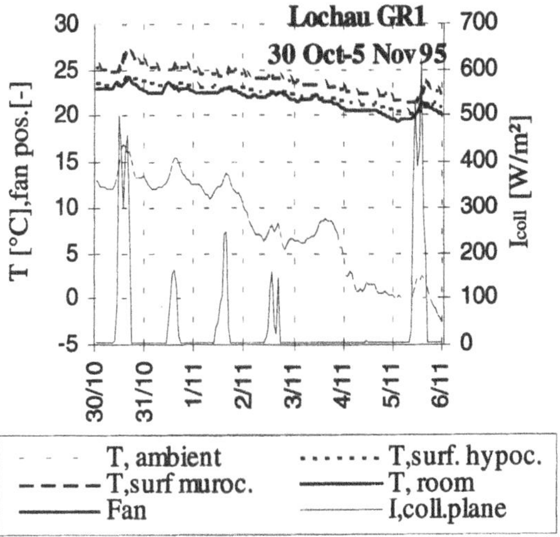


*Figure IV.3.8. Monthly solar energy gains from the collectors (measured and simulated) and total energy from the heating surfaces to the room (simulated), monitored weather data.*

the small pipe diameters (8 cm) and high air velocities in the hypocausts (1.5–3 m/s) lead to such a high heat transfer between the solar-heated air and the air ducts in the concrete that about two thirds of the collector energy gains is transferred to the murocausts. Thus, the high storage capacity and the large area of the hypocausts are not used efficiently.

To determine the system behaviour without occupant influences, the unoccupied kindergarten was monitored for two weeks during the summer holidays. Doors and windows were closed, the sun space was left unshaded and the rooms were completely shaded. Figure IV.3.10 shows the system performance on an typical summer day, with the fan turned on between 12:00 and 18:00 (heating) and between 21:00 and 8:00 (night cooling). Between the start of the fan and the maximum surface temperatures there is a time delay of 6 h for the murocausts and 8.3 h for the hypocausts. During the cooling program, a system cooling power of about 30 to 50 $W/m_{coll}^2$ was measured.

## TRNSYS simulation analysis – parametric runs

Parametric runs were performed for mean weather data at Lochau for the time period from October until April. $Q_{aux}$ is the auxiliary energy demand, $Q_{coll}$ is the active solar energy from the collectors and $Q_{pass}$ represents the passive solar energy gains entering through the windows and the sun space; $T_{grmax}$ is the maximum room air temperature in Group room 1 (October–April) and $T_{smin}/T_{smax}$ is the minimum/maximum room air temperature in the sun space (October–April)

*Collector area*

Figure IV.3.11 shows the decrease of collector energy with a smaller collector area (air volume flow rates per collector area are kept constant). There is no corresponding rise in the auxiliary energy demand, because

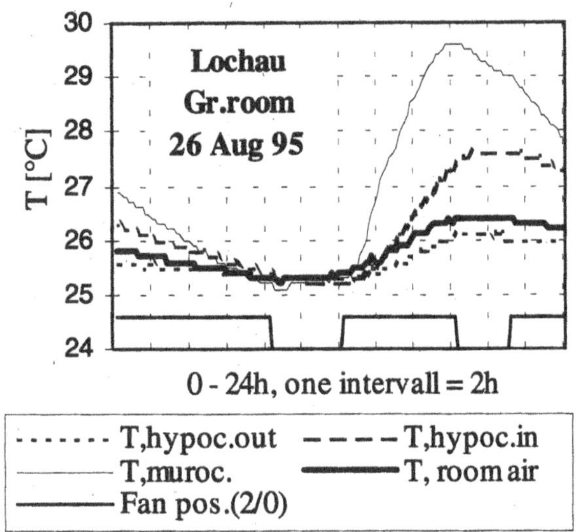


*Figure IV.3.10. Temperatures of the room air and the surfaces during one day in the heating–cooling experiment in summer*

the high solar passive gains partly compensate for the decreased collector gains. Finally, from October to April the auxiliary energy demand for the system without collector is about 8 MWh higher than the auxiliary energy demand of the existing system. 31% of the difference between the collector gains from 90 m² collector area and the system without a collector are compensated by a better usage of the passive solar energy gains.

## The glazed roof area of the sun space and comfort in the sun space

To determine the influences of the passive solar gains, the percentage of the glazed roof area of the sun space was varied (the non-glazed roof area was replaced by the roof construction used for the other rooms). Figure IV.3.12 shows the increase in the collector energy gains and the decrease in the passive solar energy gains with a decreasing area of glazed roof. Minor passive solar energy gains cause the fans to reach the upper turn-off temperature limit less frequently. As expected, the sun-space temperatures stay within a more comfortable range if a smaller area of the sun space is glazed.

*System variations*

Various configurations of the original system have been analysed. Figure IV.3.13 shows the details and results of these system variations

Although the collector energy gains decrease with a lower heating area, there is only a 0.3 MWh (1.3%) increase in the auxiliary energy demand with system SYS2 (without murocausts). The auxiliary energy demand for the system SYS3 (without hypocausts) rises by 1.4 MWh (+6.1 %).This because the high storage capacities cannot be used efficiently (see also Figure IV.3.10). On the other hand, solar passive gains compensate for the diminished collector energy gains. Besides, with system SYS2, mean floor temperatures up to 28°C have been calculated.

Without an upper limit to the room temperature for switching off the solar system fan, the active collector energy gains from October until April rise to 15 MWh. However, room air temperatures rise to 30°C. A collector inclination of 60° instead of 16° would increase the collector energy gains by 3.4 MWh (29%) and reduce the auxiliary energy demand by 2.8 MWh or 12 %.

*Collector energy – upper limit*

For systems SYS5 and SYS6 a specific collector energy gain of about 170 kWh/$m_{coll}^2$ has been calculated. This value may be considered as an upper limit for systems of this kind. Moreover, solar heat from the air collector might serve for hot-water production via an air-to-

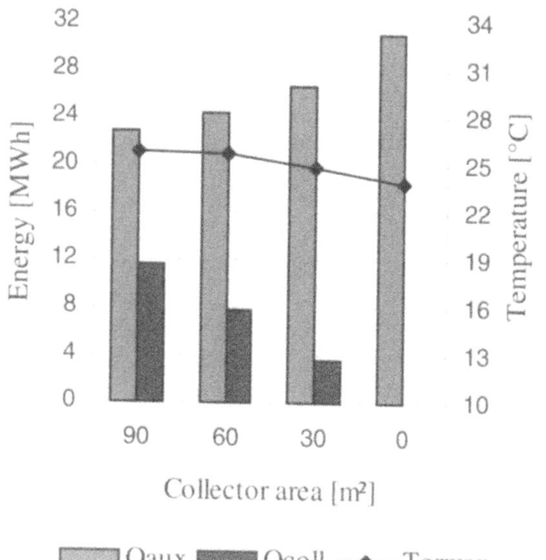


*Figure IV.3.11. Variation of the size of the collector area for simulation period October to April*

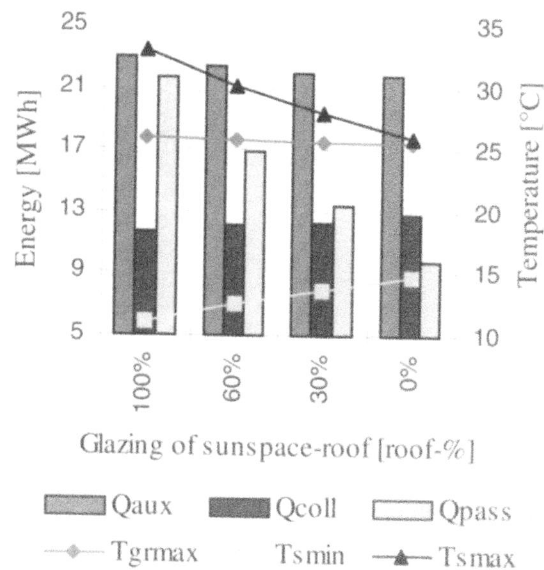


*Figure IV.3.12. Variation of the glazed roof area of the sun space for simulation period October to April*

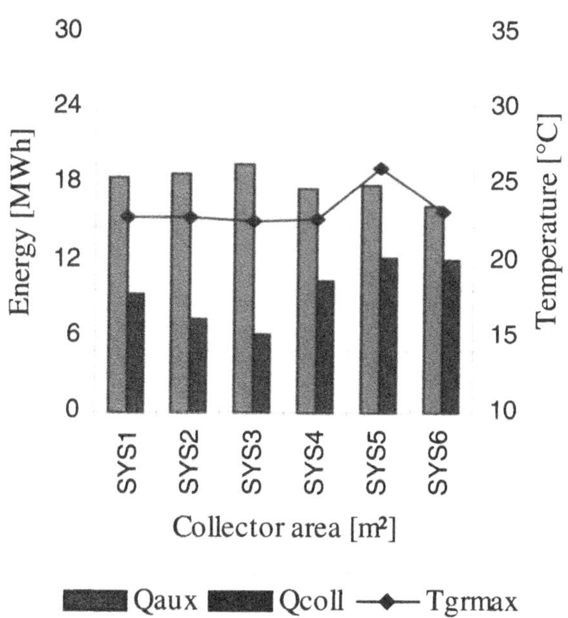


*Figure IV.3.13. System variations (simulation period October to April): SYS1 original system; SYS2 system without the murocausts; SYS3 system without the hypocausts; SYS4 system without the glazed roof of the sun space; SYS5 system without upper room-temperature limit for the fan to be turned off; SYS6 system with a collector inclination of 60°.*
*Tgrmax = group room air temperature*

water heat exchanger, particularly if there is a demand for hot water in summer.

## REMARKS

The closed-loop solar air heating system in the Lochau Kindergarten is clear and simple. It is ideal for rooms with large heating areas and it provides comfortable heated surfaces. Thermal storage within the hypocausts, murocausts and concrete ceiling provides a significant amount of solar space heating, even during overcast periods after a succession of sunny days. It also ensures a sufficient time delay between passive and active solar energy gains.

From the monitoring and simulation results the following recommendations are made for future systems of this type:

- Active and passive solar energy gains should be optimized together for buildings with important passive solar features. Thus a better solar usage (active and passive) can be achieved with optimum comfort. With large glazed spaces, use of the solar-heated air for ventilation of other building zones should be considered.
- Unnecessary duct bends (see Figure IV.3.1(b)) should be avoided by integrating the hypocausts and the murocausts into the structural engineering.
- The thickness of the hypocausts and of the murocausts can be reduced without much effect on the thermal storage capacity. The system works well with smaller heating areas, too. Larger diameters for the hypocaust and murocaust ducts (actual 8 cm; recommended 12–15 cm) would improve distribution of active solar gains. Also, the pressure drop would be less and thus a lower fan power would be required.
- Clear information about the working of the system should be provided to the users and allowances made for user control. For example, a switch between full operation (October–March), reduced operation (September, April) and no operation (summer) could be provided.

## ACKNOWLEDGEMENTS

Architect: Popelka + Schwärzler, Bregenz, Austria
Solar engineering: Kanzlei Dr. Bruck, Vienna; Christof Drexel Solarlufttechnik, Bregenz, Austria
Measurements: Eckhart Drössler, Düns; Energiesparverein Vorarlberg, Dornbirn, Austria
Simulations: Christoph Muss, Kanzlei Dr. Bruck, Vienna, Austria
Chapter author: Christoph Muss, Kanzlei Dr. Bruck, Vienna, Austria

# IV.4 Schopfloch Kindergarten

## Leonberg-Ezach, Germany

*Leonberg-Ezach*

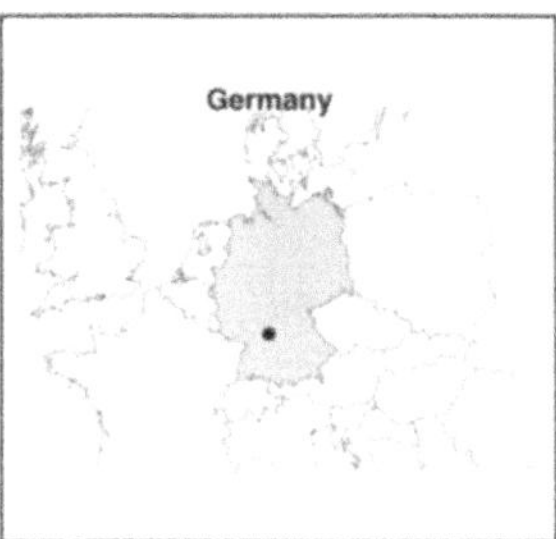


*System type 4*

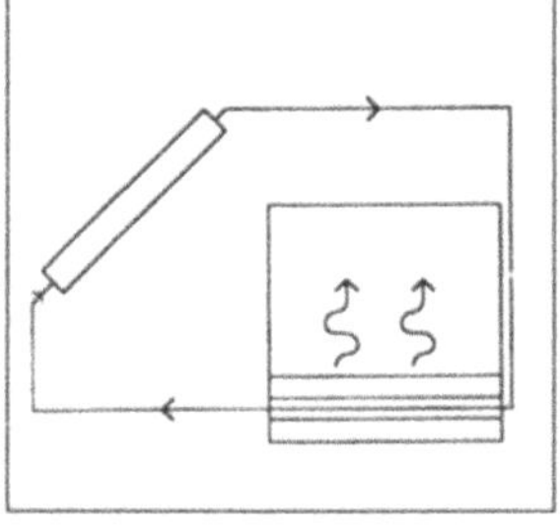

## PROJECT SUMMARY

This two-classroom kindergarten with a gym was optimized with regard to both the total solar contribution from passive solar gains through a large glazed area facing south and active solar air heating.

The active system consists of a collector, hypocaust-floor storage elements and a rock-bed storage. It offers the advantages of low heating temperatures without the problems, such as dry air or air movements, that occur with a direct air heating system. The 20 m$^3$ rock bed stores heat from a 38 m$^2$ solar air collector for the night and the next morning.

Experience since occupancy in 1988 has shown that this system works well, provided that the control system is well adjusted. As might have been anticipated, the light building construction does result in some overheating.

### Summary statistics

| | |
|---|---|
| System type: | type 4 |
| Collector type: | opaque, roof-integrated |
| Collector area: | total 38 $m_{coll}^2$ |
| Storage type: | rock bed, 20 m$^3$ |
| Total heating load: | 169 kWh/m$^2$a |
| Auxiliary heating load: | 106 kWh/m$^2$a |
| Basis: | monitored |
| Savings (active system): | 45.6 kWh/$m_{coll}^2$ |
| Floor area, volume: | 270 m$^2$, 845 m$^3$ |
| Year solar system built: | 1988 |

## SITE DESCRIPTION

The Schopfloch kindergarten is located in Leonberg-Ezach, Germany, a city located about 15 km to the west of Stuttgart:

| | |
|---|---|
| latitude | 49°N |
| longitude | 9°E |
| altitude | 400 m above sea level. |

## BUILDING PRESENTATION

The kindergarten is a single-storey flat-roofed building with one double pitched roof, ridged east–west, over the whole length of the building. The two group rooms are located on the east and west sides of the gym. The ancillary facilities are situated in the north part of the building. The rock-bed heat storage is located at the centre. Thus, the losses from the two rock-bed chambers passively heat the surrounding rooms. The ground plan and two sections are given in Figure IV.4.1.

The building has a steel skeleton with prefabricated wall elements of a wooden construction with 80 mm insulation. The pitched roof is a self-supporting steel construction with a low-*e* glazing on the northern side.

The sloped north-facing roof has about 40 m$^2$ glazing, while a white diffusely reflecting inner surface on the southern slope supplies daylight to the rooms without creating overheating problems.

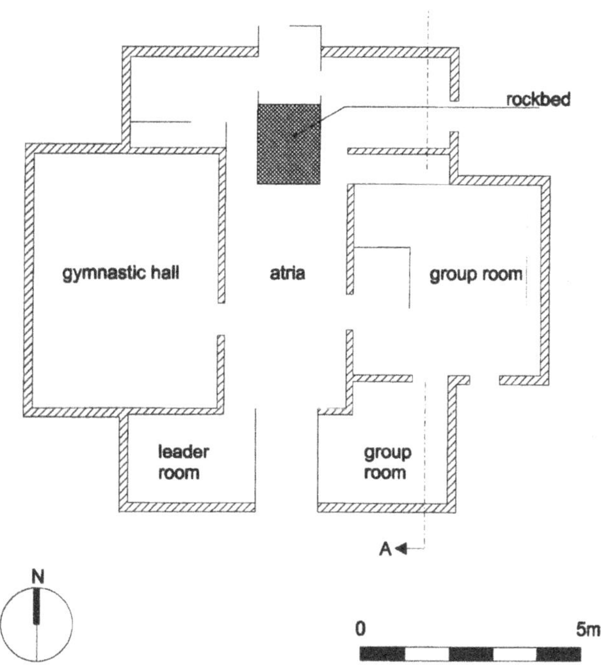


*(a) Ground plan*

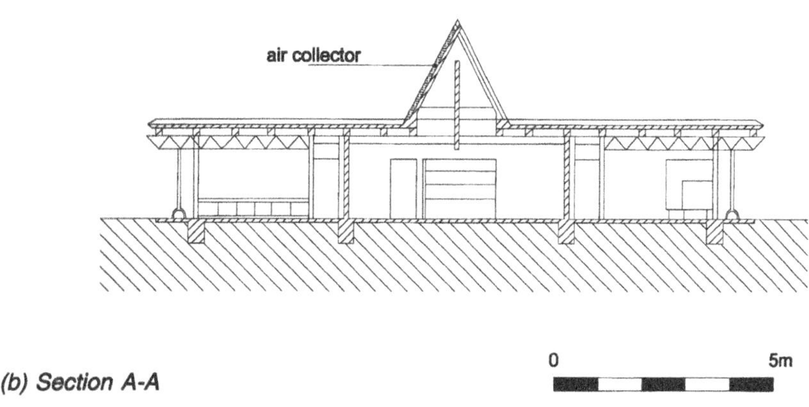


*(b) Section A-A*

*Figure IV.4.1. (a) Ground plan, (b) east–west section and (c) north–south section of the kindergarten*

The overhang is about 1.2 m and partially shades the large glazed wall areas on the south, west and east. The aim of using the overhang is to avoid additional shading devices that only work for the south orientation (see Figure IV.4.1[b]).

The building statistics are given in Table IV.4.1, while a summary of the energy demand is given in Table IV.4.2.

*Table IV.4.1. Building statistics*

| | |
|---|---|
| Gross heated floor area | 257 $m^2$ |
| Gross heated volume | 805 $m^3$ |
| Window *U* value | 3.0 $W/m^2K$ |
| Window area | 157 $m^2$ |
| Roof *U* value | 0.34 $W/m^2K$ |
| Floor *U* value | 0.59 $W/m^2K$ |
| Wall *U* value | 0.45 $W/m^2K$ |

*Table IV.4.2. Energy demand*

| | |
|---|---|
| Building total heat loss coefficient | 615 W/K |
| Auxiliary space heating | 106 $kWh/m^2a$ measured |
| Appliances and lighting | 7500 kWh/a |

## SOLAR SYSTEM

The kindergarten is heated by the concrete floor and walls, which are heated by air flowing through channels

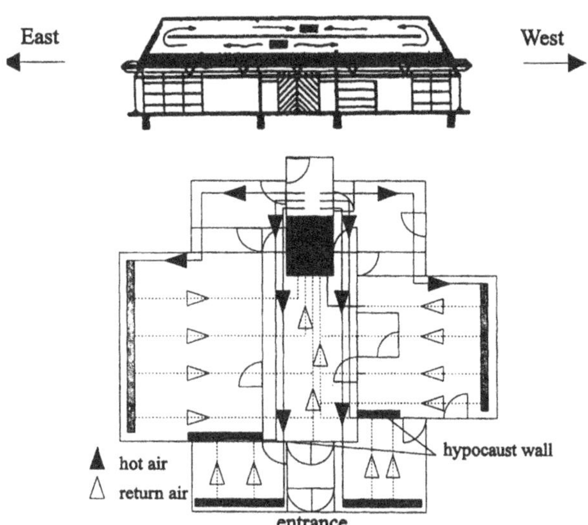


*Figure IV.4.2. Air flow in the collector and the system*

from the solar collector (Figure IV.4.2). Thus, a direct heating mode without the rock bed is possible. If the collector output is lower than the heat demand, a gas-fired auxiliary heater supplies the heat to a water heat exchanger. An additional water-to-air heat exchanger allows overheating of the wall surfaces in order to heat up the rooms quickly in the morning. Air-tight flaps in the charging and discharging ducts of the storage prevent uncontrolled discharge of the collector or the hypocaust elements. Distribution of the hot air is via the hypocaust floor and the wall elements. The two hypocaust walls are connected separately to provide for an overheated inlet. The incoming hot-air ducts are near the external walls. Return air is drawn from the sun space. Surplus solar gains from the collector are stored in the rock-bed storage. Figure IV.4.3 shows the active solar system.

### Control strategy

The whole heating and active solar system is controlled by a programmable control device. The auxiliary heating system is used if there is no heat available from the air collector or from the storage.

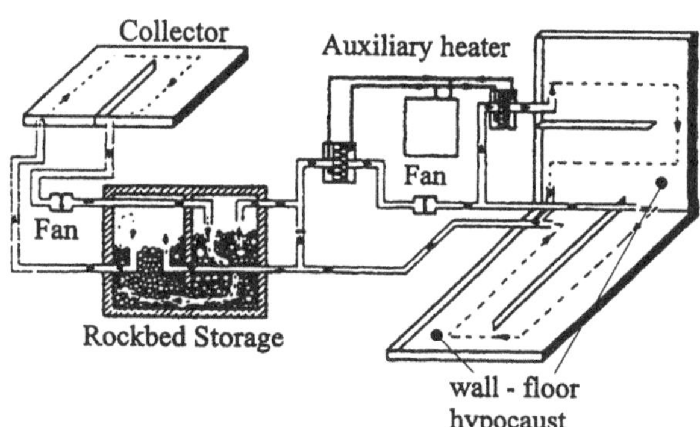


*Figure IV.4.3. The active solar system*

If the collector temperature is 8 K higher than the storage temperature, the collector fan is activated. If the temperature difference is lower than 3 K, the fan is switched off. The hot air delivered by the collector or the storage is heated if the actual inlet temperature is lower than the set point. This is influenced by the ambient temperature. The additional heater for the wall inlet air is activated if the room temperature drops below 19.5°C.

A sensor in the outlet duct of the floor heating system measures the air temperature of the outlet flow. The controller opens a bypass flap if this temperature is higher than the maximum storage temperature. Manually controlled ventilation flaps in the northern slope of the pitched roof allow the users to have individual control during the summer.

### Collector and storage

The solar air collector has a total area of 38 m$^2$ and a slope of 60°. The air flow rate is 35 m$^3$/m$^2$(collector)h. The collector fan has a rating of 0.25 kW, while the two-speed fan for distribution from the storage through the hypocaust is rated at 0.5 kW (step 1) and 2.0 kW (step 2).

Storage is in the rock bed. The capacity, including the concrete walls, is 11 kWh/K, the volume is 20 m$^3$ and the material is serpentinit rock, 50 mm in diameter. Cross-sections of the collector and floor and wall elements are given in Figure IV.4.4.

## PERFORMANCE

The site-built collector also serves as a roof covering. The absorber is a black foil, with a solar absorption of $a$ = 0.97 glued on the underlying construction. The covering is two PTFE films (Hostaflon, thickness 50 and 150 mm) stretched over an aluminium grid structure. The air circulates through an air gap of 150 mm. The transmission of the solar radiation of the films (short wave) is about $t$ = 0.9. The coefficient of solar energy, which passes through the cover and is absorbed, was calculated as $ta$ = 0.75. This value includes a decrease of the transmission due to pollution and ageing. Figure IV.4.5 shows the collector efficiency versus the temperature difference between inlet and ambient air. The collector efficiency curve shown was obtained using a linear regression.

The intercept with the $y$ axis has a value of 0.61. It is defined as $F_R(ta)$. The collector efficiency factor $F_R$ can be calculated as $F_R = 0.61/0.75 = 0.8$. This rather poor value is caused by the quite poor heat transfer between the absorber and the flowing air. The linear regression yields an $F_R k_{eff}$ = 5.4 W/m$^2$K. Thus, the value of the heat loss factor is $k_{eff}$ = 6.8 W/m$^2$K. This poor value for the double cover is explained by the relaxation of the films. An investigation after 12 months showed that the films touch each other. Consequently, a high percentage of the collector area effectively only had a single film.

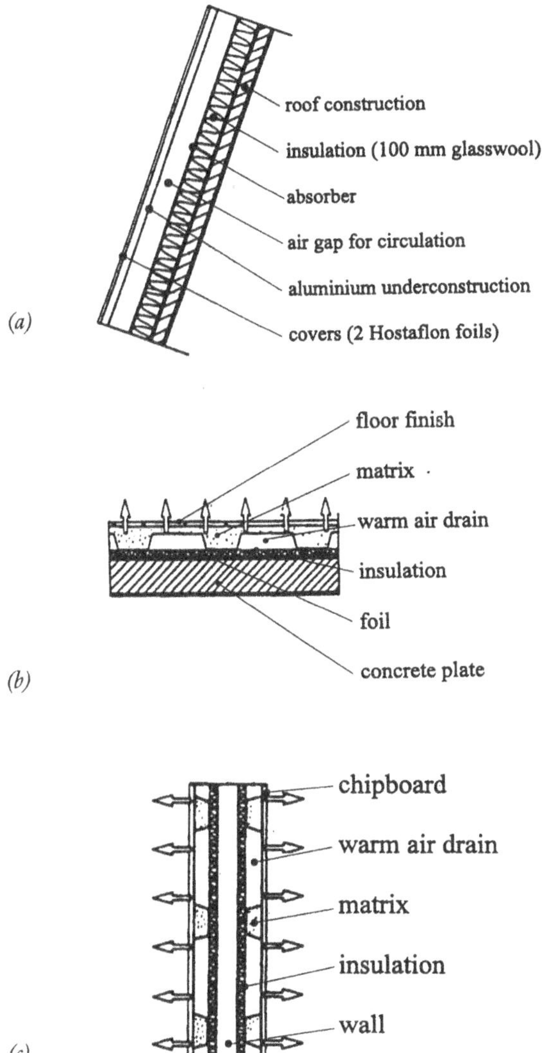


*Figure IV.4.4. Cross-sections of (a) the collector, (b) a hypocaust floor element and (c) a hypocaust wall element*

From the measurements a leakage rate of 13% from the site-built collector could be calculated. This shows that it is necessary to operate air collectors in an extraction mode.

Airflow through the collector and the storage also occurred at night owing to thermosiphoning effect. The combination of the temperature difference between the collector and the storage and the quite loose air flaps at the collector has resulted in a considerable reduction in the system efficiency. Readjustment of the air flaps substantially improved system performance.

High summer stagnation temperatures destroyed the glue bond between the absorber foil and the underlying construction. Back in operation in the autumn, the underpressure in the collector caused the foil to be 'blown out' (like a balloon). The foil then touched the films leading an additional losses and a pressure drop in the collector.

## REMARKS

The active solar system, as installed, had to be rebuilt almost immediately after installation because an air duct had been installed incorrectly. The company responsible had had no prior experience with solar systems and obviously did not fully understand this system. This meant that the storage could be charged, but not discharged. During the reconstruction, extensive measuring points were also installed. Detailed measurements were then performed during the heating periods in 1988/89 and 1989/90. Because there were comfort problems in winter, a second gas furnace and radiator system were added as the main heating system in 1996.

## CONCLUSIONS

The main cause of discomfort in winter was the long delay time of the hypocaust, which the control system could not take into account. The system performance could be improved, but the main problem, the long response time of the heating system, is an inherent

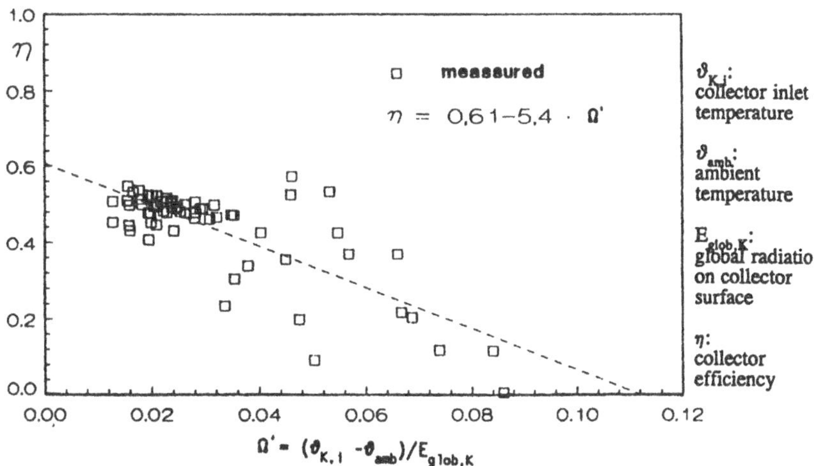


*Figure IV.4.5. Collector efficiency curve; air flow 35 m³/h*

characteristic of hypocausts. The system now functions, but its thermal efficiency still needs to be improved in order to reduce the auxiliary heating consumption of 106 kWh/m$^2$. Each component of a solar air system has to be designed carefully.

Possible improvements include:

- temporary insulation of the large glazing area in order to reduce the heat losses (the temporary insulation could also act as shading device);
- improvement of aspects of the control system, to give better control, for example, of:
  - the status and temperature of the gas boiler;
  - room temperature;
  - the speed control fan (the energy used to distribute the heat reaches 25% of the total heating energy);
- air-tight flaps
- regulation of the heating zones separately;
- ensuring that all materials used for the collector construction will withstand the stagnation temperature (the collector showed that foils without any regulation of tension are not suitable for solar air collectors);
- a two-path airflow around the absorber to increase the collector efficiency;
- insulation of all the collector ducts.

## ACKNOWLEDGEMENTS

Architect: Bela Bambek, Toblacherstrasse 30, 73773 Aichwald, Germany

Chapter author: Matthias Schuler, Helmut Meyer Transsolar Energie Technik GmbH, Nobelstrasse 15, D-70569 Stuttgart; Email schuler@transsolar.com

# IV.5 Green Park School

## Newport Pagnell, UK

*Newport Pagnell*

*System type 1*

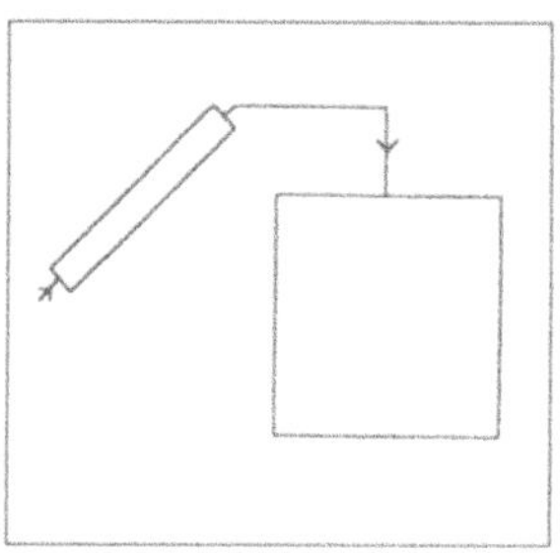

## PROJECT SUMMARY

This school for 5–12 year olds is built around a courtyard to provide good levels of daylight, solar gains and natural ventilation. Glazing on three south-facing pitched roofs converts these roof spaces into solar collectors, sometimes called a hot attic. Because the collectors are unoccupied, high operating temperatures are possible. Outside air enters through the eaves into the glazed attic collectors, is sun-warmed then further heated by the gas-fired warm air auxiliary heating system.

The solar air system is built from conventional building components and so is easy to design, construct, operate and maintain. The collector glazing forms the weather barrier and has minimal impact on the architectural design. Most importantly, the roof-space collectors are not shaded by the surroundings.

A thermal simulation program predicted that the system designed for this school would reduce auxiliary heating energy use by about 6%. Further optimization could have increased the savings to around 14%.

### Summary statistics

| | |
|---|---|
| Solar system type: | type 1 |
| Collector type: | roof space |
| Collector area: | 112m$^2$ |
| Solar system output: | 3.9 kWh/m$^2$a |
| Total space heating load: | 68 kWh/m$^2$a |
| Auxiliary heating load: | 64 kWh/m$^2$a |
| Basis: | calculated |
| Heated floor area, volume: | 1400m$^2$, 4600m$^3$ |
| Year solar system built: | 1990 |

## SITE DESCRIPTION

The school is located in Newport Pagnell, a small town on the eastern fringe of Milton Keynes, about 100 km north-west of London. Although in a suburban residential area, there is no over-shading of the south-facing facades:

| | |
|---|---|
| latitude | 52°N |
| longitude | 1°W |
| altitude | 100 m above sea level. |

The site has a temperate maritime European climate with average daily temperatures ranging from 2°C in January to 22°C in July. To the standard UK base of 15.5°C, there are 2310 heating degree days. The total annual horizontal solar irradiance is 3400 MJ/m$^2$ (945 kWh/m$^2$) and the prevailing wind is from the south-east.

## BUILDING PRESENTATION

The school was designed for 360 children aged 5 to 12 years and completed in August 1990. It is a single-storey structure with communal facilities (shared teaching areas, school hall, library, activity room, etc.) ar-

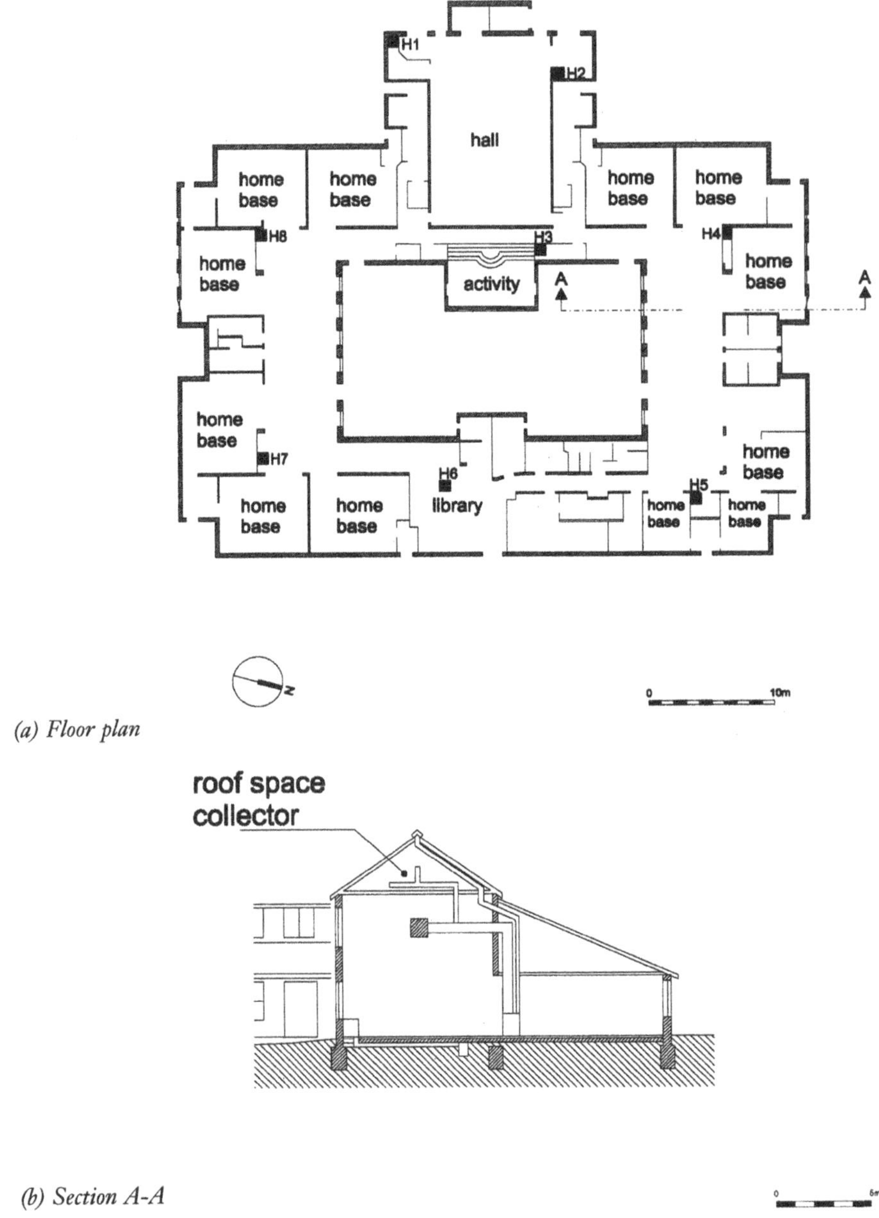


*Figure IV.5.1. (a) Floor plan, showing the location of heaters, and (b) a section of the school*

ranged around the central courtyard. Three individual teaching spaces are grouped together in the four corners at the outside perimeter. The children can enter these 'home bases' directly through the cloakrooms, which also double as draught lobbies. This can be seen in Figure IV.5.1.

The school is built on a 125 mm thick reinforced-concrete strip foundation with perimeter insulation. The outer cavity wall is brick, 25 mm extruded polystyrene and 150 mm blockwork; all internal walls are dense concrete block or load-bearing brickwork; and ceilings have 150 mm of mineral wool insulation.

The windows are double-glazed and have top-hinged wooden frames with built-in trickle vents. All the spaces can also be naturally ventilated simply by opening windows (and doors).

The lower roofs have 22.5° pitch trussed rafters with concrete interlocking tiles on felt and battens. The higher-level roofs have a 35° pitch to give improved solar radiation collection (Figure IV.5.2).

The *U*-values of some of the building elements are given in Table IV.5.1. The total heat-loss rate is 2.4 kW/K. Auxiliary heating is by means of a gas-fired warm-air heating system.

Direct solar gains through the south-facing openings make a contribution to winter-time heating. Internal blinds and generous roof overhangs help to control solar gain, particularly in summer. Light-

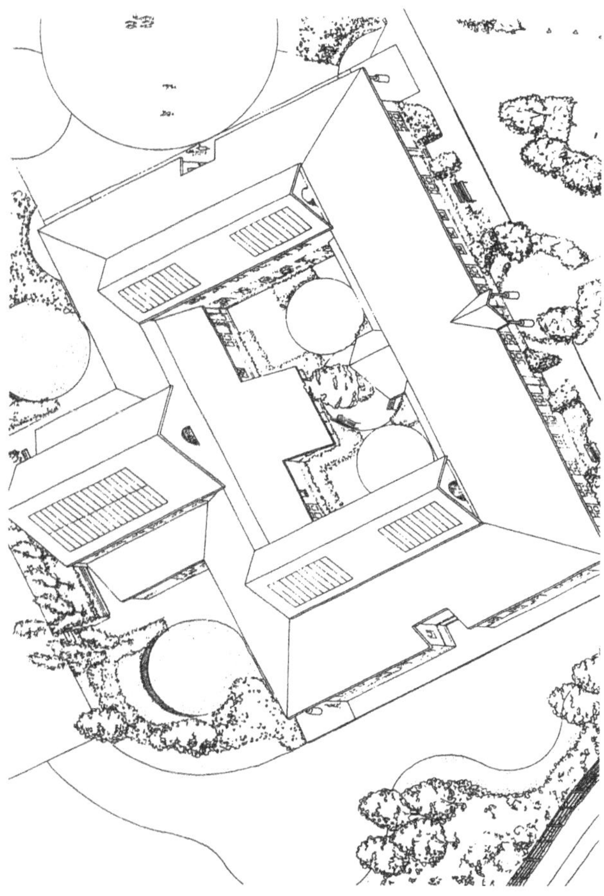

*Figure IV.5.2. Axonometric view of the school*

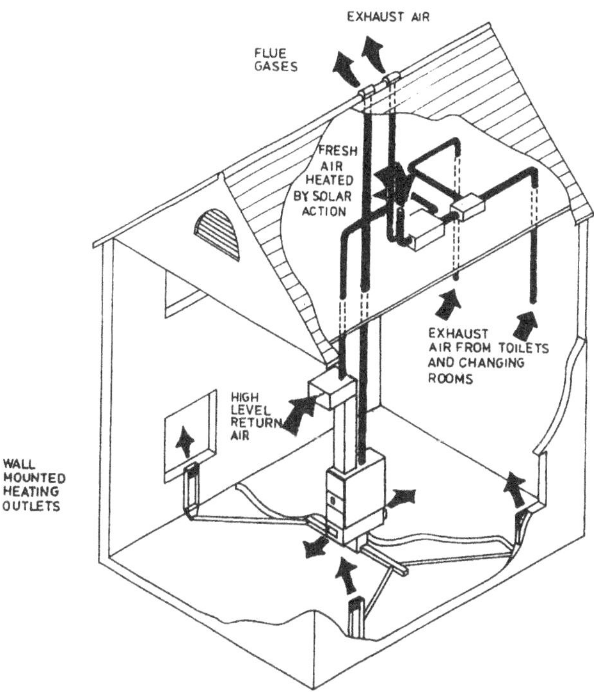


*Figure IV.5.3. Schematic diagram of the solar air system*

coloured paving helps to bounce daylight into the building.

The glazing on the south-facing roofs has the effect of converting the roof spaces into five solar air collectors, which pre-heat fresh air for the warm-air, gas-fired auxiliary heating system.

The heating system is deliberately simple. Gas-fired warm-air heaters draw in recirculated air from the teaching spaces and fresh air from the roof-space collectors (Figure IV.5.3). The air is then redistributed directly to the core or, via under-floor ducts, to the perimeter (Figure IV.5.4). A heat exchanger reclaims heat from the air that is exhausted from toilets, changing rooms and cloakrooms. The recovered heat is added to the fresh warm air from the collectors before it is fed to the heaters.

Summer comfort is improved by the incorporation of high levels of internal mass in the building to help even out peak temperatures. The aim has been to produce a building where 27°C is exceeded in the habitable spaces no more than 10 days a year.

*Table IV.5.1. Building elements*

| | *U*-value (W/m²K) |
|---|---|
| External walls | 0.42 |
| Roof | 0.28 |
| Floor | 0.38 |
| Windows | 3.0 |
| Collector 'polycarbonate' | 3.1 |

## SOLAR SYSTEM

The heating system uses eight conventional gas-fired warm-air heaters (Figure IV.5.1[a]), sized to ensure that teaching spaces can be kept at 18°C and the hall 14°C when the external temperature is –1°C. An energy management system controls the set-point temperature and the morning pre-heat is optimized based on external ambient temperature.

The warm air is distributed by rectangular insulated ducts located below the floor slab with circular distribution ducts feeding low-level outlet grills located on exterior walls (Figure IV.5.4).

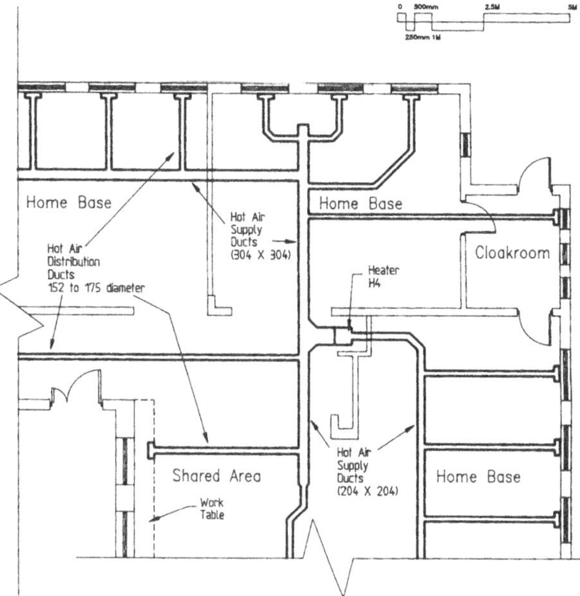


*Figure IV.5.4. Warm air ducts from heater H4*

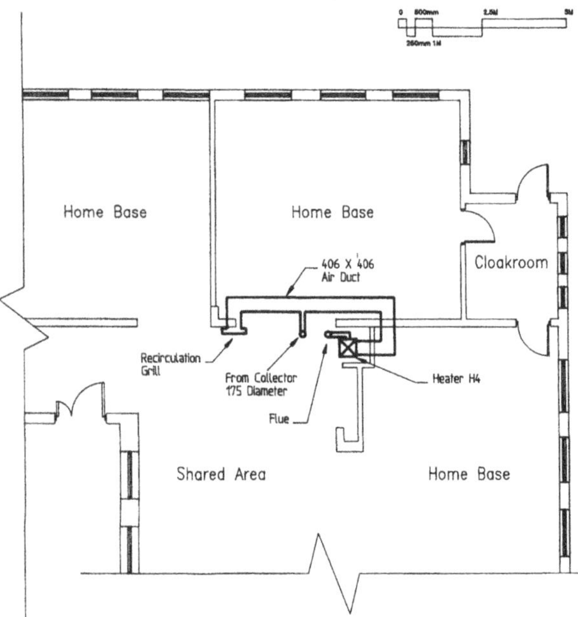


*Figure IV.5.5. Fresh air and exhaust flues from heater H4*

Air is extracted from the spaces at high level and recirculated through the heaters. The intention was that two-thirds of the total air to the heater should be from this source with one-third being fresh air from the solar collectors. In principle, all the air could be recirculated, in particular during the morning pre-heat periods (Figure IV.5.5).

Air is mechanically extracted from the showers, cloakrooms, changing rooms and toilets adjacent to the hall and passed through a (J & S HR200) cross-flow plate heat exchanger, before being exhausted. The exchanger recovers 65% of the heat to further warm the fresh air supplied to the heater.

The novelty of the system is that the fresh air supplied directly to the heaters (H4, H5, H7, H8) or through the heat exchanger to the heaters (H1 and H2) is pre-heated in glazed roof-space collectors (RSCs) and so is warmer than ambient air (Figure IV.5.6). There is no solar pre-heat to the smaller heater H3 or to the heater on the east side (H6). The largest collector, over the hall, supplies the heaters on the west side (H1 and H2) and the other collectors supply the heaters on the south and north sides. The total volume of the collectors is 440 m$^3$. The roof space above the south and north areas is all one volume. Foil-backed fire-resistant plasterboard lines the ceilings of the spaces below.

The fresh air enters the collectors at eaves level through standard ventilation slots and through semi-circular vents in the gable ends (Figure IV.5.3). These, of course, incorporate insect mesh. The collectors are built like a conventional roof, with concrete tiles on softwood battens over roofing felt. The insides of the collectors are lined with dark-brown building paper and there is 150 mm of insulation above the ceilings. To produce efficient collectors, however, some of the south-facing roofing was replaced by Röhm Makrolon SDP10, which is 16 mm double-skin polycarbonate glazing. It has a $U$-value of 3.1 W/m$^2$K and a daylight transmission of 80%, which is expected to degrade by 5% over 20 years. The polycarbonate sheets 0.6 × 2.4 m are fixed in aluminium frames, which are mounted side-by-side and then fixed to the roof trusses. This system has weight and cost advantages compared to glass systems with similar thermal performance.

Concern that high temperatures in the collectors could cause decomposition of the roof fabric led to the installation of dampered louvres and heat-activated fans in the gable ends of the collectors. These are set to expel air when the temperature in the collectors exceeds 40°C. The use of external blinds or other techniques to reduce peak temperatures was rejected as being too complicated and costly.

## PERFORMANCE

### Thermal simulation predictions

The school's design was developed with the aid of the program SERIRES. Annual simulations demonstrated

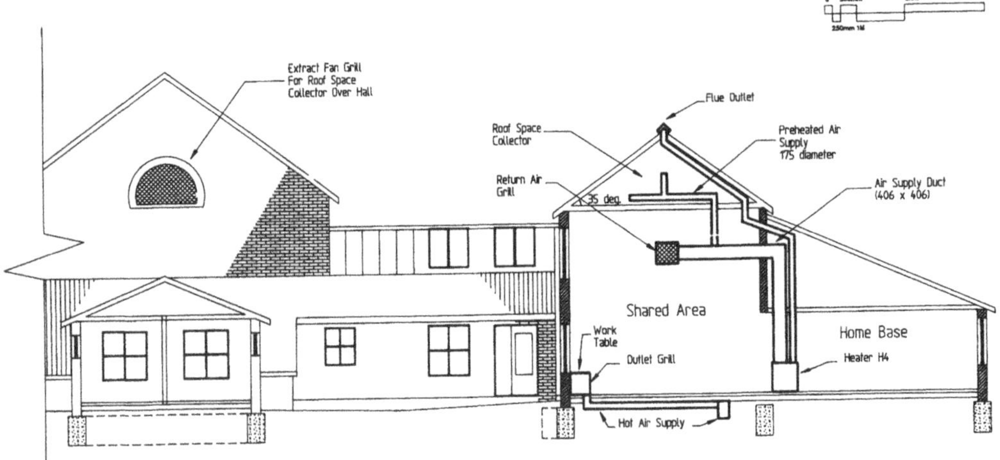


*Figure IV.5.6. Section showing the ductwork in north-east corner*

*Table IV.25.2. Energy demands of the school with and without the solar air heating system*

| | Without roof-space collector Delivered (kWh/annum) | Fraction of total (%) | With basic collectors Delivered (kWh/annum) | Fraction of total (%) |
|---|---|---|---|---|
| Auxiliary heating | 97,000 | 56 | 91,400 | 53 |
| Electric lighting | 19,400 | 11 | 19,400 | 11 |
| Solar gains* | 10,800 | 6 | 16,400 | 9 |
| Useful casual gains | 46,500 | 27 | 46,500 | 27 |

* Passive gains plus roof-space collector contributions

that the collectors reduced the already low auxiliary space-heating demands from 97,000 kWh/annum to 91,400 kWh/annum (Table IV.5.2). Even without the collectors, the demand is well below UK target values.

Predictions show that solar gains can make the most useful contribution during the autumn and spring (Figure IV.5.7). In fact, in September and June, solar heating is such that virtually no auxiliary heating is needed.

Sensitivity analyses indicated that:

- If the inside of the collector were painted black and if the insulation above the ceiling were increased, the auxiliary heating energy could have been reduced by a further 4,300 kWh per annum (i.e. to 87,100 kWh per annum).
- By increasing the proportion of fresh air which is supplied by the solar air system (rather than by infiltration) the auxiliary heating energy could be reduced. In the basic design 10m³/h per child (or about one third of all fresh air) came from the system. By increasing this to 20m³/h per child (or about two thirds of all fresh air), further savings of 5,050 kWh per annum could be achieved.

(Together these two refinements could save 14,750 kWh per annum or about 15% of the total.)

- The area of glazing in the roof space could be reduced by 30% without compromising the auxiliary heating energy demands (which would increase from 87,100 kWh/annum to 88,400 kWh/annum).
- Changing the collector orientation would not improve performance since the orientation was already near-optimal.

## Preliminary measurements

Measurements were made shortly after the school was built. During a cold period with high solar gain (e.g. 800 W/m²) the collectors reached temperatures of 27°C to 37 °C (Figure IV.5.8). On this day the classrooms achieved their desired temperature (of 18°C) by the start of school on the Monday morning.

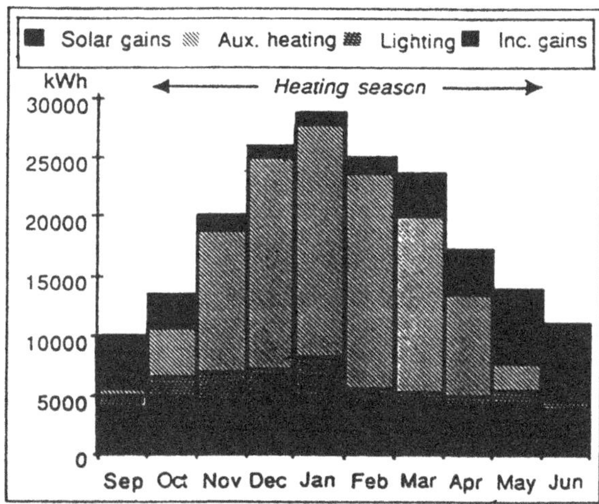


*Figure IV.5.7. Monthly energy inputs for design with basic collector*

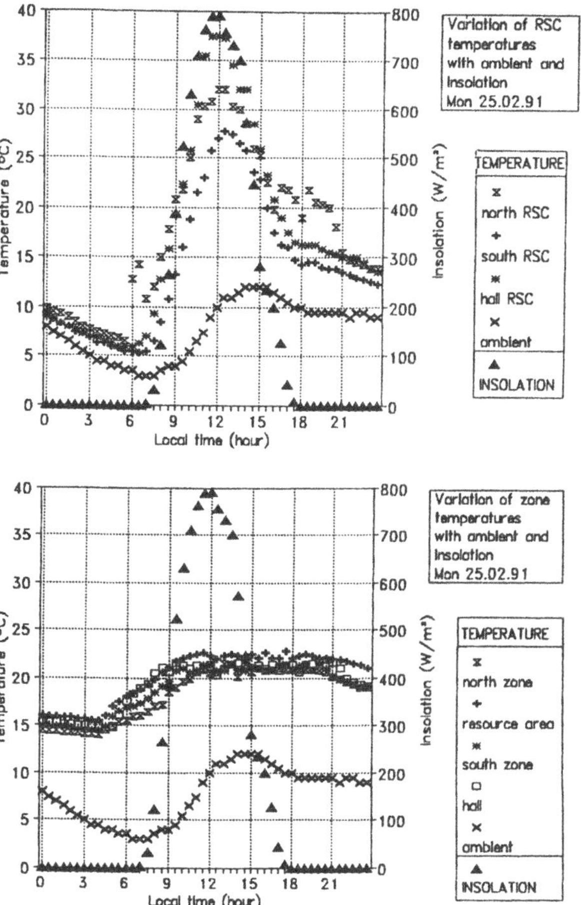


*Figure IV.5.8. Variation of the roof-space collector (RSC) temperatures and the interior temperatures for a sunny but cold Monday with the school occupied*

Unfortunately, the thermostatic control only initiated fan operation, to recover air from the collectors, when the collector air temperature exceeded 25°C. Thus, the collectors only produced useful heat for a small part of the heating season. The intention was that they should provide ventilation pre-heat by being operational at all times when the school was in use. This control problem also meant that heat was not recovered from the air exhausted from showers etc. Thus, long-term performance was well below expectations. It is anticipated that this problem will be rectified so that the predicted savings are achieved.

## REMARKS

This solar-air system is conceptually very simple. Fresh air, warmed by roof-space solar air collectors, and recovered heat from exhaust air, feeds conventional gas-fired warm-air heaters. Because schools are occupied during the daytime, there is a good match between availability of warm air and the heating periods. Therefore, thermal mass storage is not needed.

The 'glazed' area lies in the plane of the roof pitch and also acts as the weather barrier. Because roof collectors are not occupied, the air within them can be allowed to rise to higher temperatures than in conservatories and atria. Because they are mounted high up, site shading is less likely to be a problem. For optimal operation, the roof pitch may need to be greater than normal.

The collectors do not affect the construction of the occupied part of the building. Thus 'normal' construction details, building methods and site supervision arrangements can be employed. Passive solar daylighting and direct-gain heating can be designed in to work alongside the roof-space collectors.

Such collectors have also been incorporated in the Perronet Thompson School in the UK.

## ACKNOWLEDGEMENTS

Architect: Paul Markcrow, County Architect, Buckinghamshire County Council, UK

Building contractor: Llewellyn Construction Ltd., Milton Keynes, Bucks, UK.

Energy consultant: Brian Norton, PROBE, University of Ulster, Newtownabbey, Northern Ireland, UK.

Thermal analysis: Principia Mechanica Ltd, London, UK.

Chapter material: Patrick Waterfield, PROBE, and Jeremy Wagge, Bucks County Council.

Chapter author: Kevin Lomas, Institute of Energy and Sustainable Development, De Montfort University, Leicester, UK.

## BIBLIOGRAPHY

*Green Park Combined School* (1990). *Building 2000 EC series, Issue S3*, Commission of the European Communities, DG XII, December 1990, 12 pp.

Waterfield AP, Norton B (1991). Design and Performance Monitoring of Green Park Combined School. *Proceedings of the Congress of the International Solar Energy Society, Denver, USA*, **3**, (1), pp. 2676–2681.

Williams A., Bower J., Ratcliffe-Springall C., Thomas D. and Nex R. (1989). Perronet Thompson School. *Building*, January, pp 45–52.

# V Sports halls

# V.1 Introduction

## BUILDING CHARACTERISTICS

Sports halls are typified by large external facades with few windows (to avoid glare) and large flat roofs. These extensive opaque exterior surfaces are ideal for building-integrated solar air systems. Furthermore, a carefully designed, modular solar air system can improve the appearance of these structures, which are often bulky and clumsy.

## TEMPERATURE AND VENTILATION REQUIREMENTS

Athletes produce heat! To compensate for this, hall temperatures can be low, normally about 15°C. The need for space heating is hence less than normal, but high levels of activity require strong ventilation to exhaust humidity and odours. Solar-heated air can serve to heat ventilation air and pre-heat water for showers (DHW). Hot-water demand is relatively constant from day to day and the sporting season stretches throughout the year. Hot-water storage for only to two or three days has proven technically sensible and economical.

## SYSTEM DIMENSIONING

Obtaining realistic statistics for hot-water demand is imperative. The basis must be the number of showers to be taken, rather than the number of users, since experience has shown many persons do not shower after a game. Incorrect figures lead to over-dimensioned solar systems, including the solar storage tanks, with correspondingly poor system performance.

## CONCLUSIONS

Solar air systems with heat exchangers can effectively provide hot water for showers and help meet space-heating demand. Several economic solutions have performed well over the last decade, some with payback times of less than five years.

## ACKNOWLEDGEMENTS

Chapter author:
Harald Røstvik

# V.2 Karl High School Gymnasium

Munich, Germany

*Munich*

*System type 6*

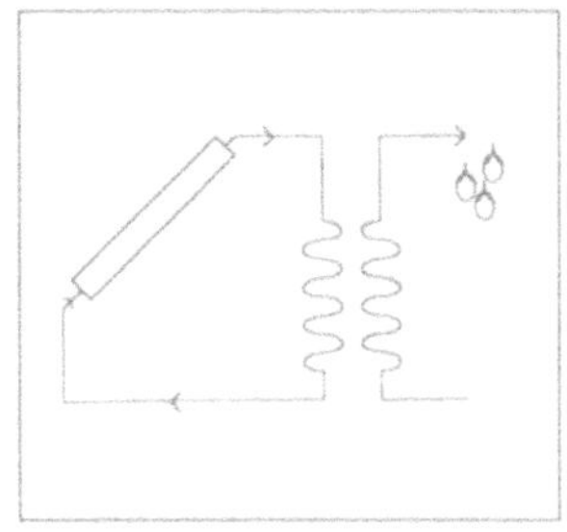

## PROJECT SUMMARY

A solar air system for space heating and domestic hot water has been added to a gymnasium built in 1972. As there already was an existing air heating and ventilation system which worked well, the investment for the solar plant was modest. Only the installation of the collectors and the domestic hot water (DHW) system were necessary.

The collector array mounted on the top of a flat roof is connected to the existing ventilation and heating system. Preheating fresh air, direct space heating and DHW generation enables the use of the collectors throughout the year. Because the applications have different operating temperatures, the solar energy gained can be maximized. In addition, the summer operation for hot-water generation allows the conventional heating system to be switched off during most of the summer. This avoids low-efficiency operation of the extremely oversized gas furnace. Accordingly, the savings are much higher than just the solar energy gained. The aim of this retrofit was also to determine effective control strategies and to gain insights for future installations.

### Summary statistics

| | |
|---|---|
| System type: | type 6 |
| Collector type and area: | opaque roof-mounted, 190 m$^2$ |
| Storage type, volume, capacity: | water, 1800 litres, 2.1 kWh/K (only DHW) |
| Annual contribution of the solar air system: | 40.5 kWh/m$^2_{floor}$ |
| Annual auxiliary heat consumption: | 165 kWh/m$^2_{floor}$ |
| Basis: | monitored |
| Heated floor area: | 1302 m$^2$ |
| Year solar system installed: | 1991 |

## SITE DESCRIPTION

The Karl High School is located in Munich Pasing:

| | |
|---|---|
| latitude | 48°N |
| longitude | 11°E |
| altitude | 515 m above sea level. |

The mean temperature is 8°C, ranging from a maximum of 36°C to a minimum of –32°C. There are 4020 degree days and the mean wind speed is 1.9 m/s.

## BUILDING PRESENTATION

The building, consisting of two gymnasia, changing rooms, showers and toilets, had a poor thermal standard at the time of construction. Only the roof was

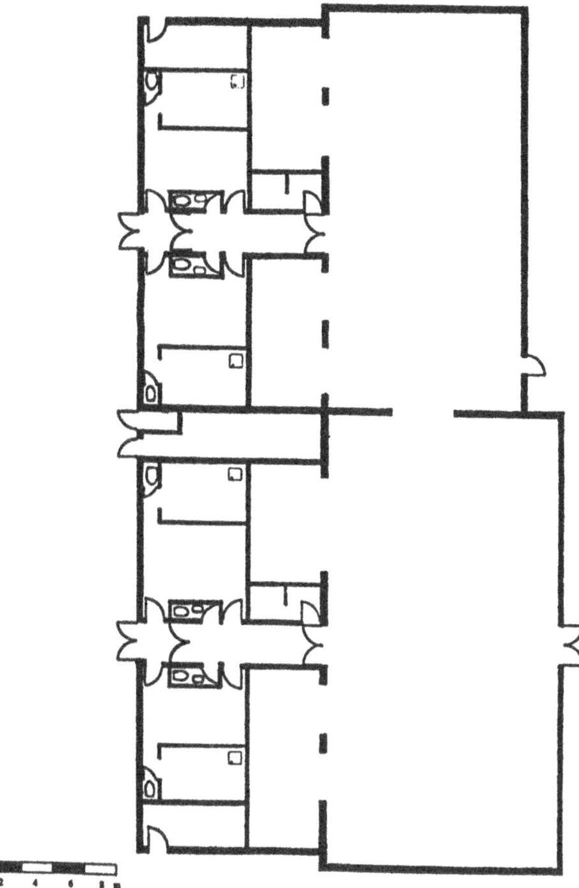

*Figure V.2.1. Plot of the building*

improved thermally when the solar retrofit was done in 1991. The building statistics are given in Table V.2.1 and details of construction in Table V.2.2.

A floor plan and section are given in Figure V.2.1.

### Building heating coefficients

Since the solar retrofit, all the data presented have been measured, except passive gains and infiltration, which were calculated with TRNSYS. The consumption of natural gas before renovation was measured and other data from that time were calculated. The gas demand before renovation was corrected in terms of the number of degree days so that there is the same climatic basis for comparison between different years. Data are given in Table V.2.3.

## SOLAR SYSTEM

The original heating system was complemented by solar air collectors installed on the roof of the building. Because the two parts of the gymnasium are different sizes, the collector array was adapted to thc different energy demands. Two open collector loops with a common air inlet for the air exchange to the ambient were installed. The amount of ambient air is 24% of the total mass flow.

The outlet of each collector field is connected to the conventional air heating system located at the front of the building. The collector loops are open inside the building. The collectors are under low pressure. Figure V.2.2 shows the collector array, Figure V.2.3 the scheme. During summer and when the temperature inside the gymnasium exceeds a fixed limit, the solar system is shifted over to providing the building with hot water. In this case the collector loop for the air heating is closed and the system is a closed loop with an integrated heat exchanger (see the scheme for water heating in Figure V.2.4 ). The inlet duct for air heating, which is built as twin duct, is then used as an inlet and outlet duct. Additional motor-driven flaps close and open the different parts of the system. The gas furnace, which provides both air heating and DHW is then switched off. By avoiding low efficiency operation and stand-by losses, the energy savings are higher than the solar energy gained. Even the new modern furnace still has 74% stand-by losses when it is in operation for DHW only (the losses mainly depend on the size and capacity of the whole heating circuit).

*Table V.2.1. Building statistics*

| | |
|---|---|
| Year of construction | 1972 |
| Year of retrofit | 1991 |
| Orientation | Azimuth – 10° |
| Gross heated floor area (gfa) | 1302 m² |
| Heated volume | 4863 m³ |

*Table V.2.3. Heating data*

| | |
|---|---|
| Conduction load | 131 $kWh/m_{gfa}^2a$ |
| Air change load | 59 $kWh/m_{gfa}^2a$ |
| Passive gains | 37 $kWh/m_{gfa}^2a$ |
| Solar gains | 41 $kWh/m_{gfa}^2a$ |
| Savings | 46% |

*Table V.2.2. Details of construction*

| | | |
|---|---|---|
| Walls | $U$ = 1.85 W/m²K | Concrete panel construction 24 cm |
| Windows | $U$ = 3.3 W/m²K | Metal frame in the changing room |
| | $U$ = 3.5 W/m²K | Glass bricks |
| Roof | $U$ = 0.33 W/m²K | Flat roof, extruded concrete, styrol insulation |
| Total $U$-value | 1.81 W/m²K | |
| Heating system | Atmospheric gas furnace, 360 kW (replaced 1994 with 190 kW) | |
| | Changing rooms, radiators | |
| | Gymnasium, two air heaters | |
| Ventilation | Five exhaust ventilation systems | |

*Figure V.2.2. View of the collector array*

### Distribution

The existing air heating system was not changed, except that two flaps were installed to open or close the collector loop and to open or close the existing air intake. The power of the fan was raised only slightly to increase the fan speed.

### Controls

The solar air system is controlled by two differential thermostats that open the flaps of the collector field for air heating or changing to DHW mode. The fans of the air heating system are controlled by the DDC (direct digital control) of the building because they are components of the existing heating system. The DDC receives two signals from the thermostats of the collector, ON or OFF and DHW-Mode ON or OFF. The DDC additionally switches the fan speed depending on solar radiation (two-stage fans) and controls the back-up heating or the inlet temperature with the existing heat exchangers.

Set points are:

$\Delta T_{collector}{}^{3}$ 2 K for air heating (maximum temperature of gymnasium ≥ 20.5°C)
$\Delta T_{collector}{}^{3}$ 10 K for water heating when $T_{gym}$ exceeds maximum; maximum storage temperature ≥ 75°C

### Construction details

To avoid leaks through the flat roof, the roof membrane was not penetrated and the collectors are fixed to stainless steel tracks with anchors at the eves to withstand wind pressure.

### Components and data of the solar system

The components are given in Table V.2.4.

## PERFORMANCE

The building, the solar system and the weather conditions were monitored from 1992 until June 1996. For performance analysis and parametric studies additional TRNSYS simulations, validated by measurements, have been carried out.

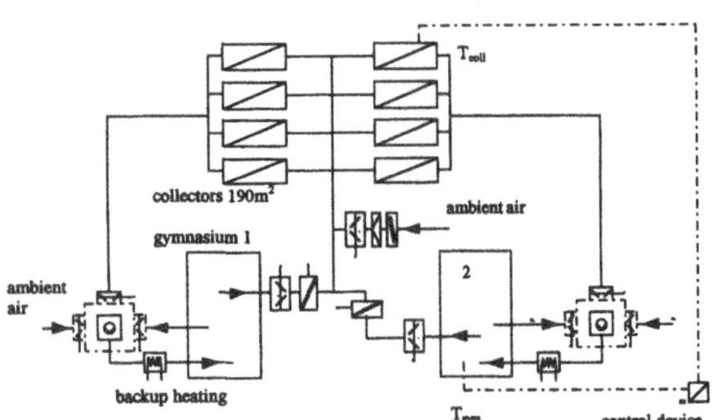


*Figure V.2.3. Scheme of the system when used for air heating*

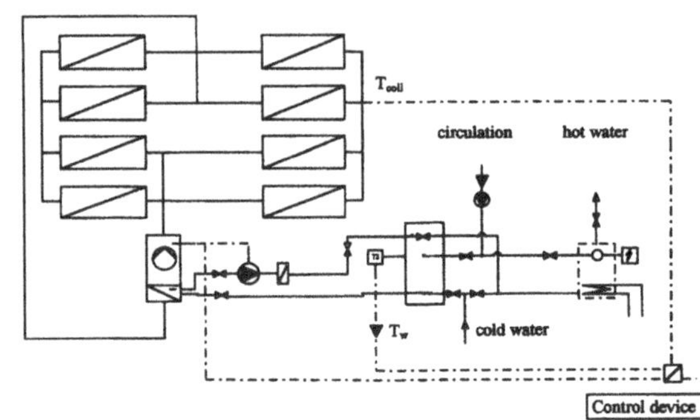


*Figure V.2.4. Scheme of the system when used for water heating*

*Table V.2.4. System components*

| | |
|---|---|
| Collector | 190 $m^2$ Grammer solar air panel, net area 174.8 $m^2$ |
| Operation modes | Direct space heating<br>DHW heating |
| Absorber | Black and selective surface |
| Setup | Two arrays with common inlet<br>Eight rows of collectors<br>76 modular collectors, each with a 2.3 $m^2$ aperture<br>Collectors: four modules per row |
| Solar contribution | 283 $kWh/m_{coll}^2$ |
| Flow rate | air heating: 29 to 44 $m^3/m_{coll}^2h$ ( two stages)<br>DHW: 22 $m^3/m_{coll}^2h$ |
| Control | Differential thermostats connected to a DDC |
| Back-up heating | Water-to-air heat exchanger |
| Storage | 1000 litres and 800 litres |
| Heat exchanger | Six rows of stainless steel tubes with aluminium lamellae, 20 kW |
| Distribution | air heating: two fans (existing) 1.5/4 kW each, switched to higher speed when insolation $^3$ 400 $W/m^2$<br>water heating: 1.4 kW |
| Consumption of fans | additional consumption: 1770 kWh/a, $COP_{air}$* = 23.5<br>DHW: 1120 kWh/a $COP_{water}$* = 8.3 |

* COP is defined as the ratio of solar gains to consumed fan power. If the efficiency of the furnace is taken into account, the total savings are much higher, which leads to higher COPs.

The performance of the solar system is excellent because of the high gas savings. These savings are even higher than the received insolation, because of avoided low-efficiency operation of the gas furnace when it is switched off during the summer.

Another reason for the high savings is that the air intake for fresh air has been moved to the collector inlet. Before renovation (Figure V.2.5) it was located at the point of lowest pressure in the system. This led to high leakage rates in spite of closed flaps and additional heating requirements for fresh air that was not needed.

Some design parameters of the system were checked by TRNSYS calculations. The original design was done by empirical methods. The influence of collector area on gained heat for air heating is given in Figure V.2.6. The selection of the collector area depends not only on the collector output. Additional savings can be achieved by switching off the conventional heating system during summer season for DHW heating. The collector system in this case is slightly oversized.

The influence of the amount of fresh ambient air mixed with the air flow from the collector output in the case of air heating is given in Figure V.2.7. This shows the potential for solar air heating of gymnasia.

## REMARKS

The heating systems of gymnasia are oversized during most periods of operation and therefore provide a good application for solar air heating systems. A high fuel-saving potential can be realized by avoiding low furnace efficiency and stand-by losses. Because conventional air heating systems often exist and can be used for this application, the investment is low.

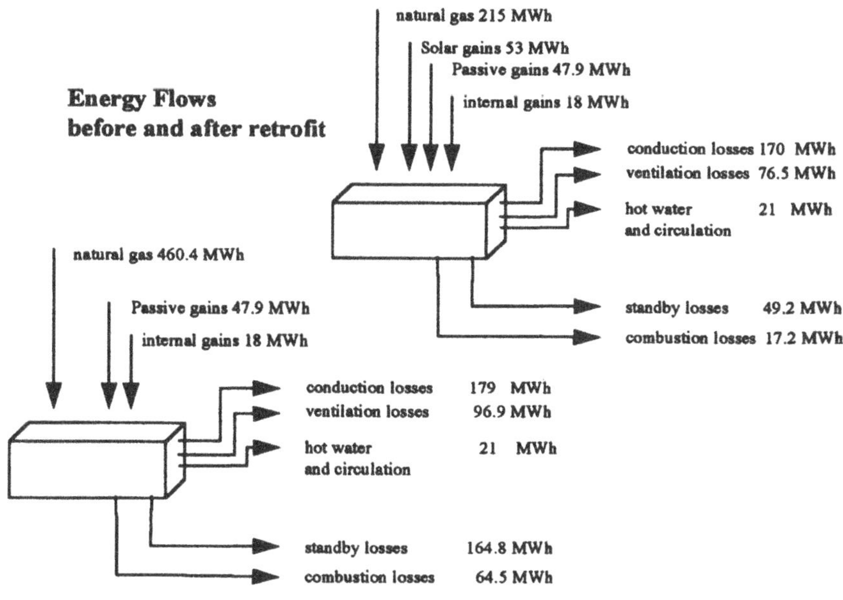


*Figure V.2.5. Energy flows to and from the building before and after renovation*

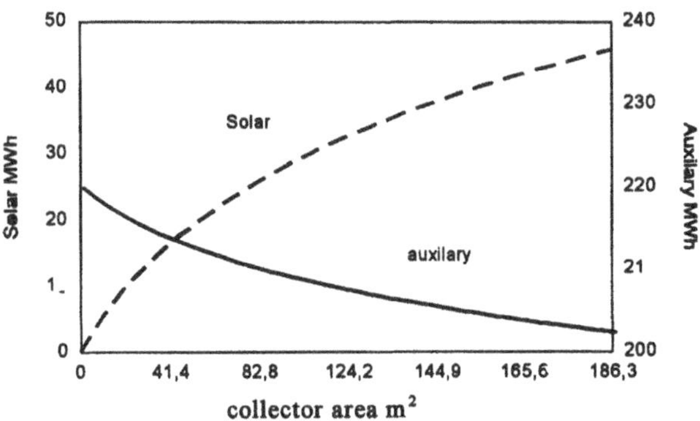


*Figure V.2.6. Usable heat from collector and auxiliary energy demand as a function of the collector area for air heating.*

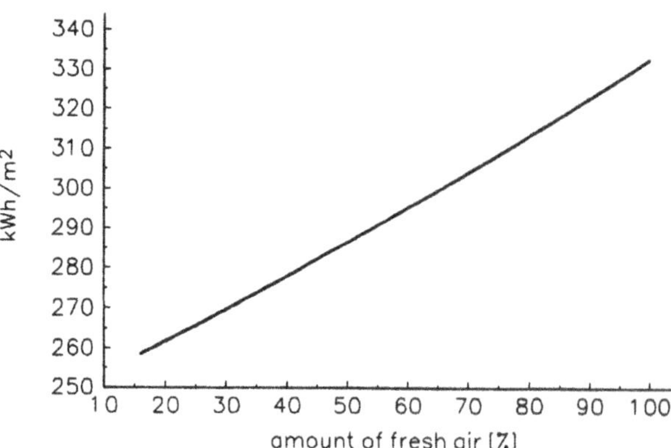


*Figure V.2.7. The energy gain of the collector as a function of the percentage of fresh air mixed with the air flow from the collector when the system is being used to heat the gymnasium (air heating); all the other parameters used are as given in Table V.2.4*

## ACKNOWLEDGEMENTS

Manufacturers of solar components:
Collector: Grammer SKT, Werner v. Braun Strasse 6, D-92224 Amberg, Germany
Control: Resol GmbH, Fänkenstrasse 26, D 45549 Sprockhövel, Germany
Architect: W. Landherr, Blütenburgstrasse 25, D-80636 München, Germany
Energy consultants: H. Barthel, Werner v. Braun Strasse 6, D-92224 Amberg and J. & J. Morhenne, Schülkestrasse 10, D-42277 Wuppertal, Germany
Owner: Community of Munich
Chapter author: J. Morhenne, Schülkestrasse 10, D-42277 Wuppertal, Germany

## BIBLIOGRAPHY

Morhenne J, Barthel H (1997). *Solare Luftkollektoren für kommunale Gebäude am Beispiel von Turnhallen.* Abschlussbericht BMFT 329016C, Bundesministerie für Forschung u. Technologie D-52428 Jülich.

# V.3 Athletics Hall, Odenwald School

## Heppenheim, Germany

*Heppen-heim*

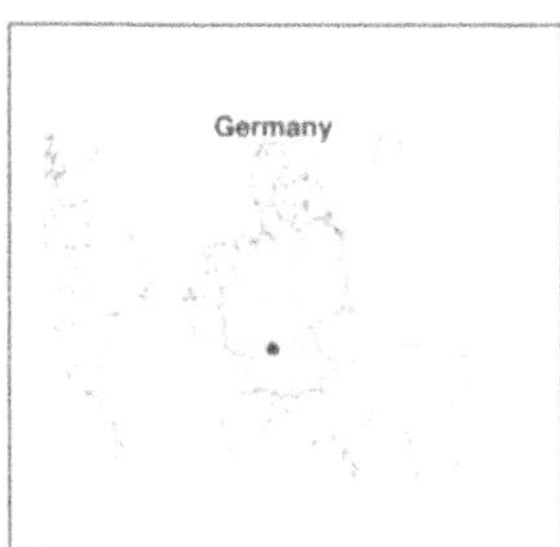


*System type 1*

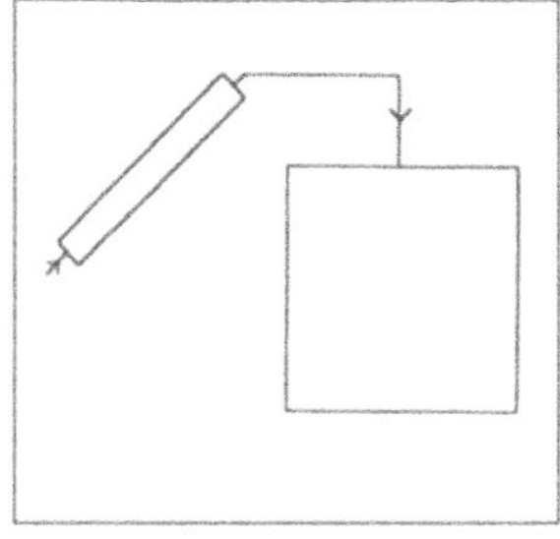

### PROJECT SUMMARY

This building, completed in 1995, is a good example of how to use a glazed foyer, not only as a climatic buffer zone, but also for preheating the inlet air by solar gains. The completely glazed west-oriented foyer is used as a huge air collector to preheat ventilation air during the heating period.

The glass superstructure across the hall stores a movable curtain, serves as a skylight and enhances the natural ventilation of the hall due to the chimney effect. The stiffening ribs of the floor are also used as an air duct to the hall and as an installation duct. Photovoltaic-powered fans are used to move solar preheated air into the hall.

#### Summary statistics

| | |
|---|---|
| System type: | type 1, |
| Collector type and area: | spatial collector, 256 m$^2$ |
| Storage type, volume, capacity: | building structure, 18 m$^3$, 10.4 kWh/K |
| Annual contribution of the solar air system: | 4 kWh / m$^2_{floor}$ |
| Annual auxiliary heat consumption: | 102 kWh / m$^2_{floor}$ (calculated) |
| Annual auxiliary heat consumption: | 103 kWh/m$^2_{floor}$ (measured) |
| Basis: | monitored and calculated |
| Heated floor area: | 1095 m$^2$ |
| Year solar system built: | 1995 |

### SITE DESCRIPTION

The Odenwald school is a private School located near Heppenheim, a small city about 50 km south-east of Frankfurt am Main:

| | |
|---|---|
| latitude | 49°N |
| longitude | 9°E |
| altitude | 113 m above sea level. |

### BUILDING PRESENTATION

The athletics hall of the Odenwald School is built into the hillside to the south-east of an existing building, as shown in Figure V.3.1. Thanks to the slope of the site,

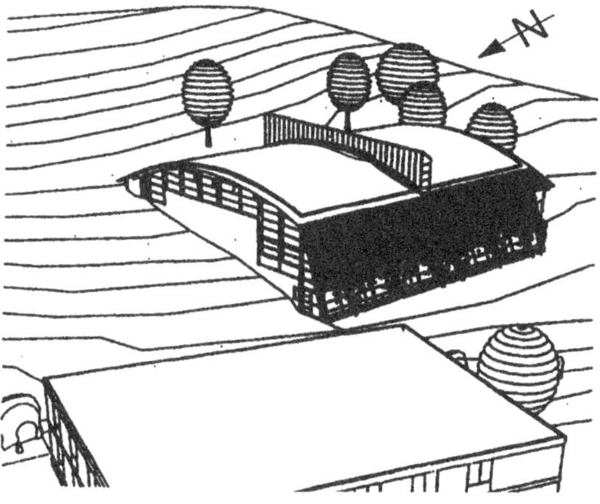


*Figure V.3.1. An isometric view*

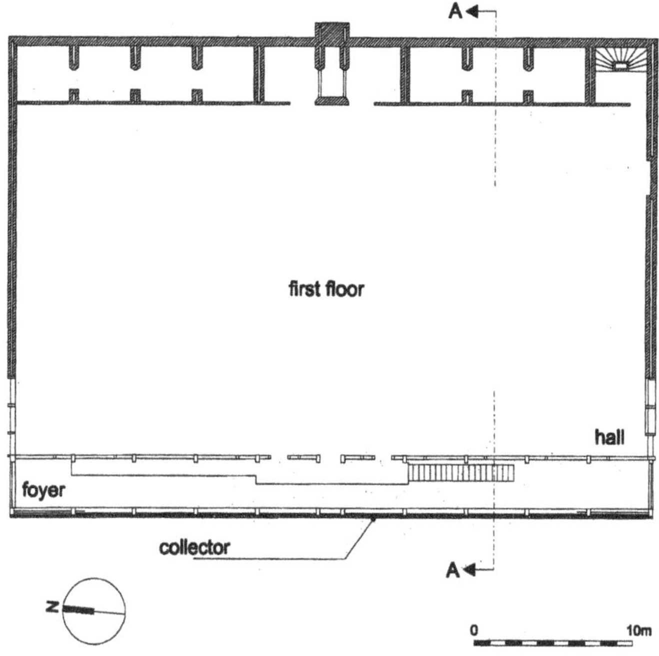


*(a) Floor plan*

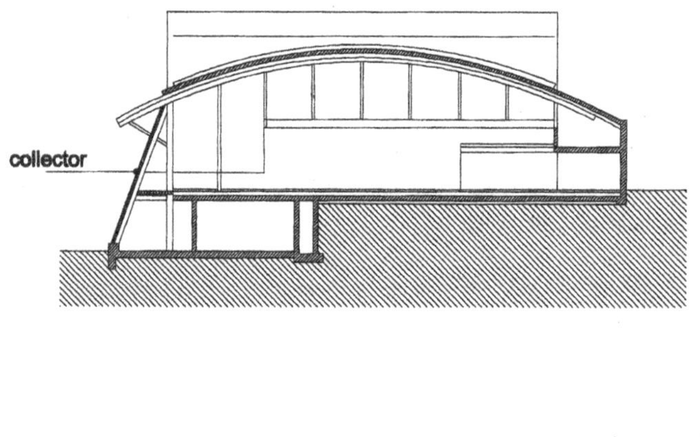


*(b) Section*

*Figure V.3.2. (a) Plan and (b) section*

the locker rooms and auxiliary facilities have been placed below the hall floor level, thus minimizing the footprint and volume of the entire complex.

The large vaulted roof is spanned with a light-weight structure consisting of slim, glue-jointed trusses and a structural shell of 27 mm laminated wood. The fixed shading devices on the south-west facade were added mainly for architectural and stylistic reasons. The entire roof is planted, thus compensating the loss of grassland occupied by the building. A glass superstructure running across the entire hall houses a movable curtain, which contributes to the daylighting of the hall and also enhances natural ventilation due to its chimney effect. The end walls of the hall have extensive glazing with wooden frames. Shading of the south facade can be controlled by means of adjustable blinds.

A plan and section of the hall are shown in Figure V.3.2.

For the 8 m high glazed conservatory, which serves as both a solar collector and a heated buffer zone, fixed shading devices are attached

The hall is heated by the existing furnace of another building. The domestic hot water system is supple-

*Table V.3.1. U-values*

| | $U$ (W/m²K) |
|---|---|
| Glazing hall – outside | 1.3 |
| Glazing hall – foyer | 5.0 |
| Glazing foyer – inclined | 1.7 |
| Glazing foyer | 1.3 |
| Roof | 0.27 |
| Floor of hall | 0.26 |
| Floor of foyer | 0.56 |
| Wall – outside | 0.42 |
| Wall – earth | 0.57 |

*Table V.3.2. Structural and insulation data*

| | |
|---|---|
| Roof | 27 mm laminated wood |
| | 100 mm polystyrol |
| | earth |
| Floor of hall | 70 mm lime |
| | 140 mm insulation and floor heating element |
| | 250 mm concrete |
| Floor of foyer | 50 mm lime |
| | 65 mm insulation |
| | 200 mm concrete |
| Wall | 25 mm wood |
| | 85 mm insulation |
| | 200 mm concrete |
| Wall to earth | 200 mm concrete |
| | 60 mm insulation |
| Building heating load by conduction losses | 66 kWh/m²a (calculated) |
| Building heating load by ventilation losses | 52 kWh/m²a (calculated) |
| Auxiliary space heating | 102 kWh/m²a (calculated) |
| Appliances and lighting | 7800 kWh |

mented by some solar water collectors on the south facade.

### Building statistics

The gross heated floor area is 1094.5 m² and the heated volume 6943.4 m³. $U$-values are given in Table V.3.1 and structural and insulation data in Table V.3.2.

## SOLAR SYSTEM

Figure V.3.3 shows the solar system on a sunny winter day with the foyer working as a huge air collector. Sunlight absorbed by the massive floor warms up the air. Two PV-powered fans at the top of the wall between the foyer and the hall blow the preheated air into the hall. If there is no irradiation, the inlet air is preconditioned by an earth channel and heated by a radiator.

In summer the fans are deactivated. A fixed shading device, good ventilation and the natural shading by special plants reduce the risk of the collector overheating.

The system is controlled by three temperature sensors: outside, at the top of the foyer and in the hall. The fans have no controls; they are regulated by the actual irradiation, which varies the PV-power.

## PERFORMANCE

The two fans in the wall between the foyer and the hall work at 12 V, with a peak power input of 56 W. Each has a flow rate of 1500 m³/h. The maximum power output of the two photovoltaic panels is 100 W and the maximum total flow rate is about 3000 m³/h.

### The collector

The collector is a west-oriented spatial collector, with a glazed facade area of 256 m² a frame ratio of 15% and a volume of 660 m³.

### Savings

By using the foyer as an air collector with inlet air from the earth channel (see Figure V.3.3), up to 4.3 MWh/a, i.e. 4 kWh/m² are saved. The relatively low savings from the solar preheating are mainly caused by the west orientation of the glazing, a result of architectural considerations.

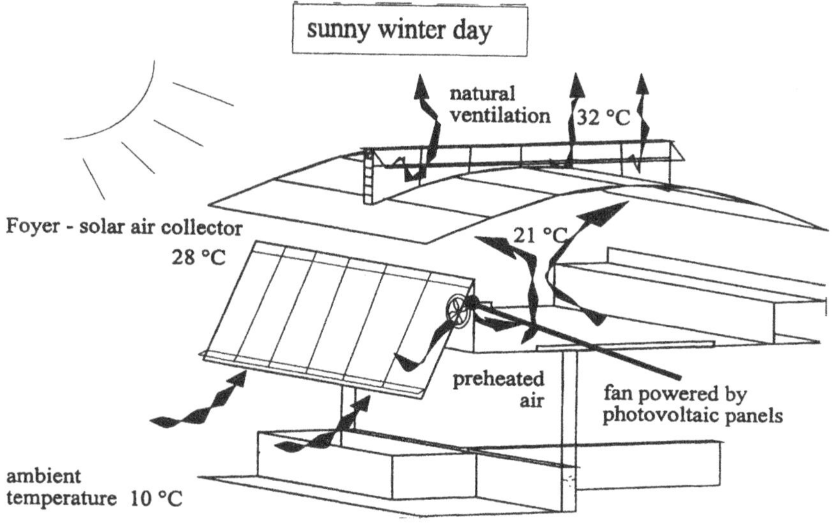


*Figure V.3.3. The solar system on a sunny day*

## REMARKS

The calculated heating-energy consumption of the building is about 118 kWh/m$^2$a with a heating set temperature of 20°C.

The energy savings due to the earth duct and the preheated air from the foyer total 4.3 MWh/a, leaving a remaining heating demand of about 114 kWh/m$^2$a. A night set-back (set temperature 15°C) reduces this to 102 kWh/m$^2$a.

The quite high calculated values for the heating demand are a result of having a heating set temperature of 20°C and a high occupancy rate from 8:00 to 22:00 six days a week.

Problems occurred during the erection of the building because communication between the planning engineers was not adequate. Important information was not available to the craftsmen. The following are a few examples:

- The opening from the channel to the hall was completely filled with wiring by the electrician. Ventilation through the earth duct into the hall was therefore hardly possible.
- The electrical engineer selected a fan that could never work with the installed photovoltaic panel, because the starting power exceeded the maximum output of the PV panels.
- The device to control the system was planned by an engineer who did not know exactly how the whole system was supposed to work. With the installed control devices, it is not possible to control the system as necessary. The changes to the control system are expensive, because it is built of analogue components.

## CONCLUSIONS

It should be guaranteed that all involved parties understand how the whole building and its systems function, including the craftsmen on the building site.

Using digital components for control systems, instead of a hard-wired analogue components, would greatly simplify future changes.

## ACKNOWLEDGEMENTS

Architect: Buero plus + Professor Peter Hübner; Project Architect Martin Grün, Goethestrasse 44, Pf 107, D-72654 Neckartenzlingen; Tel. + 49 7127 9207 0; Fax + 49 7127 9207 90

Energy Concept: Transsolar; Project Engineer Mathias Schuler, Helmut Meyer Nobelstrasse 15, D-70569 Stuttgart; Tel. + 49 711 679760; Fax + 49 711 6797611

# V.4 Stavanger Squash Centre

Norway

*Stavanger*

*System type 6*

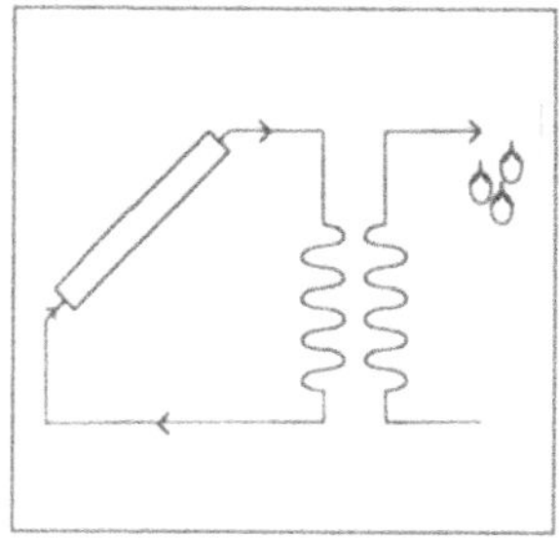

## PROJECT SUMMARY

Although Stavanger is the technological and financial oil-capital of Norway, the Stavanger Squash Centre was until recently the largest solar building in Norway with 120 m$^2$ of collectors.

The active, building-integrated, solar air collector in the 45° roof facing 15° east of due south, has now been delivering solar-heated hot water for the showers for 15 years. The solar system consists of several standard products put together in a new way. Monitoring has shown that the system produced 18,000 kWh/m$^2$ a (150 kWh/m$_{coll}{}^2$a). If operated as planned, it could have had a solar contribution of 45,000 kWh/a (375 kWh/m$_{coll}{}^2$a), resulting in a 19% solar fraction of total demand.

### Summary statistics

| | |
|---|---|
| System type: | type 6 |
| Collector type and area: | roof-integrated air collector, 120 m$^2$ |
| Storage type, volume: | water tank, 6000 litres |
| Annual contribution of solar air system: | 150 kWh/K$_{coll}$ |
| Basis: | monitored |
| Heated floor area, | 1400 m$^2$, 8500 m$^3$ |
| Year solar system built: | 1981 |

## SITE DESCRIPTION

The sports building is located on the outskirts of Stavanger – 3 km from the centre of the city – next to a sports arena with an unobstructed view to the south:

| | |
|---|---|
| latitude | 58°N |
| longitude | 6°E |
| altitude | 50 m above sea level. |

There are 1727 sunshine hours per year. 50% of the 850 kWh/m$^2$a of annual solar radiation in Stavanger comes from an overcast sky. The mean outside temperature is 7°C. The heating season is 230 days (27 September to 14 May). There are 3,000 degree days (temperature base 9°C).

## BUILDING PRESENTATION

The building was constructed in only seven months at a cost of only US $570/m$^2$ (excluding VAT). A site plan is shown in Figure V.4.1 and a plan and section of the building in Figure V.4.2. Figure V.4.3 shows how the snow slides off the collector.

Other energy-related features in the building are:

- earth-sheltering;
- heat recovery from shower water;
- heat recovery on exhaust ventilation.

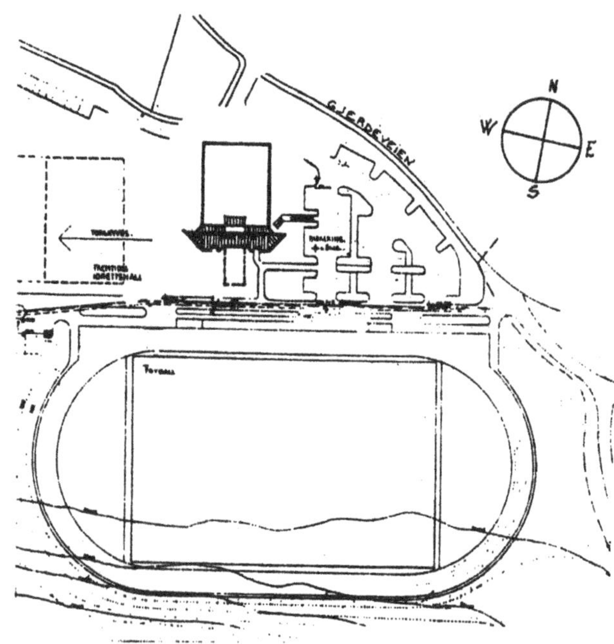


Figure V.4.1. Site plan

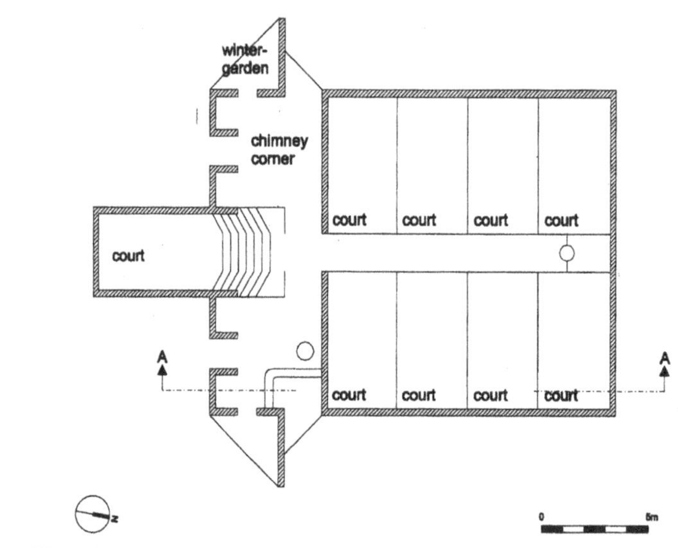


(a) Floor plan

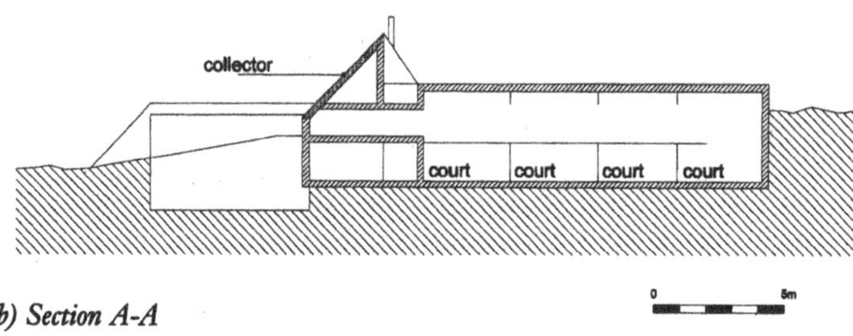


(b) Section A-A

*Figure V.4.3. View of the building, showing how, in winter, the snow slides off the collector*

The building was designed to require 250,000 kWh/a (180 kWh/m²a)or 35% of the energy used annually in similar, but conventional, buildings without the same energy-saving features and without a solar system (700,000 kWh = 535 kWh/m²).

The 1995 economic value of the annual energy saving of 450,000 kWh in Norway, for both oil and electricity (hydropower) is US $38,000 (0.09/kWh). The planned energy uses in the building were as shown in Table V.4.1. The traditional cooling system has been avoided, resulting in capital savings of US $25,000 plus reduced running costs for the owner.

## SOLAR SYSTEM

The heating system provides:

*Table V.4.1. Planned energy use*

| | |
|---|---|
| Hot water, mainly showers, | 90,000 kWh<br>140,000 kWh with<br>50,000 kWh from solar |
| Ventilation and space heating | 100,000 kWh |
| Lighting (turning into heat) | 60,000 kWh |
| Total | 250,000 kWh |

- heat for ventilation air;
- hot water for the showers;
- hot water for the kitchen;
- hot water for the underfloor heating.

The double-shell electric hot water tank with capacity 1,500 litres (inner tank) has an additional outer magazine of 300 litres. A 75 kW electric heating element heats the inner tank. Of this 25 kW is back-up in case of solar-system breakdown.

Cold water from the public water utility enters the building at a low temperature of 7 to 10°C. Normally the water is first preheated by passing it through a pipe in the 'grey-water' tank for heat recovery from waste water. From here the water enters the 6000 litre solar tank in a closed loop. If necessary, the solar-heated water is then further heated to the needed temperature in the electric hot-water tank. The system is illustrated in Figure V.4.4. The solar water heating system for showers and kitchen is illustrated in Figure V.4.5.

### The collector

The collector is glazed with a single transparent fibreglass cover. A corrugated aluminium sheet with selective coating serves as the absorber. Originally a 4 mm glass covered the collector, but this had to be replaced after a few years. Poor workmanship in fixing the glass to the collector resulted in tension and glass breakage.

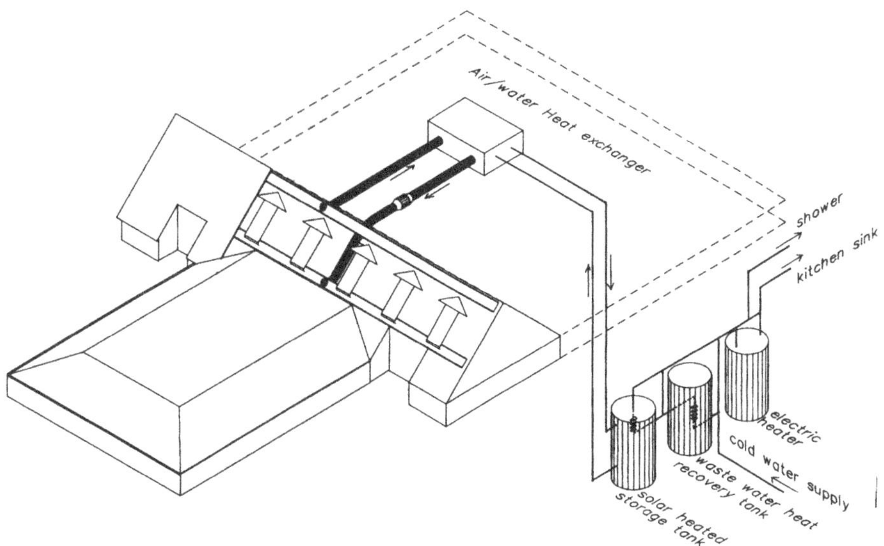


*Figure V.4.4. Solar heated air circulated in a closed loop through the air-to-water heat exchanger behind the top of the collector*

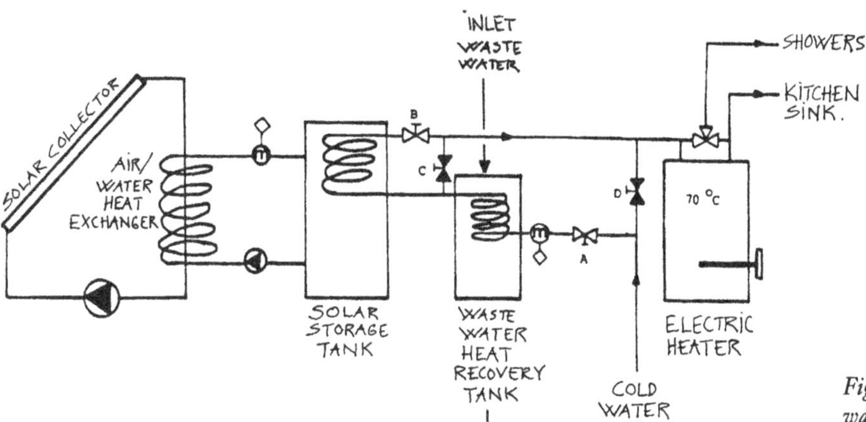


*Figure V.4.5. Principles of the solar water heating system for the showers and the kitchen*

Behind, a 10 cm space is created by timber battens on pressed non-toxic timber boards. In this space the solar-heated air circulates. From a distribution channel – a deepening of the collector at its lower part – air is distributed from the centre of the 25 m wide roof out to each side. As the air heats up, it rises and is collected in a similar distribution, a deepening of the collector at the top, leading to the main channel, delivering the warmed air to the air-to-water heat exchanger. Since 50% of the annual solar radiation in Norway is from an overcast sky and since the weather varies rapidly from overcast to clear and back to overcast, it was necessary to design a collector with a low heat capacity so that it could react quickly to weather changes. In addition, low cost was a must, because of the very low prices of traditional energy in energy-rich Norway. Hence, the collector is constructed of standard building materials.

The collector provides good insulation for the roof since it contains either warm air or static air. This results in less need for insulation in the roof construction under the collector. A slimmer construction could be designed, saving construction costs. The collector was designed for an air flow of 3,790 $m^3/h$ (25 $m^3/hm_{coll}^2$).

## The heat exchanger

The air-to-water heat exchanger is positioned just behind the solar collector in a small north-facing room on the roof. The fan that forces the air to circulate through the collector is also positioned here. The heated air is passed through the heat exchanger, where the solar heat is transferred to water and transported, via insulated water pipes, to the solar storage tank in the technical room in the basement. The circulation of solar-heated water from the air-to-water heat exchanger to the storage tank was dimensioned for 96 litres/min.

## Storage

The 6000 litre water storage tank contains a water-to-water heat exchanger for transferring the solar-heated water.

## Distribution

From the storage tank, the solar-heated water is transported, via the electric heater tank, to the showers and washbasins. The assumed need for hot water for showers etc. was 10–15$m^3$/day.

## The controls

A standard computerized controller regulates all functions. The solar collector circuit is regulated by a temperature differential thermostat. This starts the solar collector fan and the water pump in the air-to-water heat-exchanger circuit when the temperature at the top of the collector is more than 10 K higher than the temperature in the bottom of the storage tank.

The storage tank is equipped with a small open expansion tank to take up the expansion in the 'passive' solar-heated water in the storage tank as it expands when heated and contracts when cooled (tapped).

The closed water-pipe circuit transporting water from the air-to-water heat exchanger to the storage tank is equipped with a closed expansion tank for the same purpose.

## PERFORMANCE

In the planning stages, the data in Table V.4.2 were used. Based on this, the solar contribution for showers was calculated to be 58,000 kWh/year. It turned out, however, that only 120$m^2$ of the collector was in operation. This reduced the circulating air in the collector to 25 $m^3$/ $hm^2$.

*Table V.4.2. Data used in the planning stages*

| | |
|---|---|
| Hot water for showers | 15,000 litres per day. |
| Solar collector size | 120 $m^2$ |
| Circulating air in collector | 25 $m^3/hm^2$. |
| Insulation on collector | 1,250 $kWh/m^2a$ (tilt 45 ) |

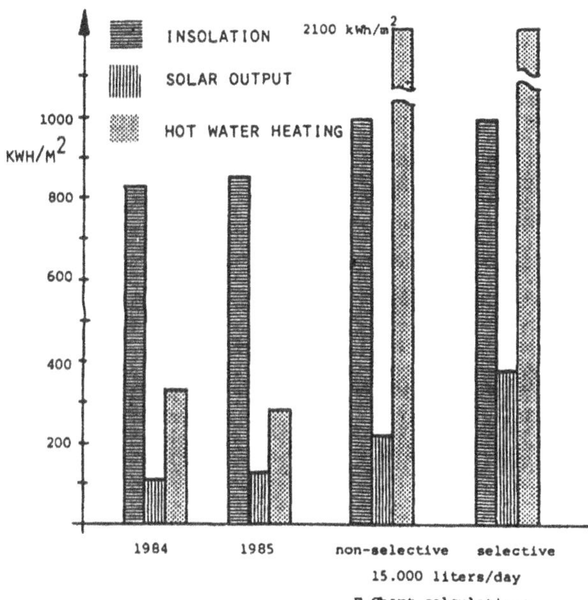


*Figure V.4.6. Energy consumption comparisons*

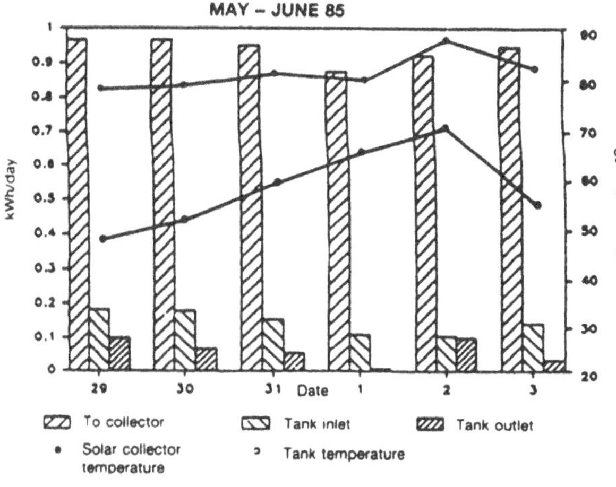


*(a)*

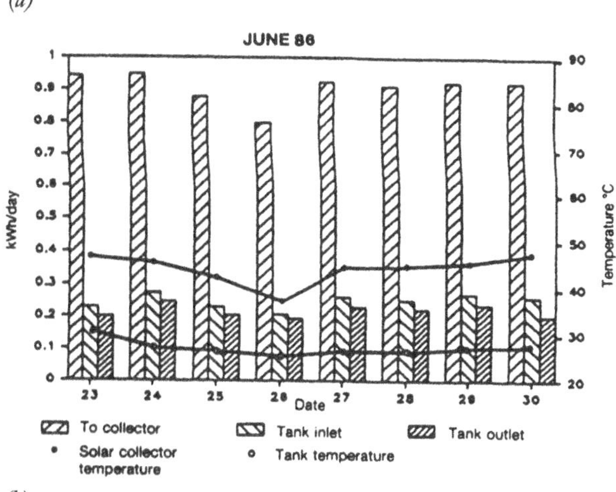


*(b)*

*Figure V.4.7. Monitoring results from (a) 3000 litres/day hot-water use and (b) 20,000 litres/day forced hot water use.*

This resulted in a calculated expected annual solar contribution of: 43,500 kWh/year = 362 kWh/m²(solar collector). The extra cost of the system beyond normal construction was US $10,000 (1981).

The monitoring of the 120m² collector has, however, shown that the water usage in the building has been very low – only 20% of planned use (Figure V.4.6).

A combination of monitoring and forced water use tests (opening up shower taps to let hot water out) was performed to establish the system efficiency (Figure V.4.7). All costs have been adjusted to 1986 prices, when the monitoring took place. Energy prices resulting from the solar contribution are calculated on the basis of a 7% interest rate and a system life span of 20 years.

After adjusting for the normal insulation year (30 years average), conclusions can be drawn and related to the planned annual solar contribution from the 120 m² collector of 43,500 kWh/year (362 kWh/m²(solar collector)).

Because much less water was used than planned, the solar system has an annual contribution of 18,000 kWh/year (150 kWh/m²(solar collector)).

The heat recovery of the shower water ('grey water') was not included in the initial planning calculations. Since the heat recovery 'competes' with the solar contribution and reduces the annual output from solar, an adjusted calculation excluding heat recovery was done to compensate for this, as follows:

> 20,400 kWh/year = 170 kWh/m² solar collector (120 m² solar collector).

Adjusted to water volume usage as planned, the figure changes to:

> 27,000 kWh/year = 225 kWh/m² solar collector (120 m² solar collector).

Finally, when all comparable data are considered, a direct comparison between planned and actual output can be made, including the effects of a selective coating (the selective surface was ineffective due to dust covering it). The end results confirm that the design tool gave reasonable results: 45,000 kWh/year = 375 kWh/m²(solar collector) year.

## The economics

The solar energy cost can be calculated as follows (necessary investment, 1986):

> Monitored US $14,000: 18,000 kWh/year = US $0.77/(annually produced kWh).
> Planned US $14,000: 45,000 kWh/year = US $0.31/(annually produced kWh).

The figures for planned output resulted in a long-term limit cost of US $0.03/kWh. In 1986, the average consumer of energy in Norway (electricity from hydropower) paid US $0.06/kWh. New Norwegian hydropower projects, now in development, are pricing their energy at a minimum of US $0.09/kWh.

## REMARKS

Three main experiences should be noted:

- The effectiveness of a selective coating is quickly lost when dust covers the absorber. In the case of Stavanger Squash Centre, this was caused by glass breakages that resulted in sand 'storms' filling the collector.
- Closing an in-situ built collector should take place during calm weather if the building site is dusty (as most building sites are). If breakages occur during operation, repairs should be made immediately or the broken area should be covered up. If dust and sand enter the collector, the absorber should be vacuumed cleaned, not washed down. Washing makes the situation worse.
- The efficiency of a solar system in a sports building depends on proper design and dimensioning with regard to water use in the building. Over-dimensioned collector area results in reduced return on investment.
  - Most consultants have rules of thumb for water use in sports buildings. These should be questioned since it often appears that less water is used than expected. (Many people turn up in the sports building in sports clothes and return home to shower.) Hence, the number of participants is very misleading for planning water consumption.
- Heat recovery from waste water (grey water) 'competes' with solar heat. The investment needed for both heat recovery and the solar system should therefore be considered carefully.

## ACKNOWLEDGEMENTS

System supplier: ABB Miljø AS, Norway
Solar system and building design: Civil architect MNAL Harald N. Røstvik, Sun Lab., Norway
Monitoring agent: SINTEF, Trondheim, Norway

# VI Industrial buildings

# VI.1 Introduction

## BUILDING CHARACTERISTICS

One of the best solar applications is air heating for industrial buildings. Their typical one-storey geometry with an expansive, often poorly insulated roof leads to substantial space heating needs. In addition, heating fresh air to replace exhausted air can often be all that the space heating requires. Many solar air heaters to supply tempered fresh air have proved cost-effective today.

## VENTILATION AIR HEATING

A solar heating system in which fresh air is drawn through the collector and discharged into the building achieves maximum efficiency. If building air is recirculated through the solar collector, it can only operate when the panel temperature exceeds room temperature. When outside air is heated, any temperature rise is useful, so the system can even function on a cloudy day; a temperature rise of 5 to 20 K above ambient air temperature is normally sufficient. If the fresh air is heated above room temperature, the solar system also contributes to the space heating.

Two types of solar ventilation system are presented here, glazed and unglazed. The latter in its simplest form can cost less than a recirculating space heating system when many costly items of conventional solar systems, such as thermal storage, separate support structures for the collectors and collector glazing, are omitted. These cost savings allow paybacks in the range of two to five years.

## DESTRATIFICATION ENERGY SAVINGS

Distribution ducting below the ceiling can be added to supply fresh air throughout the entire building and in the process will destratify any ceiling heat that may accumulate, achieving additional energy savings. These additional energy savings from destratification can equal the solar savings. The Bombardier factory and the USA Army hangar are excellent examples where this has been applied.

## SUMMARY

Whenever an industrial application is being considered, the solar system should be designed first to heat the required fresh air. An existing surface such as the south wall should be used as the mounting surface for the solar collectors. The case histories in this section illustrate the significant savings and benefits that are possible for industrial buildings.

## ACKNOWLEDGEMENTS

Chapter author: John Hollick

# VI.2 Mätzler Garage

## Vorarlberg, Austria

*Vorarlberg*

*System type 5*

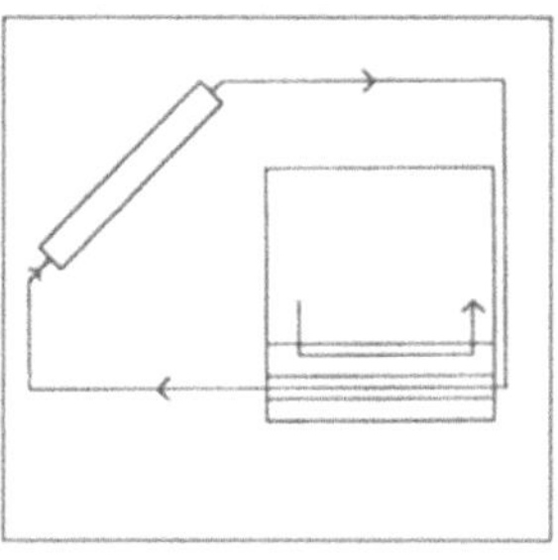

### PROJECT SUMMARY

The Mätzler Garage, a small workshop for a private motorcycle collection, incorporates a solar air collector covering the south facade of the building. The heat delivered by the collector is circulated into a hypocaust and subsequently mixed with the garage air before returning to the collector. This provides sufficient comfort in the workshop without auxiliary heating for most of the heating season. Accordingly, no permanent auxiliary heating system has been installed. The building is a simple rectangular, 6 × 9 m, wooden construction with a flat roof. A skylight dome of polycarbonate provides ample daylighting. The potential of this system for use in larger commercial workshops has been analysed by simulation. As an alternative, solar-heated air can be circulated directly into the garage by natural convection.

#### Summary statistics

| | |
|---|---|
| System type: | type 5 |
| Collector type and area: | opaque, wall-integrated, 19.8 m$^2$ |
| Storage type, capacity: | hypocaust m$^3$, 14 kWh/K |
| Annual contribution of the solar air system: | 47 KWh/m$^2$a |
| Annual auxiliary heat consumption: | 92 KWh/m$^2$a |
| Basis: | monitored/calculated |
| Heated floor area/volume | 55 m$^2$/210 m$^3$ |
| Year solar system built: | 1993 |

### SITE DESCRIPTION

The workshop is located in the Rhine Valley in Vorarlberg, Austria. This region, at the upper end of Lake Constance, is plagued by autumn fog. There is no major shadowing south of the building:

| | |
|---|---|
| latitude | 47°N |
| longitude | 9°E |
| altitude: | 410 m above sea level. |

The seasonal heat degree day average is 3530, with reference to the Austrian base of 20/12°C. With reference to the same standard the number of heating days is 219.

### BUILDING PRESENTATION

The building is a lightweight wooden construction with an earth-covered flat roof. The walls are thermally insulated with 12 cm of Rockwool, the roof with 18 cm. The building's thermally effective mass is restricted to the hypocaust in the floor. A plan and section are shown in Figure VI.2.1. The building statistics are given in Table VI.2.1 and the *U*-values of the building elements in Table VI.2.2.

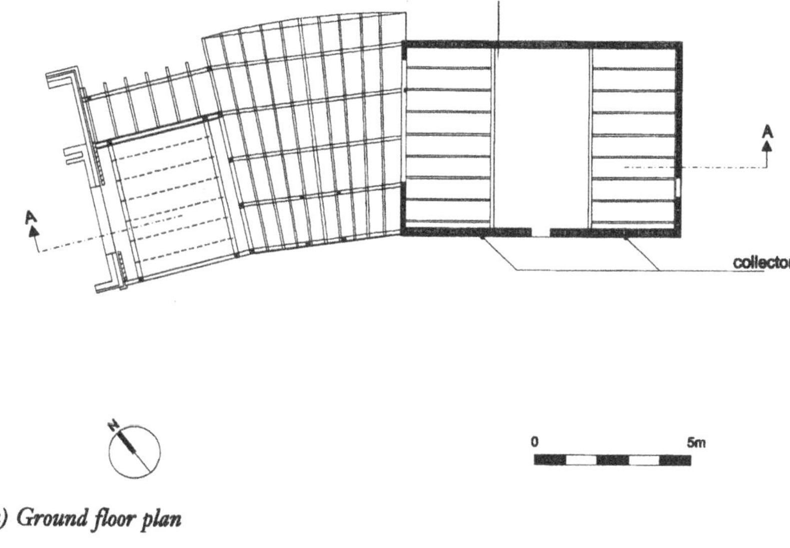


*(a) Ground floor plan*

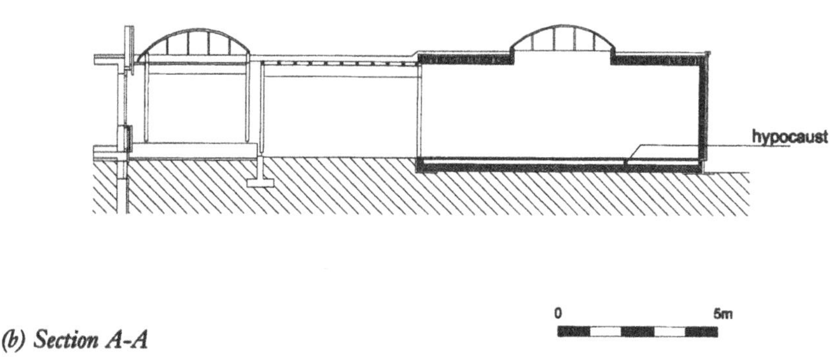


*(b) Section A-A*

*Figure VI.2.1. (a) Ground floor plan and (b) section*

Table VI.2.2. *U*-values of building elements

| Element | *U*-value (W/m$^2$K) |
|---|---|
| Walls: | |
| north, east, west | 0.32 |
| south with collector | 0.30 |
| garage door | 0.95 |
| roof | 0.23 |
| skylight (polycarbonate) | 3.30 |
| windows | 1.3 |
| floor to ground | 1.99 |

## SOLAR SYSTEM

The active open-loop system with hypocaust storage works in two modes (Figure VI.2.2).

In the main mode a fan blows solar-heated air from the air collector into the hypocaust floor and subsequently into the room, from where it is drawn back to the collector.

In the alternative mode solar-heated air thermosiphons directly into the room. This gives the user the option of heating the room more quickly. A one-way vent flap between the collector and the fan allows solar-heated air into the room when the fan is deactivated. When the fan is in operation, the one-way vent flap closes. To prevent cold air sinking from the collector and entering the room at night, the lower opening, between the collector and room air, also has a one-way vent flap.

### The collector

The air collector is single-glazed with air underflow. It has a corrugated aluminium absorber and a selective surface. The net collector area is 19.8 m$^2$, the tilt is 90° and the orientation 15° west of south. The air flow rate through the collector is 47.5 m$^3$/m$^2$(collector area)h.

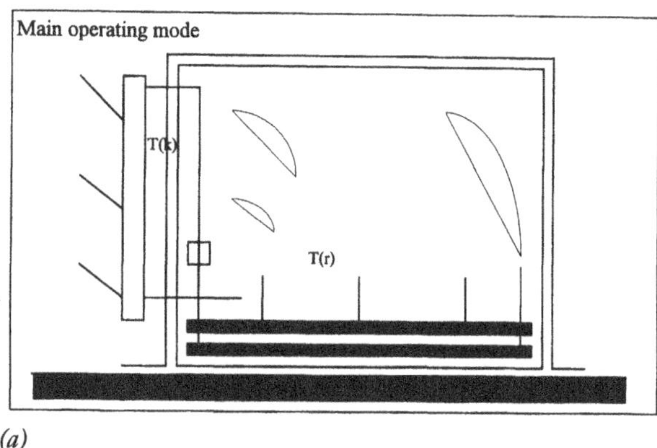


*(a)*

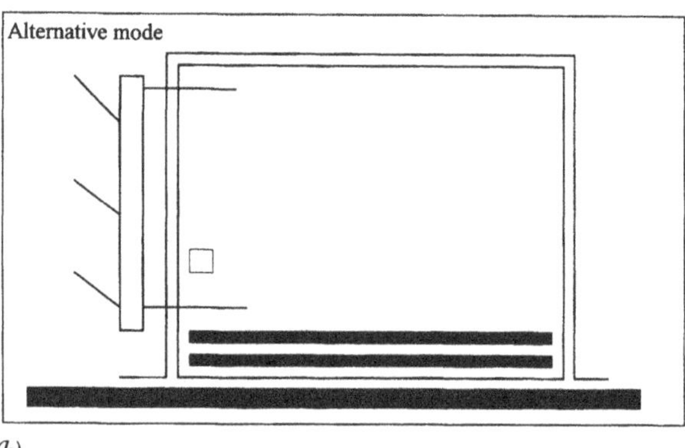


*(b)*

*Figure VI.2.2. (a) Main operating mode and (b) alternative mode*

The collector covers the whole roof of the building, for practical, economic and aesthetic reasons. In this way, the constructional details of the surrounding building components are simpler and cheaper. Site assembly proved to be simple, thanks to a standardized mounting procedure and partial prefabrication. A section of the air collector is shown in Figure VI.2.3(a).

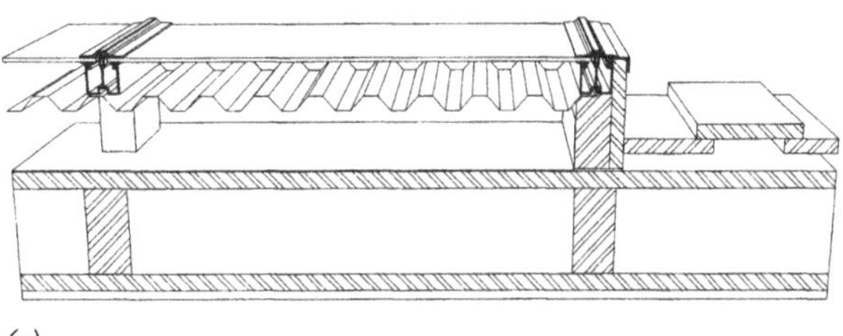

*(a)*

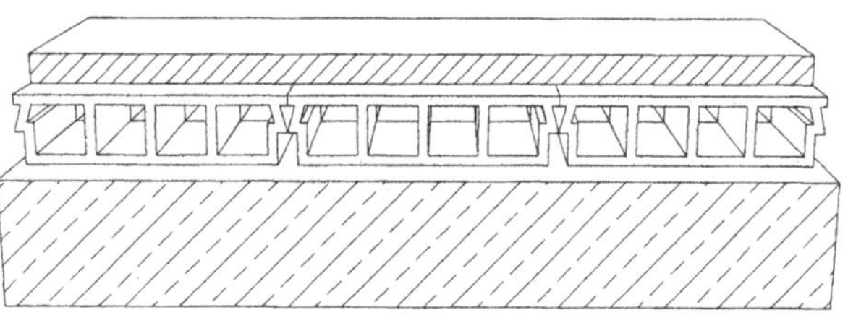

*(b)*

*Figure VI.2.3. Detail sections of (a) the air collector and (b) the hypocaust*

### Storage

The hollow core floor consists of a cement screed above hollow bricks (Figure VI.2.3[b]) on top of a concrete slab foundation, which is not insulated from the ground below. (This saved on cost but reduces performance.) The heat capacity of the concrete foundation slab and the hollow core floor is at 14 kWh/K.

### Distribution

During collector operation hot air passes through the hypocaust floor and is blown into the room. The heat stored in the hypocaust radiates into the room. In the alternative mode the air enters the room directly.

### Controls

The fan is controlled according to room and collector temperatures. A number of back-draught dampers are self-operating and need no controls. The thermosiphon mode is manually shut off in summer, when there is an option of venting collector air to the outside while, at the same time, air is drawn out of the room.

## PERFORMANCE

### General considerations

The system effectively provides a solar-tempered climate appropriate for a workshop. A building such as this, with a single large room, has no problems with heat distribution and the costs for such a solar installations were modest.

### Design considerations

A suitable design of such a system depends on the required indoor temperature. This system is suitable for workshops where relatively low temperatures are acceptable. With lower room temperatures the collector is more efficient. For cool indoor temperatures (just frost-free) thermal insulation under the floor is not essential.

### Where did the idea come from?

The design of the Mätzler Garage was inspired by the workshop of a concrete supplier in Ohio, USA. The building consisted of a cheap lightweight shed with additional thermal insulation and an open loop thermosiphoning air collector. The only thermal storage was the contents of the building itself (the heavy steel trucks, sand, cement and water). The main requirement was merely to keep the shed above freezing during severely cold periods. Thus the insulation under the floor was unnecessary.

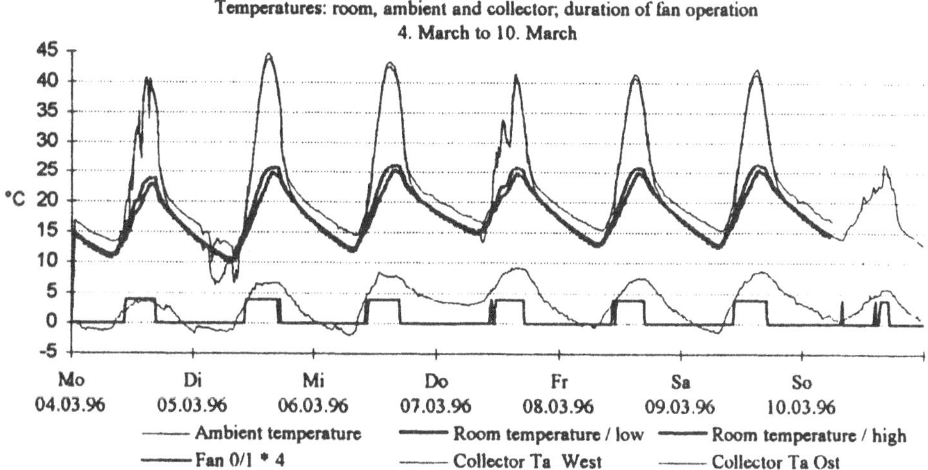


*Figure VI.2.4. Thermal behaviour*

*Table VI.2.3. Buildings analysed*

| Building | Floor area ($m^2$) | Height (m) | Door area ($m^2$) | Occupants | Collector area ($m^2$) | Hypocaust area ($m^2$) |
|---|---|---|---|---|---|---|
| Building 1 | 54 | 3 | 9 | 2 | 20 | 46 |
| Building 2 | 200 | 4.5 | 24 | 4 | 60 | 140 |
| Building 3 | 800 | 6 | 48 | 8 | 190 | 400 |

## Thermal behaviour (measured)

The first week of March 1996 illustrates the thermal behaviour of the building (Figure VI.2.4). The outdoor temperatures varied between −1°C and +9°C. The indoor temperatures varied between +10°C at night and +25°C during the operation of the collector.

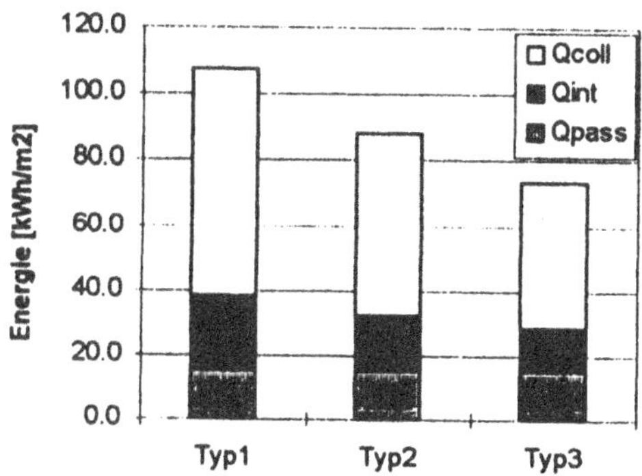


(a)

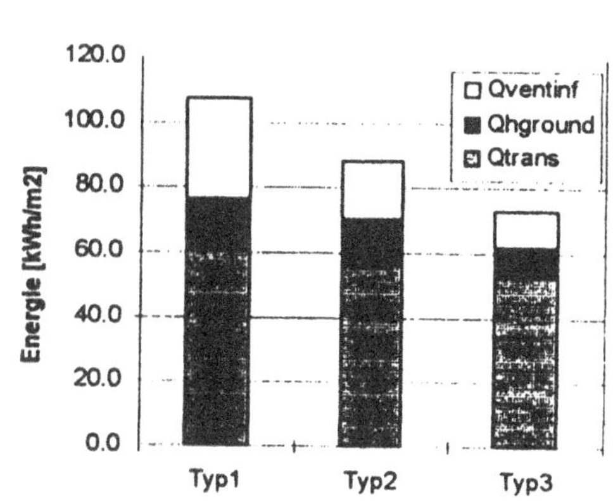


(b)

*Figure VI.2.5. Yearly specific heat gains (a) and losses (b)*

## Simulations and sensitivity studies

To assess the suitability of this solar air system for larger workshops, the effects of variations of building size, room temperature and insulation were analysed.

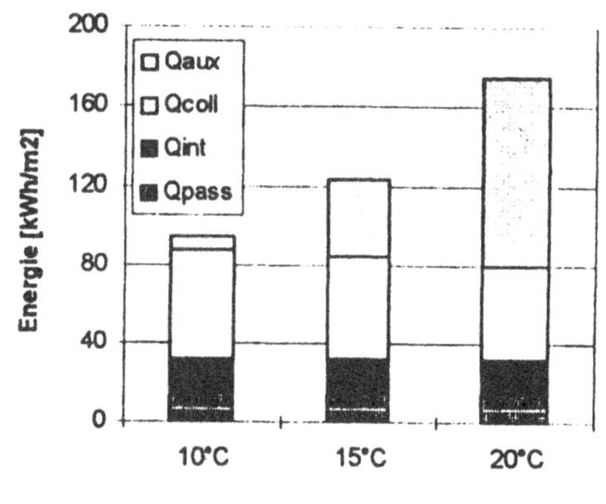


(a)

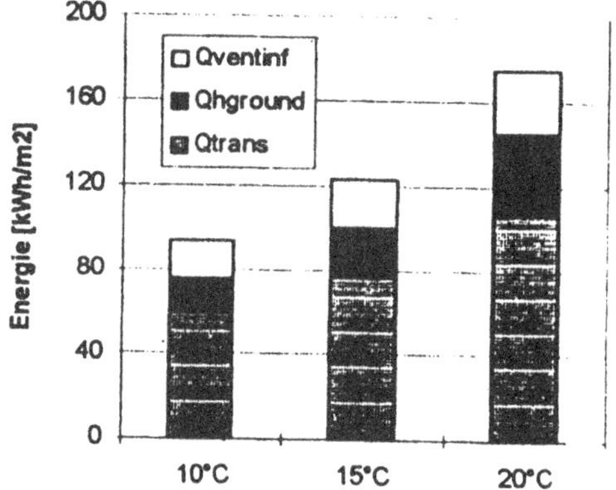


(b)

*Figure VI.2.6. Yearly specific heat gains (a) and losses (b) for Building 2 with different set points*

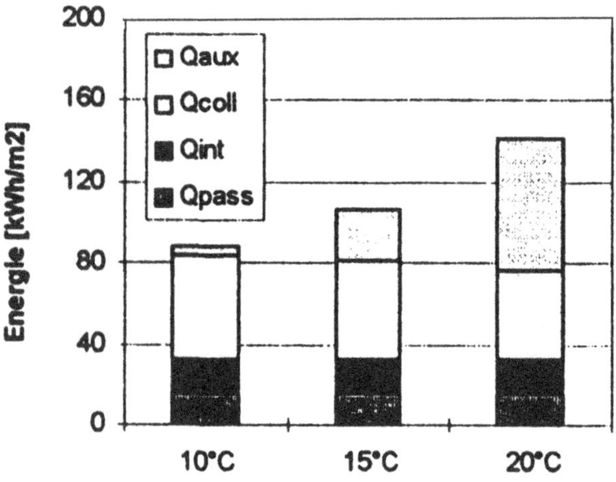


(a)

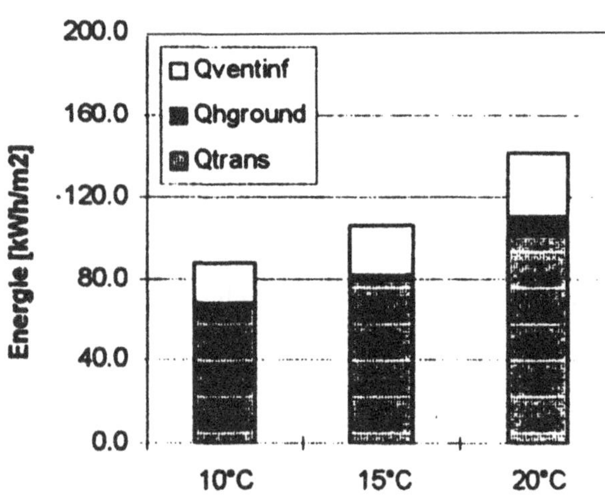


(b)

*Figure VI.2.7. Yearly specific heat gains (a) and losses (b) for Building 2 with different set points and 6 cm of ground insulation*

The building construction and their level of thermal insulation were the same as in the Mätzler Garage. Instead of the large skylight arch, north-facing skylights with a tilt of 70°, with an area of 15% of the roof and a *U*-value of 3 W/m$^2$K were considered.

Three building sizes were analysed (Buildings 1–3; Table VI.2.3). Building 2 was simulated for different heating set points. Auxiliary heating was applied for temperatures (10, 15 and 20°C). Three levels of thermal insulation to the ground (0, 3 and 6 cm of extruded polystyrene under the concrete slab foundation) were compared.

## Results of simulations

*Building size*

The three buildings were simulated without auxiliary heat and without any additional ground insulation. The specific heat gains of the three buildings are shown in Figure VI.2.5(a) and the losses in Figure VI.2.5(b).

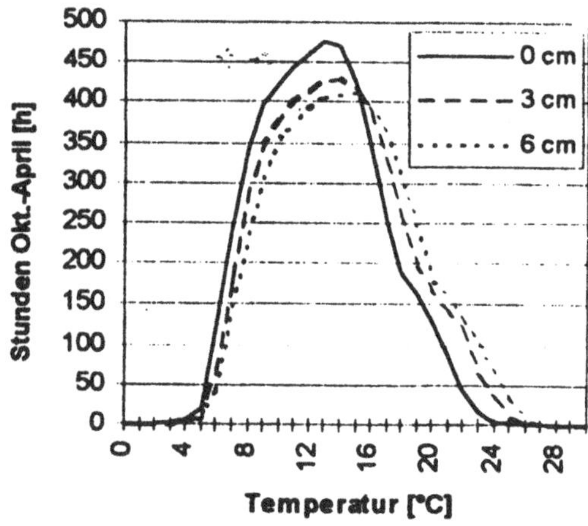


*Figure VI.2.8. Room temperature distribution of Building 2 with different thicknesses of ground insulation*

*Auxiliary heating at different temperatures*

To show the suitability of this system for different thermal comfort levels, Building 2 was assumed to be heated with the set points 10, 15 and 20°C. The results in Figure VI.2.6 show that the need for auxiliary heating is quite low at temperature levels up to 15°. However, for higher room temperatures, insulation from the ground becomes imperative.

*Ground insulation*

The effect of ground insulation is shown in Figure VI.2.7 from the same simulation as in Figure VI.2.6, but with 6 cm of ground insulation. The annual energy savings at 20°C were 30 kWh/m$^2$.

*The effect of ground insulation on thermal performance*

The room temperature distribution (October–April) using different thicknesses of ground insulation is shown in Figure VI.2.8. It is clear that ground insulation is appropriate for higher temperature levels. For lower temperature levels such as 10°C the thermal inertia of the ground may help to sustain the comfort level.

## REMARKS

The use of this system seems promising for larger workshops with only low temperature requirements. For higher temperature levels the same system requires ground insulation. The positive effect of substantially increasing the size of the building did not meet the expectations of the author.

## ACKNOWLEDGEMENTS

Architect, system designer, chapter author: Sture Larsen, Hörbranz

Monitoring and simulations: Manfred Bruck, Kanzlei Dr. Bruck, Vienna; Christoph Muss, Kanzlei Dr. Bruck, Vienna; Eckhard Drössler, Düns; Josef Burtscher, Energiesparverein Vorarlberg, Dornbirn, Austria

Manufacturers of the selective absorber surface: TeknoTerm Energi AB, Askim, Sweden

# VI.3 Kägi Steel Warehouse

## Winterthur, Switzerland

*Winterthur*

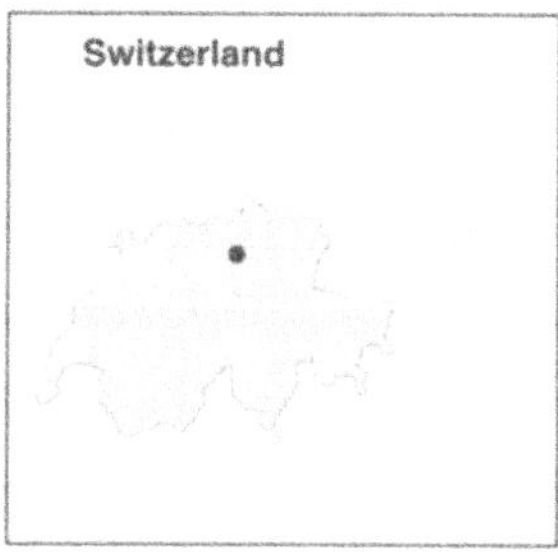


*System type 2*

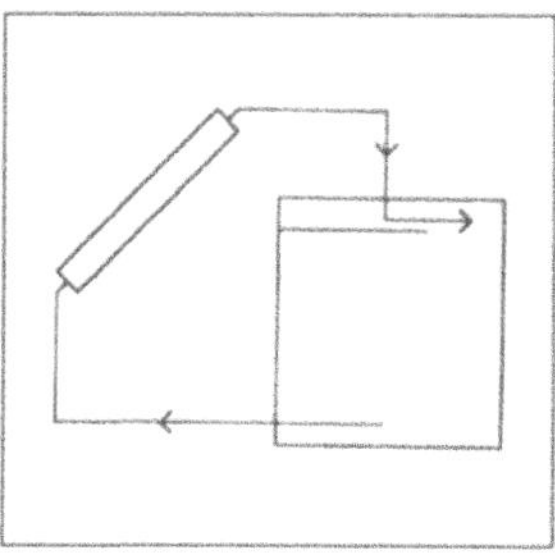

## PROJECT SUMMARY

The steel warehouse of the Kägi company in Winterthur near Zürich is a two-hall complex. The older (left) hall (3,000 m$^2$, 30,000 m$^3$) was completely uninsulated and has an open basement (6,500 m$^3$) to store infrequently-used items. To prevent condensation on the cold steel surfaces in spring and early summer and to provide a smaller temperature difference between the hall and the basement for workers in the summer (and avoid productivity loss due to illness) as well as to raise temperatures in the winter, the following measures were taken in a step-by-step approach:

- A ventilation system was installed (1986) to circulate sun-warmed air from under the roof to the basement. It was shown that, under the given circumstances, such a system worked well and was cost-effective in comparison with a conventional heating system.
- Complete external insulation was added to the hall (1994) and a roof-integrated 350 m$^2$ air collector installed so that the circulating air could be warmed.
- Construction of a separate adjoining new hall (1996) profited from all the above experience. The air collector is integrated in the facade facing south-east.

The first stage (1986) was monitored in detail, while only rough measurements were made for the modified system (1994). The 1996 system for the new hall was analysed by computer simulation.

### Summary statistics

| | |
|---|---|
| System type: | type 2 |
| Collector type and area: | roof integrated, 350 m$^2$ |
| Storage type, volume, capacity: | basement, 6700 m$^3$, 930 kWh/K |
| Annual contribution of the solar air system: | 16.6 kWh/m$^2_{floor}$ |
| Basis: | monitored |
| Heated floor area: | 3000 m$^2$ |
| Year solar system installed: | 1994 |

## SITE DESCRIPTION

The buildings are situated in a small industrial area outside the city of Winterthur, about 20 km north-east of Zürich. The site lies in a shallow valley running east–west and profits from good solar exposure. The photograph shows the complex of the two connected warehouse halls as seen approximately from the south ('sun view' in January):

| | |
|---|---|
| latitude | 48°N |
| longitude | 9°E |
| altitude | 440 m above sea level. |

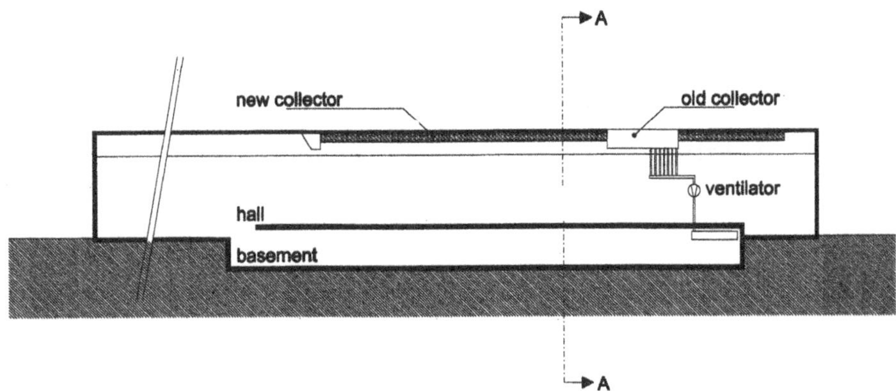


*(a) Longitudinal section*

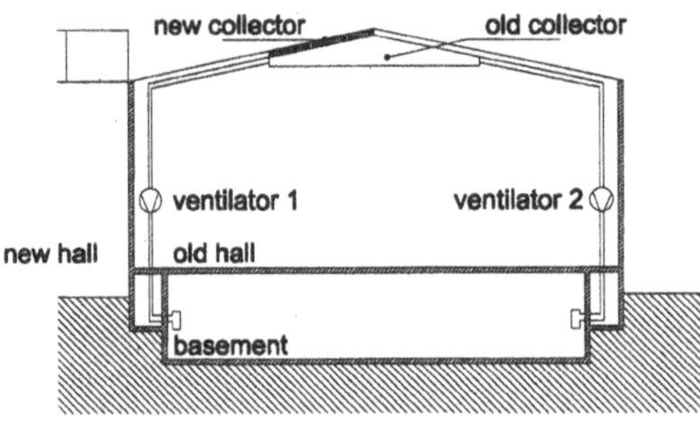


*(b) Section A-A*

*Figure VI.3.1. (a) Longitudinal section through the warehouse hall and (b) cross-section A–A*

The climate is a typical 'central European climate' (i.e. overcast and foggy in winter).

## BUILDING PRESENTATION

The complex consists of two connected halls (2 × 3000 m$^2$). Under the old hall (on the left in the photograph) there is a basement with a heavy concrete foundation and ceiling. The roof surfaces are tilted by 15° and oriented south–east and north–west. Sections are shown in Figure VI.3.1.

The primary function of the building is to store steel pipes and, when the halls were open, there were severe problems with extreme temperature differences and condensation, particularly in the basement. In 1986 the halls were closed in; automatic doors for the trucks were installed and the air ventilation system transporting sun-warmed air to the basement was added.

The old hall has no heating system other than the 'air collector'. The temperature closely follows the ambient temperature in its daily and seasonal fluctuations. However, the fluctuations can be more extreme than those of the ambient air, because, mainly in summer, the thin, dark, uninsulated roof acts as a collector and thus is a big additional radiating heat source, which affects the people working there. At night and in the winter the roof loses heat to the cold sky by radiation. The cooled air sinks from the roof to the open basement, with temperatures dropping to nearly 0°C, whereas in a normal (closed) basement they would stay much closer to the soil temperature (5° to 10°C).

*Table VI.3.1. The building elements and their U-values*

| Element | *U*-values ($W/m^2K$) before 1994 | U-values ($W/m^2K$) after 1994 |
|---|---|---|
| Walls | 5.5 | 0.6 |
| Roof | 5.5 | 0.6 |
| Windows/skylights | 6.0 | 3.1 |

*Table VI.3.2. Characteristics of the basement*

| | |
|---|---|
| Volume | 6,700 $m^3$ |
| Surface area | 3,300 $m^2$ |
| Mass | 2,800 tonnes |
| Stored steel | 500 tonnes |
| Heat capacity | 930 kWh/K |

The hall is of lightweight steel-frame construction and is built on a 3 m high lime–sand brick basement foundation. The upper part is made of corrugated fibre panels with a band of single glazed windows (glass or polycarbonate). The roof also consists of corrugated fibre boards with skylights. In 1994 the hall was completely insulated with 6 cm of rockwool. The building elements and their *U*-values before and after 1994 are given in Table VI.3.1. Even though there are many tonnes of steel pipes stored within it, the above-ground hall, with its huge volume of 30,000 $m^3$, must be considered as an extremely light construction.

On the other hand, the basement is a very heavy concrete construction (uninsulated from the ground with the following characteristics given in Table VI.3.2. The total heat storage capacity of nearly 1 MWh/K is very impressive. Figure VI.3.2 shows an inner view of the hall including the air collector.

## SOLAR SYSTEM

As a result of natural temperature stratification and little air exchange in the basement, temperatures stay below 10°C up to spring and early summer. Because the moisture content of the air is considerably higher in spring and summer than in winter, frequent condensation on the cold steel surfaces used to occur, decreasing the value of the stored material. Furthermore the workers, clothed lightly for the warm environment in the hall (temperatures around 30°C), often caught cold when working temporarily in this cold and damp basement.

*Figure VI.3.3. The open basement underneath the hall; in the background is one of the air outlets*

To solve this problem, a simple open ventilation system was installed in 1986 to blow sun-warmed air (9,400 $m^3/h$) from under the roof into the basement (Figure VI.3.3). For this purpose a 'suction box' was installed under the roof (Figure VI.3.2 middle). It consisted of a very light ceiling made of 5 cm rockwool panels and open on both sides. This simple installation worked very successfully. Figure VI.3.3 shows a picture taken from the entrance of the basement towards the rear, where the air outlets can be seen. Figure VI.3.4 illustrates the collector construction. In 1994 the hall was completely insulated and therefore a new solution had to be found to gain sun-warmed air. The 'suction box' was closed and 350 $m^2$ of air collectors were integrated into the roof. The goal was, as before, to prevent:

- extremely low temperatures in winter;
- extreme temperature differences (between the hall and the basement) in spring and summer;
- loss of productivity due to illness;
- condensation on the steel surfaces.

*Figure VI.3.2. Inner view of the hall including the air collector. The old air collection box, which is now closed at both ends, can be seen in the middle*

*Figure VI.3.4. The collector during the construction phase*

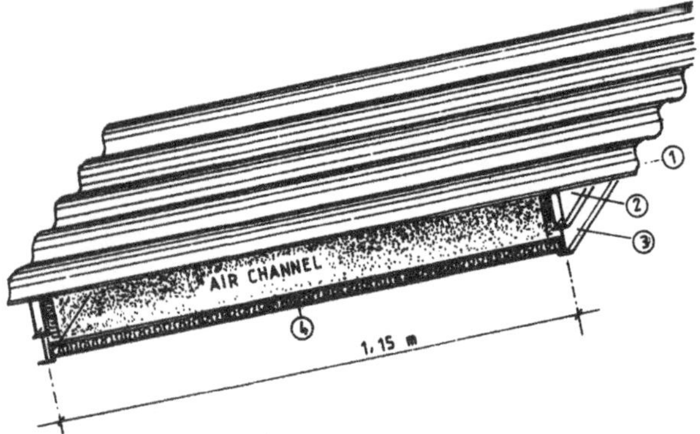


*Figure VI.3.5. The collector construction: (1) outside cover of corrugated polycarbonate; (4) inner cover of rockwool plates, covered white to the hall and black to the collector air channels; (3) direct integration with the supporting steel structure, which, together with the spacer (2), is also insulated from the air gap*

The system is regulated electronically with two inputs: the air temperature difference between collector and basement and the relative humidity of the collector air. Only if the temperature difference is greater than 10 K and the relative humidity lower than 40% are the two ventilators switched on. Details of the collector and the 'storage system' are given in Table VI.3.3.

## PERFORMANCE

Comparison between the under-roof air extraction system (1989) and the roof-integrated collectors for the uninsulated hall (1994) shows the temperatures given in Table VI.3.4.

Of course, comfort in the hall improved slightly owing to the insulation effect. There are smaller daily and seasonal temperature fluctuations and the 'comfort feeling' has definitely increased because no hot (summer) or cold (winter) surface temperatures occurred in the roof.

The performance of the air system stayed about the same (output 50 MWh/a), but the output air temperature increased remarkably. The run times of the ventilation system decreased from 2,200 to 1,300 hours a year because of the different control setting: on = $DT$(collector – basement) > 10 K. This balances out the use of the 'better' collector, having the disadvantage of very little run time in winter. Figure VI.3.6, for a typical sunny day in April 1995, clearly shows the temperature changes in the basement as a result of the warm air input; without this there would have been no temperature change during the day. The mean collector efficiency is 25%.

*Table VI.3.3. Technical data for the collector and the 'storage system'*

| | |
|---|---|
| Overall area of the 15° tilted south-east roof | 1500 $m^2$ |
| Area of 15° tilted south-east absorber | 300 $m^2$ |
| Air volume rate/absorber area | 30 $m^3/hm^2$ |
| Fan power/absorber area | 7 $W/m^2$ |
| Storage capacity/absorber area | 3 $kWh/Km^2$ |

*Table VI.3.4. Temperatures measured in January and July before and after the insulation of the hall*

| Typical values (°C) | January | | July | |
|---|---|---|---|---|
| | 1989 | 1994 | 1989 | 1994 |
| Outside air average | 0.3 | 0.1 | 19.4 | 22.4 |
| Collector maximum | 10 | 20 | 40 | 50 |
| Hall average | 1.6 | 3.1 | 22.8 | 24.9 |
| Basement average | 4.7 | 6.5 | 20.4 | 23.0 |

### Parametric studies by computer simulations

Based on measured data for 1994–95, extrapolations were made by computer modelling (TRNSYS) to answer several additional questions. One key issue is the fact that a south-east surface, tilted by only 15°, has almost the same irradiation as a horizontal plane, as can easily be shown by using any meteorological tool ( e.g. METEONORM). This also means that there is too much gain in summer and almost none in winter. Since industrial halls tend to have only slightly tilted roofs, the only alternative would be to place the collector on a vertical wall, which in this case is also south-east oriented.

The following variations were investigated:

0. As is, (azimuth –45° from south/tilt 15° from horizontal – 45/15; see Figure VI.3.7).
1. Same as variation 0, but vertical south-east (–45/90).

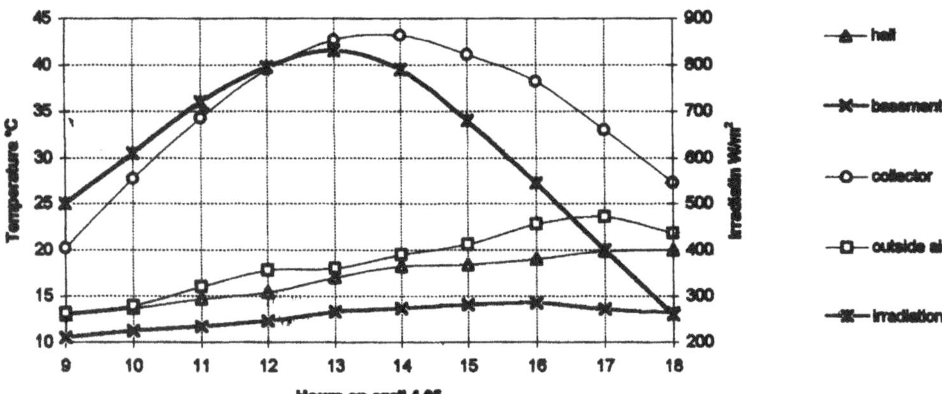


*Figure VI.3.6. The temperatures on a typical sunny April day*

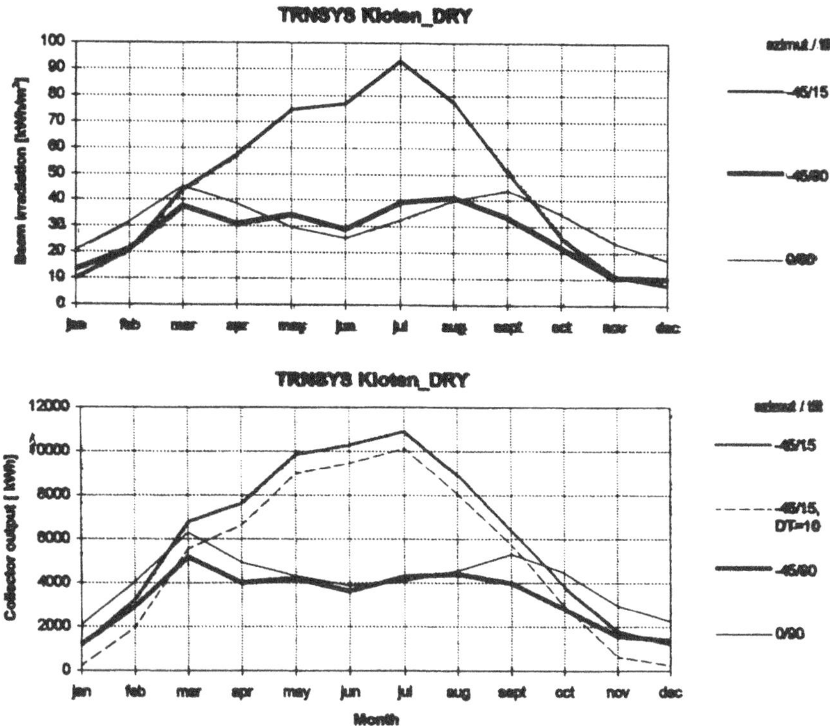


*Figure VI.3.7. Simulation outputs*

2. Same as variation 1, but vertical due south.
3. Same as variation 1, but with improved (double polycarbonate) glazing.

To simply check the advantages of collector orientation regarding irradiation input METEONORM was used. To gain insight into how the whole system (hall–collector–basement) performs, TRNSYS was used, together with an air-collector module. Figure VI.3.7 shows the beam irradiation on the collector surface according to METEONORM (top) and TRNSYS collector energy output. For comparison the original (validation) case with its higher temperature difference in the control setting is plotted.

Qualitatively, the collector performance closely follows the pattern of the direct-beam irradiation of the sun, as can be clearly seen in Figure VI.3.7. This is reasonable because collectors mainly function when the sun is shining. The output in spring is a little higher and in autumn a little lower because 'storage' temperatures (basement/hall) are different. The conclusions that can be drawn are as follows:

- Placing the collector vertically flattens out the performance, allowing year-round constant control settings without excessive output (and ventilator electricity consumption) in summer.
- Winter performance is very sensitive to control settings. A (too) large temperature difference (10 K) for fan operation results in almost zero output from November to January. Reducing that value to 5 K improves the situation drastically and was chosen as standard for all other variations shown.
- Winter performance cannot be improved substantially by placing the collector vertically, facing south-east because vertical planes are much more sensitive to deviation from due south than are tilted planes. Nevertheless, a due-south vertical plane shows excellent performance in winter, spring and autumn and would therefore be a definitive improvement.
- Temperatures in the basement do not differ much among the variations because of the very heavy construction which is uninsulated from the ground.

The benefit of better collector glazing (smooth double instead of corrugated single-sheet polycarbonate) could not be analysed with TRNSYS because the available models for air collectors were not suitable for such parametric studies. Nevertheless, the effect on collector output itself can be calculated using a comparatively simple tool for estimating collector performance based on irradiation data and outside air temperature. The equations for heat and irradiation balance were set up in a PC spreadsheet. Iterations for different conditions returned the respective values for collector performances, resulting finally in a common collector–performance curve. With this collector output was calculated and added for every hour over a year, based on weather and hall data from TRNSYS. This procedure comes very close to a 'real' simulation and is adequate to show the difference between the two cases. A better collector cover loses less heat, but on the other hand admits less solar radiation. The net effect on collector performance is an improvement in performance of an average of 40%. (VI.3.8).

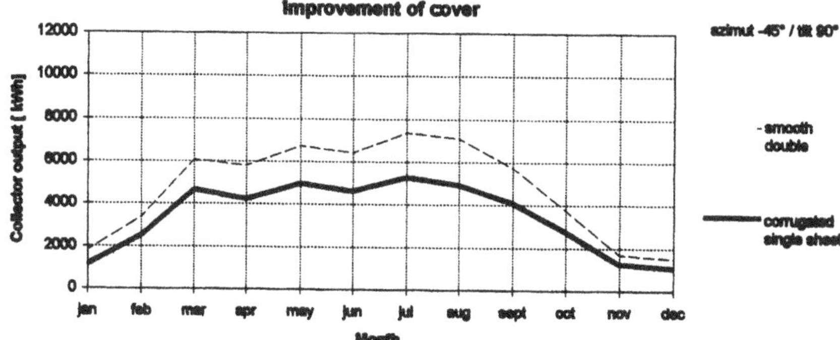


*Figure VI.3.8. Comparison of different air-collector covers (polycarbonate)*

## REMARKS

It is important to circulate warm air from warmer to colder regions in a huge warehouse in order to prevent cold, moist and unhealthy working areas, especially in basements. It is most cost-effective to simply extract the air from under an uninsulated roof. Generally, temperature stratification is as good as an insulated collector construction.

When a building such as a industrial warehouse is to be insulated, which is generally recommended for comfort reasons, integrating an air collector in a more or less south-facing roof or wall works well. A vertical collector is more sensitive to orientation but has several advantages:

- even distribution of collector energy over the year (no overheating in summer), which also allows optimal control settings;
- protection from rain and dirt;
- easier cleaning and visual inspection;
- easier mounting;
- shorter air channels, resulting in a lower pressure drop and fan electricity consumption.

A vertical air collector improves winter performance over one tilted 15° only if it is oriented more or less due south (±15°). Above an azimuth of 45°, no further improvement occurs.

Fan control settings (temperature difference) are critical for performance in winter when insolation is weak and outside temperatures low. 5 K seems to be a reasonable control setting.

This air system definitively improved basement comfort conditions, but had little influence on hall temperatures since the volume of the hall is so big. However, the somewhat higher surface temperature of the floor just above the basement improves the 'comfort feeling' there.

The time constant of the concrete storage mass (basement ceiling and foundation) can be extrapolated to be about one month.

To prevent condensation problems completely, the control system of the fans must take into account the dew point, *but dewpoint control is expensive*. The project showed that if the system is working correctly and is only controlled by temperature difference, it is possible to keep the temperature level above the dewpoint in almost any condition all year round! This would make dewpoint control superfluous. The system is then a very attractive and low-cost alternative, although a choice has to be made between certain compromises concerning the number of hours below dewpoint, the amount of heat transferred and the season of operation.

Generally, the project showed that using passively sun-warmed air is effective in modifying the temperature of a basement and keeping it dry. We believe that there is a potential for such a system for diverse basement storage areas, for many kind of goods.

## ACKNOWLEDGEMENTS

Engineer and Solar System design: Kägi AG and Amena AG
Measurements/author of paper: A. Gütermann, c/o Amena AG, Steinberggasse 2, CH-8402 Winterthur, Switzerland

## BIBLIOGRAPHY

*Luftkollektoranlage Stahllager Kägi AG in Winterthur* (1995). R+D project report Nr. 11096 (in German). HBT-Solararchitektur, ETH-Hönggerberg, CH-8093 Zürich.

*METEONORM: Meteorologische Grundlagen für die Sonnenenergienutzung* (1995). (In German). Bundesamt für Energiewirtschaft, CH-3003 Bern.

*Thermische Messungen am Stahllager Kägi & Co. In Winterthur* (1990). Monitoring project final report (in German). HBT-Solararchitektur, ETH-Hönggerberg, CH-8093 Zürich.

*Window Air Collectors – a Physical Model and its TRNSYS Implementation* (undated). Basler & Hofmann, Forchstrasse 395, CH-8029 Zürich.

# VI.4 Wewer's Brickyard

Helsinge, Denmark

*Helsinge*

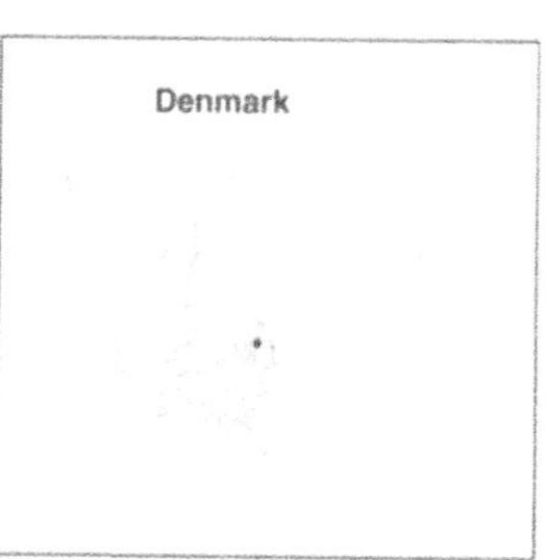


*System type 1*

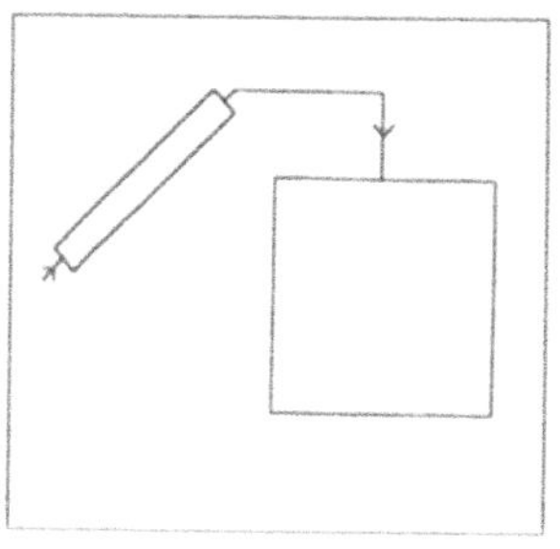

## PROJECT SUMMARY

Wewer's Brickyard is a demonstration project with 126 $m^2$ of multi-function solar energy panels (MF-panels), built in the northern part of Zealand, Denmark. The MF-panels preheat the ventilation air to the building. The potential yearly contribution from an MF-panel under Danish climatic conditions may exceed 900 kWh/$m^2$. The extra cost compared to a traditional wall or roof is small.

The high performance of the MF-panel system is due to the MF-panel being not only a solar collector, but also a heat exchanger (heat recovery unit) between the fresh air coming into the building and exhaust air leaving the building. Being a part of the wall or roof construction, the MF-panel further decreases the heat loss through that part of the building.

### Summary statistics

| | |
|---|---|
| System type: | type 1 |
| Collector type and area: | opaque on the roof 126 $m^2$ |
| Annual contribution from the solar air system: | 19.2 kWh/$m^2a_{floor}$ |
| Annual auxiliary heat consumption: | 286 kWh/$m^2a_{floor}$ |
| Basis: | monitored |
| Heated floor area, volume: | 500 $m^2$, 1650 $m^3$ |
| Year solar system built: | 1991 |

## SITE DESCRIPTION

The demonstration project is located just outside Helsinge in northern Zealand, Denmark:

| | |
|---|---|
| latitude | 56°N |
| longitude: | 12°E |
| altitude | 30 m above sea level. |

There is a temperate coastal climate.

## BUILDING PRESENTATION

The brickyard manufactures traditional bricks for the building industry and for export. The building, where the MF-panel is installed, contains a workshop for manufacturing brick beams and special-purpose bricks.

Figure VI.4.1 shows a drawing of the building with the approximate location of the MF-panel and a plan of the rooms of the building. The building has uninsulated brick walls and single-pane windows. The roof is poorly insulated, with only an approximately 50 mm layer of badly located mineral wool. The main entrance to the building is through a draughty sliding wooden door (the barn-door type) to room 1; the gates to room 3 and 5 are also draughty. The building has thus a very high heat loss due both to transmission and to natural infiltration. The heat demand for space heating for the heating season 1992/93 was measured to be 143,000

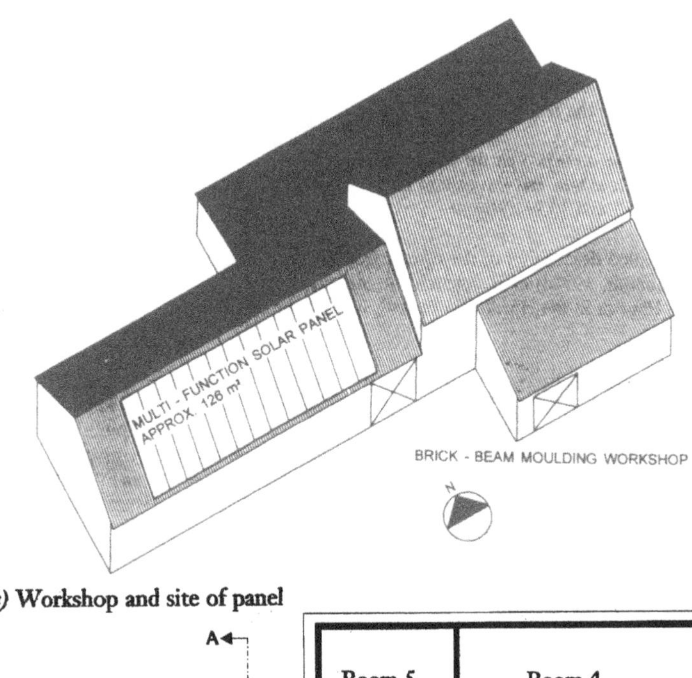


*(a)* **Workshop and site of panel**

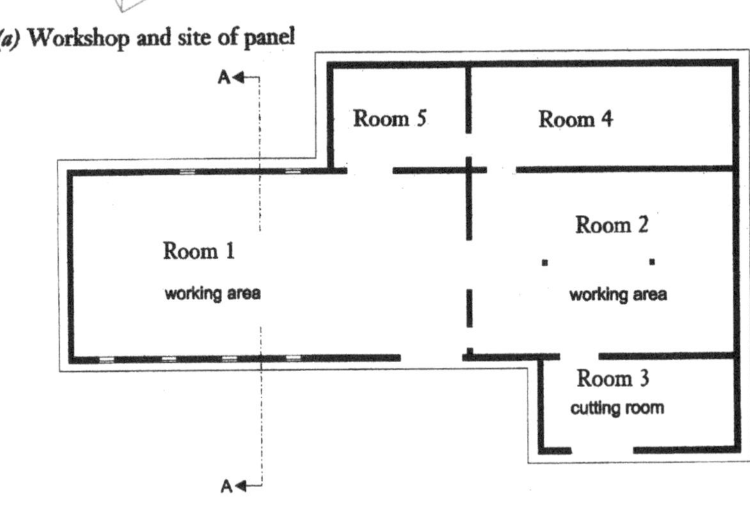

***(b) Floor plan***

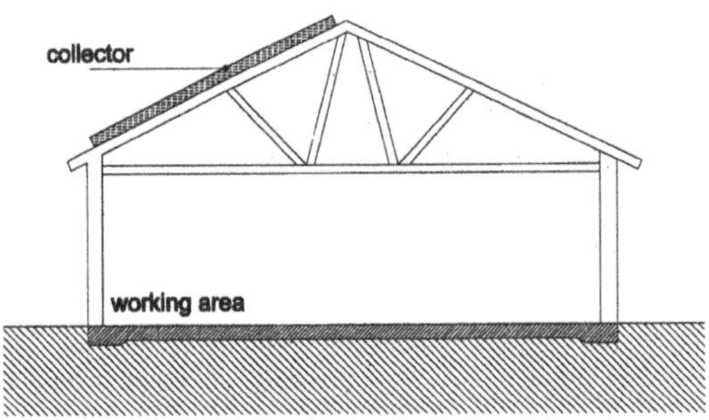


*(c) Section A-A*

*Figure VI.4.1. (a) The workshop, showing the appropriate location of the MF-panel, (b) a floor plan and (c) section*

*Figure VI.4.2. The interior of room 1; the arrows indicate some of the inlets and outlets of the MF-panel system*

kWh (286 kWh/m²a). The infiltration rate has not been measured. The building has no domestic hot water.

Brick beams are manufactured in both rooms 1 and 2. Room 3 is a cutting room, where the grooves for the steel reinforcement are cut. Room 4 serves as an office and as an area for breaks. The gas boiler is located in this room. The mixing machine is located in room 5. Figure VI.4.2 shows the factory interior.

## SOLAR SYSTEM

The MF-panel system consists of 126 m² of MF-panels connected directly to the rooms without any storage in between. 56 m² of MF-panels are connected to room 1 and 56 m² to room 2, while 14 m² is connected to room 3.

The idea behind installing the MF-system was to improve the comfort for the workers and to decrease the drying time for the brick beams. Comfort could be improved by increasing the room temperature and decreasing the humidity and this would automatically reduce the drying time of the brick beams.

### The solar collector

The MF-panels are mounted on an existing south-facing roof with a tilt of 19°.

How the MF-panel functions is shown in Figure VI.4.3. The MF-panel consists of a transparent cover, behind which a metal sheet with a trapezium-corrugated profile is located. The two plates are mounted in front of a normally insulated outer wall or roof. At the top of the panel there are box-shaped manifolds. In this way it is possible to let air into the two sets of air gaps formed by the trapezium-corrugated metal sheet. The panel can, therefore, act both as a solar collector and as a heat exchanger between fresh air coming into the building and exhaust air leaving the building. If the exhaust air is led down between the insulated wall and the trapezium-shaped plate, the heat loss through the wall will be reduced further, as the temperature on the outside of the insulated wall will be higher during the heating season than the ambient air temperature. Details of the top and bottom of the MF-panel are shown in Figure VI.4.4.

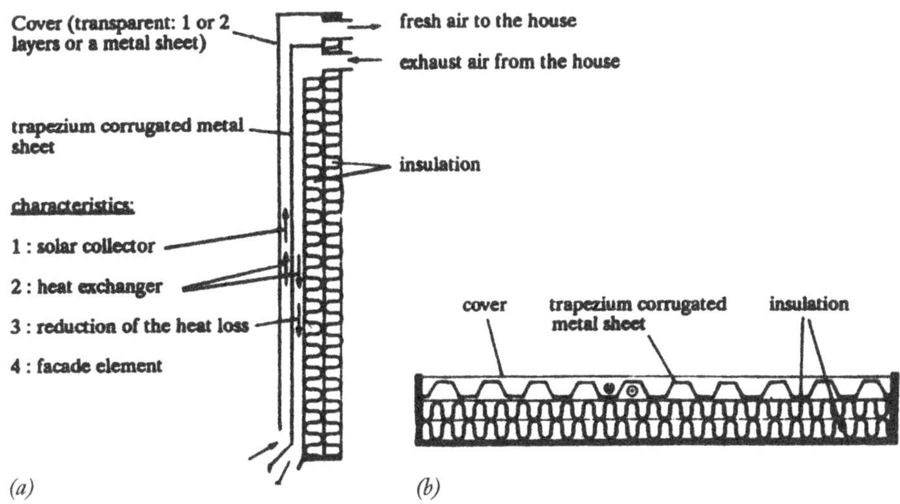


*Figure VI.4.3. (a) Vertical and (b) horizontal sections through the MF-panel*

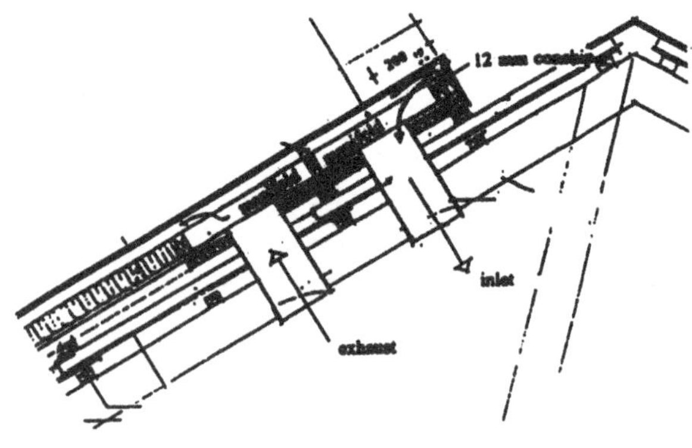


(a)

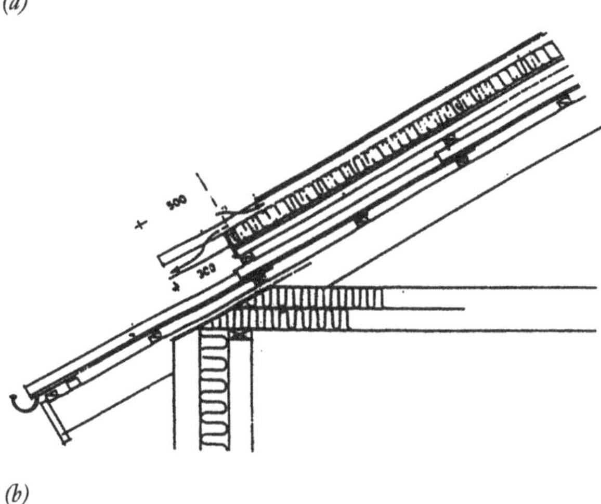

(b)

*Figure VI.4.4. Construction drawing of the (a) top and (b) the bottom of the MF-panel*

The area of the MF-panel is approximately $6 \times 21$ m$^2$ = 126 m$^2$. The cover of the MF-panel is divided into 26 sections, each with a width of 0.81 m. The trapezium plate is 0.5 m longer than the cover in order to separate fresh air from exhaust air, as seen in Figure VI.4.4(a).

The cover consists of double-walled, 10 mm thick ribbed UV-protected polycarbonate sheets. The cover is fixed to the MF-panel by means of aluminium profiles. The smallest distance between the cover and the absorber, and the absorber and the reverse-side insulation is 10 mm. The absorber of the system is made of trapezium-corrugated steel sheets with a black surface. The joints between the steel sheets have been carefully sealed in order to avoid leaks between fresh and exhaust air. The chosen profile is shown in Figure VI.4.5.

In order to create an even distribution of the incoming air over the absorber, a pressure drop was created at the top of the MF-panel, at the manifold. The trapezium plate was not stopped at the manifold but was continued almost to the top of the manifold, as shown in Figure VI.4.4(b). An air gap of only 15–20 mm was left for the air to flow into the manifold. The reverse-side insulation of the MF-panels consists of 75 mm mineral wool. For more details on the MF-panels please refer to Jensen (1990) and 1994).

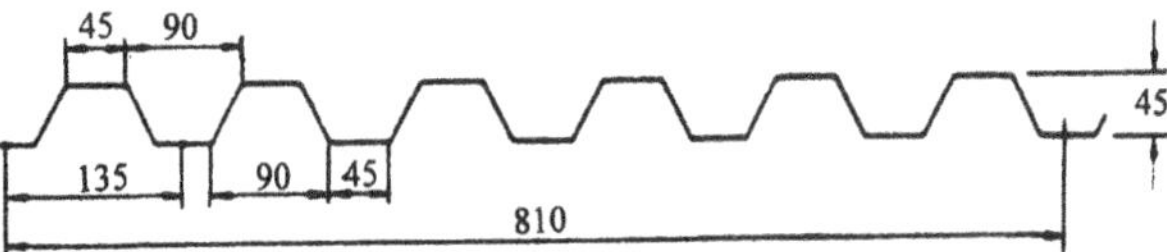


*Figure VI.4.5. The profile of the trapezium-corrugated steel sheets*

## Storage

There is no storage in the system except for the storage capacity of the building itself.

## Distribution

The MF-panel serves both as a solar air collector and as a heat exchanger between exhaust and fresh air. For that reason a pair of ducts between the MF-panel and the building is necessary.

The MF-panel is connected to the ducting system via nine pairs of channels (Figure VI.4.6). Rooms 1 and 2 have fairly similar volumes, 840 and 645 m$^3$ respectively, while the volume of room 3 is much smaller at 180 m$^3$. It was, therefore, decided to connect approximately 56 m$^2$ of absorber area, using four pairs of ducts, to each of rooms 1 and 2, while room 3 was connected to the last pair of ducts, connected to approximately 14 m$^2$ of absorber area. The gas boiler is located in room 4, so it was decided not to connect any part of the MF-panel to this room, which is heated by the heat loss from the boiler. The ducts to the rooms are insulated with 30 mm of mineral wool.

Because the MF-panel was installed on an existing roof, positioning the ducting was rather difficult. The precise location of the ducts has, to a large extent, been determined by the location of the attic rafters.

The inlet air is blown into the room through diffusers right under the ceiling, while the outlets are located along the walls, 1 m from the floor. In this way no short cut exists between the inlets and the outlets. Figure VI.4.6 shows the location of the inlets and outlets, while Figure VI.4.2 shows some of the inlets and outlets in room 1.

Room 3 is only connected to the collector part of the MF-panel; it is thus not possible to exchange heat between fresh air and exhaust air for this room. The reason for this is that, because of the many openings, the infiltration into the room is very large, so heat recovery would be pointless. The inlet to room 3 is on the back wall right under the ceiling, as indicated in Figure VI.4.6.

During the summer, when there is no heat demand, air in the MF-panel stagnates. In order to decrease the stagnation temperature three motor-driven bypasses open when the MF-panel becomes too hot, so that it is cooled by the stack effect. The location of the bypass ducts is shown in Figure VI.4.6.

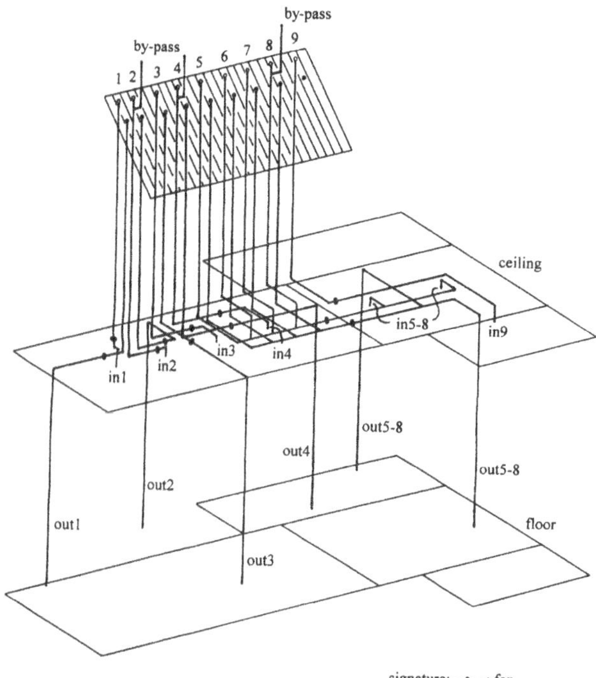


*Figure VI.4.6. Isometric drawing of the connection between the MF-panel and the rooms of the building*

## Control

A rather simple control was selected for the system as the use of the building did not require precise, sophisticated control. The system is thus controlled by interconnected thermostats and hygrostats. Two thermostats are located in the attic, to sense the temperature at the top of the MF-panel. One of these thermostats determines if the temperature of the MF-panel is high enough to deliver solar-heated air to the building. This thermostat is connected to the thermostats in the rooms. The other thermostat in the attic senses if the temperature of the MF-panel is high enough for the bypass for cooling to be opened.

The thermostat at the top of the MF-panel is connected to the thermostats in the rooms in such a way that solar-heated air from the MF-panels is only directed into the rooms if the air temperature is high enough to be utilized (above 25°C) and the room temperatures are below a certain limit. The inlet of solar-heated air is controlled separately for the three rooms.

If, however, the humidity of the room 1 or room 2 becomes too high, but the temperature of the MF-panel is too low to be utilized, the MF-panel will be switched to exchanger mode, i.e. the exhaust air will be passed through the MF-panel, for each of the two rooms separately.

When the MF-panel is not in operation the building is naturally ventilated.

The air flow in the collector mode slightly exceeds two room air volumes (28 $m^3/hm^2_{collector}$) while the flow of fresh air in exchanger mode is 1.4 times the room volume (19 $m^3/hm^2_{collector}$).

This was the intended control strategy. However, conditions and demands developed differently from expected. As a result of changes in the production in the workshop, rather high humidity in the rooms was often desired, in order that special-purpose bricks should not dry out too fast and be damaged. The MF-system was, therefore, blocked from acting in exchanger mode.

## PERFORMANCE

### Measured performance

The system was comprehensively measured for two years, starting in June 1992. The performance of the system was evaluated for the heating season 1992–93.

The mean efficiency of the MF-panel in collector mode was measured to be 39%, which is 18% higher than expected based on tests on a prototype.[1] The efficiency in exchanger mode was measured by special-purpose experiments to be 44%, which likewise was 20% higher than expected.

The energy supply from the MF-panel during the heating season has been measured to be 9,600 kWh. This gives a solar fraction just above 6%. For the same period, the fans used 480 kWh of electricity equalling 5% of the useful heat delivered from the MF-panel to the building.

The amount of energy delivered from the MF-panel was much lower than planned. This was due to the MF-panel not acting as a heat exchanger between the fresh air entering the building and the exhaust air leaving the building. The amount of saved energy for this mode of operation is much larger than for the collector mode. The MF-panel further only preheats the air if the outlet temperature of the MF-panel is higher than the room temperature.

The annual performance of the MF-panel system equals 76 $kWh/m^2_{collector}$. However, in view of the very limited operating conditions, the performance is acceptable. The performance of the MF-panel system under better circumstances is estimated in the following.

### Calculated performance

Using a calculation tool and the actual measured efficiencies of the MF-panel acting as solar collector and heat exchanger, the following scenario has been evaluated.

The air to the building is always admitted through the MF-panel, preheated by solar radiation during daytime and by the exhaust air during night-time and periods with low solar radiation. The air flow through the MF-panel is 65 $m^3/hm^2_{collector}$ in collector mode and 33 $m^3/hm^2_{collector}$ in heat-exchanger mode. The higher air flow rates increase the efficiency of the MF-panel as a solar collector, but decrease the efficiency as a heat exchanger. The efficiency of the MF-panel as collector is calculated to be 68%, while the efficiency as heat exchanger is reduced to 40%.

The yearly potential performance under Danish conditions is 910 $kWh/m^2_{collector}$, obtained as follows:

- solar collector mode: 164 $kWh/m^2a_{collector}$.
- heat exchanger mode: 723 $kWh/m^2a_{collector}$.
- saved heat loss through the construction covered by the MF-panel: 23 $kWh/m^2a_{collector}$.

The minimal reduction of heat loss through the construction covered by the MF-panel results because of the high insulation standard of Danish buildings. For less well insulated buildings this value would be considerably higher.

## REMARKS

This building is less suitable for demonstrating the MF-panel concept than originally thought. However, the project still illustrates the potential of the concept and the measurements showed higher efficiencies than expected based on tests of a prototype of the panel.

The actual MF-panel at Wewer's Brickyard was not cost-effective owing to the incorrect running of the system. However, if the MF-panel is operated as intended and the flow rates of air through the panels are correct, the simple payback time of the system will only be a few years.

The measurements showed that the by-passes for summer cooling of the MF-panels were inadequate in maintaining acceptable panel temperatures. The panels should thus be actively ventilated or, still better, constructed of materials able to withstand stagnation temperatures.

## ACKNOWLEDGEMENTS

Architect: The Architects Arne Meldgaard & Co., Hellerup, Denmark

Energy consultants: Svend Engell Nielsen, Techline, Helsinge, Denmark, and Søren Østergaard Jensen, DTI Energy, Taastrup, Denmark

Building owner: Wewer's Brickyard, Helsinge, Denmark

Chapter author: Søren Østergaard Jensen, DTI Energy, Taastrup, Denmark

## REFERENCES

Jensen SØ (1990). *Results from Tests on a Multi-Function Solar Energy Panel. The Thermal Insulation Laboratory*, Report no 213. 1990, Institute for Buildings and Energy, Building 118, DK-2800 Lyngby, Denmark.

Jensen SØ (1994). *MF-demonstration project at Wewer's Brickyard in Helsinge.* Report no 267, Technical University of Denmark, Institute for Buildings and Energy, Building 118, DK-2800 Lyngby, Denmark.

# VI.5 Bombardier Inc. Factory

Valcourt, Canada

*Valcourt*

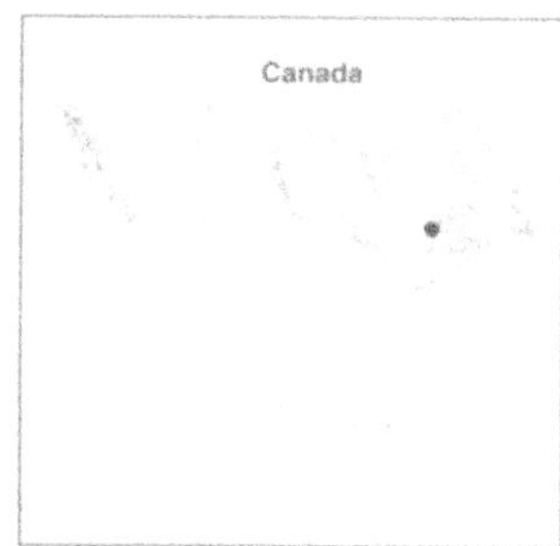


*System type 1*

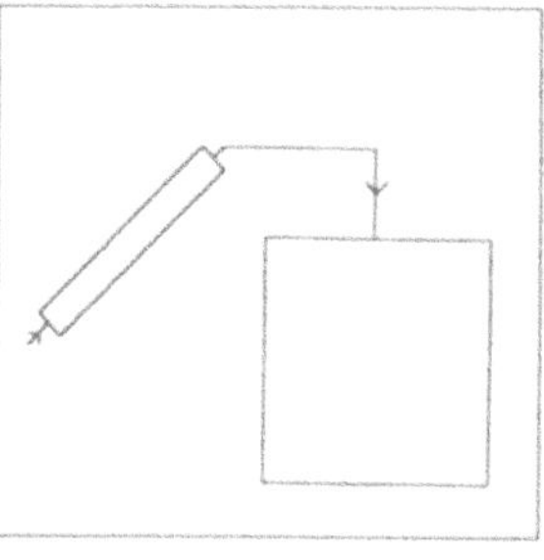

## PROJECT SUMMARY

Bombardier Inc. of Montreal, Canada has installed over 15,000 m$^2$ of unglazed solar air collectors at its various manufacturing plants in Quebec. The solar collectors decrease operating expenses and significantly improve indoor air quality. Bombardier's engineers compared the cost of this system with other methods of recladding and improving ventilation and found the costs to be the same. With no extra costs for free solar heating, the system has an immediate payback, along with other benefits of improved air quality and attractive architecturally designed walls that utilize solar energy.

Bombardier's first solar heating installation at their Sea Doo manufacturing plant has a solar wall area of 740 m$^2$. The heat-absorbing surface area is 611 m$^2$ of a custom dark olive-green colour, with the balance being the white canopy plenum along the top and vertical dividers. The entire surface of the solar panel is separated into six sections with one fan per section. Wall-mounted ventilation fans were installed to bring in a total of 71,400 m$^3$/h of heated ventilation air. Monitored solar and destratification savings for the year 1993–94 were CA $33,000 (ECU 20,440) based on energy savings of 894,000 kWh.

### Summary statistics

| | |
|---|---|
| System type: | type 1 |
| Collector type: | unglazed perforated metal |
| Collector area: | 611 m$^2$ |
| Ventilation: | 71,400 m$^3$/h (117 m$^3$/m$^2$h) |
| Solar system output: | 431,000 kWh/a (705 kWh/ma) |
| Destratification savings: | 463,000 kWh/a (758 kWh/m$^2$a) |
| Basis: | monitored |
| Year solar system installed: | 1993 |

## SITE DESCRIPTION

The Sea Doo factory is located in Valcourt, Quebec, approximately 100 km south east of Montreal. The south wall is at the front of the building and is visible to everyone entering the building. Valcourt has approximately 4670 heating degree days to base 18°C and 5200 degree days to base 20°C. Heating is required for nine months of the year:

| | |
|---|---|
| latitude | 46°N |
| longitude | 72°W |

## BUILDING PRESENTATION

The building was constructed in the 1960s and had walls made of a precast mix of concrete and foam. These walls had deteriorated badly, with cracks, peeling paint, wall separation and numerous openings along the eaves. The building also suffered from negative air pressure,

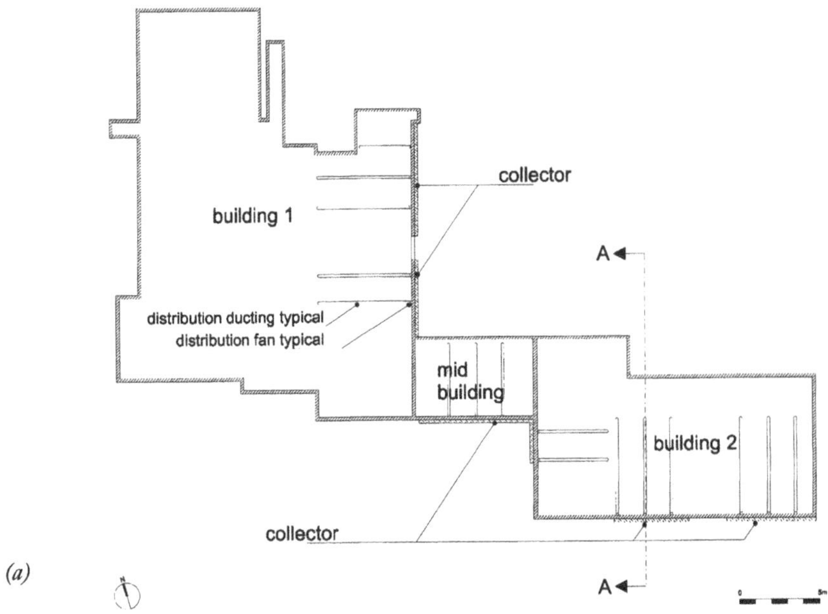


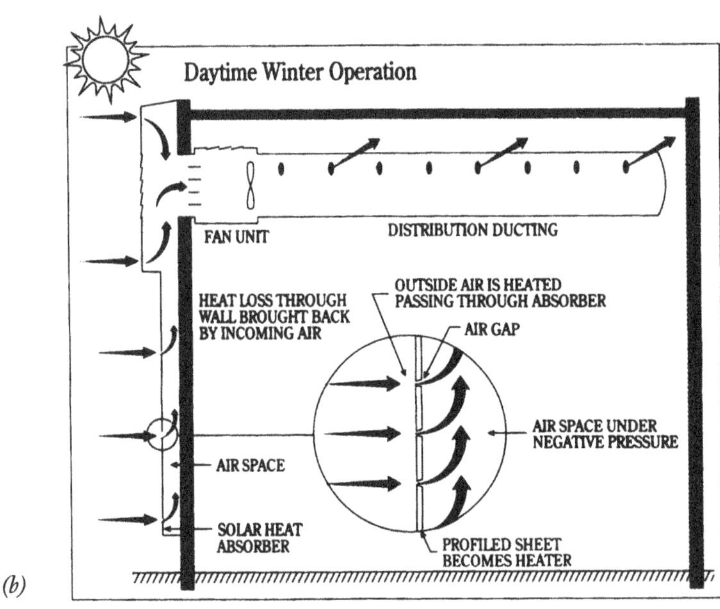


*Figure VI.5.1 (a) Plan view of the buildings with SOLARWALL panels and distribution ducting and (b) a cross-section of the solar ventilation heating system*

resulting in poor ventilation and inefficient exhaust systems, particularly on colder days. A decision was made to repair the walls and improve the ventilation and air quality inside the plant.

An architect was retained to design an attractive facade since the company's clients from around the world visiting the plant would see the solar panels. The wall is in two parts, with the office, entrance and a new glass facade for the office between the two solar-wall sections. The architect chose a dark olive-green colour to complement the colour over the office areas. A white canopy along the top and vertical white dividers approximately 17 m apart break up the long walls and create an attractive feature wall, which does not resemble a typical solar panel. Figure VI.5.1 shows a plan view and section of the buildings.

## SOLAR SYSTEM

### The system

The site-fabricated 0.9 m high white canopy runs the full length of the south wall. The solar absorber below the canopy looks like conventional metal cladding, ex-

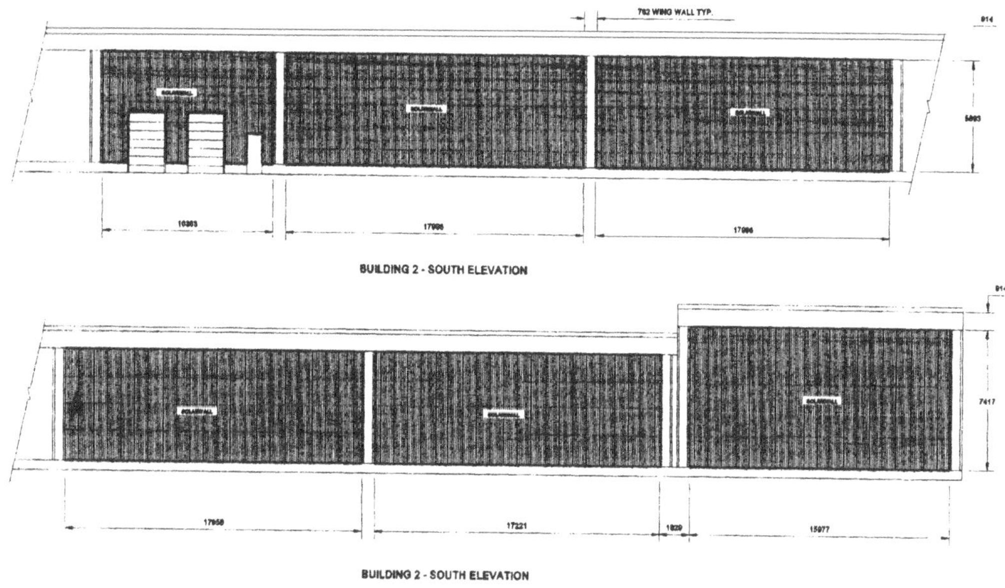


*Figure VI.5.2. South elevation drawing of the solar panels on building 2*

cept that it is mounted out from the main wall and is perforated to allow outside air to infiltrate the metal, where it picks up solar heat. A framework of steel vertical studs and horizontal Z-bars was first installed onto the existing wall, leaving an air space of 220 mm. The solar cladding was then attached to the horizontal Z-bars using colour-matched stainless-steel self-tapping screws. The air flow rate through the solar collectors is 117 m$^3$/m$^2$. Figure VI.5.2 is a drawing of the panels on building 2, while the photograph in Figure VI.5.3 shows the solar panels with the white header and white vertical architectural sections.

## Distribution

New fans on the south wall draw the solar-heated air into the building and distribute it throughout the plant through ducting at ceiling level. The fresh air mixes with stratified warm air at the ceiling and destratifies it, providing additional heating for the fresh air on days when there may be insufficient solar heat (Figure VI.5.4))

## Controls

The fans are run whenever the building is occupied. Modulating dampers between each solar panel and its associated fan self-adjust automatically to maintain a minimum temperature in the distribution ducts. If the air is too cold, the dampers modulate to mix ceiling air with outside air so as to maintain the minimum set point.

## Storage

There is no storage in the system, although the thermal mass of the wall does provide some heat to the air cavity

*Figure VI.5.3. The solar panels with white header and white vertical architectural sections*

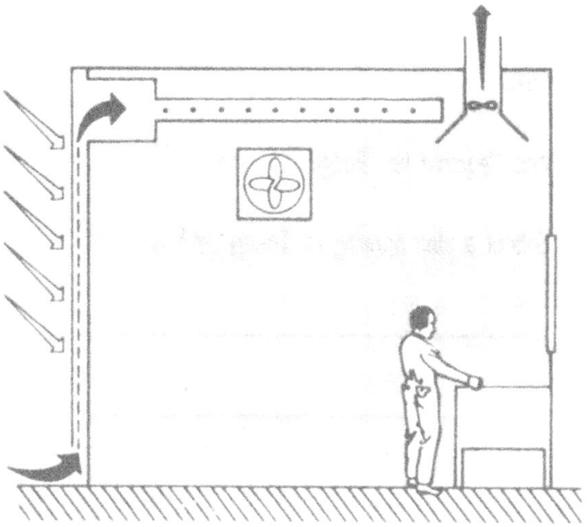

*Figure VI.5.4. The solar wall (SOLARWALL) supplies make-up air, which reclaims ceiling heat*

*Table VI.5.1. Summary of the savings*

| | | |
|---|---|---|
| Solar gain and wall heat recovery | 431,000 kWh/a | 705 kWh/m²a |
| Destratification | 463,000 kWh/a | 758 kWh/m²a |
| Total | 894,000 kWh/a | 1463 kWh/m²a |

in the evening. At night, heat loss through the wall is passed on to the incoming air in the solar collection air cavity, a form of heat recovery.

## PERFORMANCE

The first solar installation, monitored by the Quebec government, saved CA \$33,000 (ECU 20,440) between March 1993 and February 1994 based upon total energy savings of 894,000 kWh/a. A summary of the savings, which include solar gain and destratification, both absolute and per m² of collector is given in Table VI.5.1.

The freedom to choose a colour is important; going to a non-black paint only reduces the solar efficiency by a few per cent. The architect can compensate for any reduction by increasing the panel area or accept the reduction, since the appearance of the building is a primary concern.

## REMARKS

Pleased with the results of the initial installation in 1993, Bombardier selected SOLARWALL for the facade of its new administration building and for other walls of the complex. The company has won energy awards for the innovative solar heating system, which was installed for the same cost as a conventional air make-up system and wall cladding.

After the success of the Valcourt installations, Bombardier's Canadair aircraft division selected grey solar heating panels to reclad 10,000 m² of its south- and west-facing exterior walls in its Montreal building, furnishing up to 1,120,000 m³/h of make-up air. The existing exterior asbestos-tile walls offered poor insulating qualities, resulting in excessive energy costs and poor interior air quality. According to the feasibility study conducted by Enermodal Engineering in 1994, the Canadair SOLARWALL installation is expected to generate energy savings of 8,310,000 kWh/a or CA \$180,000 (ECU 111,500) annually with the system operating five days per week, 16 hours per day. The cost of the solar system is the same as the company's budget for recladding and purchasing new air make-up fans, which means that the solar project has an instant payback.

## ACKNOWLEDGEMENTS

Solar panel manufacturer: Conserval Engineering Inc., 200 Wildcat Road, Toronto, Ontario M3J 2N5, Canada; Tel. 416 661-7057; Fax 416 661-7146

Solar panel: SOLARWALL

Building owner: Bombardier, Valcourt, Quebec, Canada

Designer and installer: Matrix Inc., Kirkland, Quebec, Canada

Chapter author: J.C. Hollick, Conserval Engineering Inc., Toronto, Canada

# VI.6 JRC Research Building

## Ispra, Italy

*Ispra*

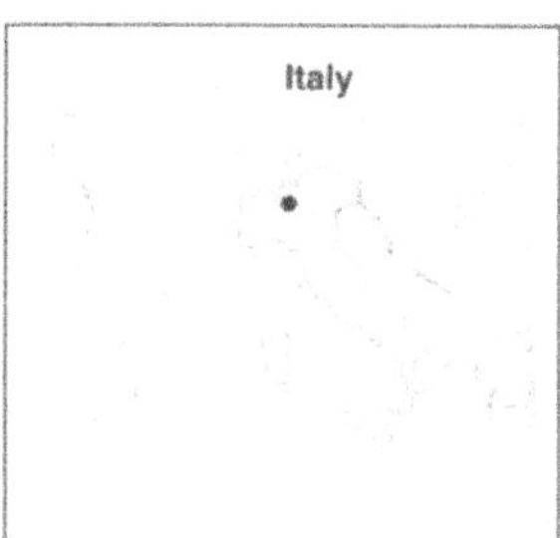


*System type 1*

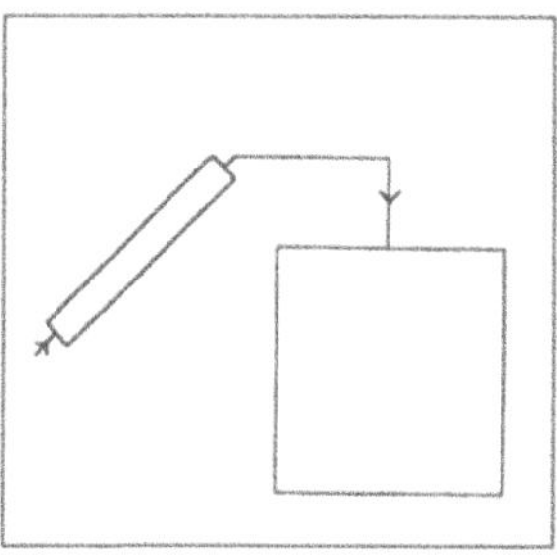

### PROJECT SUMMARY

Building 45 of the Joint Research Centre in Ispra was retrofitted with both solar air and solar hot water systems. The air panels provide the heat for the ventilation system. The panels were integrated into the south wall with a curved plenum as an architectural feature of the building. The installation was completed in 1995 and monitoring began in 1997. The unglazed solar-air panels are dark grey in colour and look like conventional metal cladding. Engineers from three countries were involved in the selection and design of the innovative system.

#### Summary statistics

| | |
|---|---|
| System type: | type 1 |
| Collector type and area: | 370 m² + 180 m² curved canopy |
| Storage type, volume, capacity: | no storage |
| Annual contribution of the solar air system: | 350 kWh/m² solar panel |
| Basis: | calculated |
| Year solar system installed: | 1995 |

### SITE DESCRIPTION

The Joint Research Centre Ispra (JRC Ispra) is located north of Milan, Italy and is owned by the European Union:

latitude 46N
longitude 9E.

### BUILDING PRESENTATION

The JRC Ispra is a research centre with numerous buildings built in the period 1958/68 to meet short-term needs in the nuclear field at a time when energy-saving considerations were of little concern. In 1993 it was decided to transform the JRC Ispra into a model of an environmentally friendly research centre. One of the first steps was to retrofit a limited number of masonry buildings to save energy and use alternative energies.

As Building 45 consumed an excessive amount of energy, it was decided to improve the energy efficiency of the building as well as improve its appearance. The specific measures that were applied are:

- The existing heating and ventilation system was modified and a new air-handling system was installed.
  A solar air heating system was added to heat the ventilation air.
- A variable air flow system and air distribution ducting were installed.
- Windows and wall and roof insulation were upgraded.
- Two new water storage tanks were added, one for the new water solar collectors and the other as a buffer for the district heating system.
- An energy management system was incorporated.

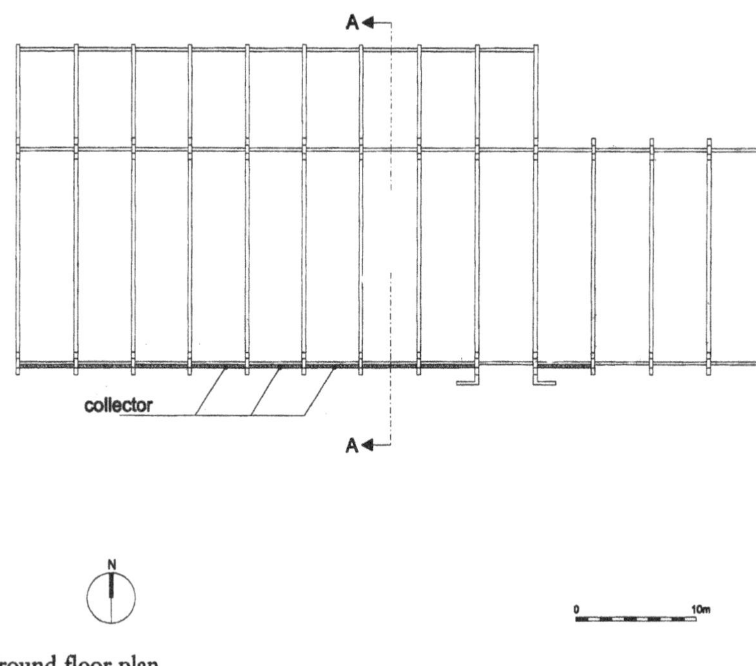


*(a)* Ground floor plan

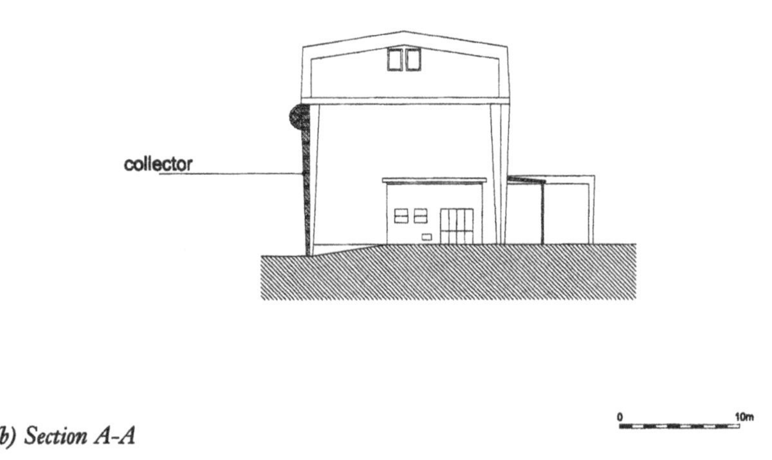


*(b) Section A-A*

*Figure VI.6.1. (a) Ground floor plan and (b) a section of the building.*

Figure VI.6.1 shows the plan of the ground floor and a section, while Figure VI.6.2 shows the location of the solar air panels on the south facade.

## SOLAR SYSTEM

### The system

Solar wall panels were installed on a portion of the south wall and are used to preheat the outside air drawn through them into the building. The solar panels are connected to the west HVAC unit, which draws in between 4,000 $m^3/h$ and 20,000 $m^3/h$ of solar-heated air. Additional heat is provided, when necessary, by hot-water coils in the HVAC unit. In summer, the solar panels are by-passed and outside air is brought in directly. The air is then cooled by circulating lake water through the coils in the HVAC unit. If additional heat or ventilation air is required, a second HVAC unit on the east wall can be used, but it is not connected to the solar heating panels.

### The collector

The main innovative aspect of the heating and ventilation system is the presence of a large all-metal air solar collector placed in front of nine bays on the south

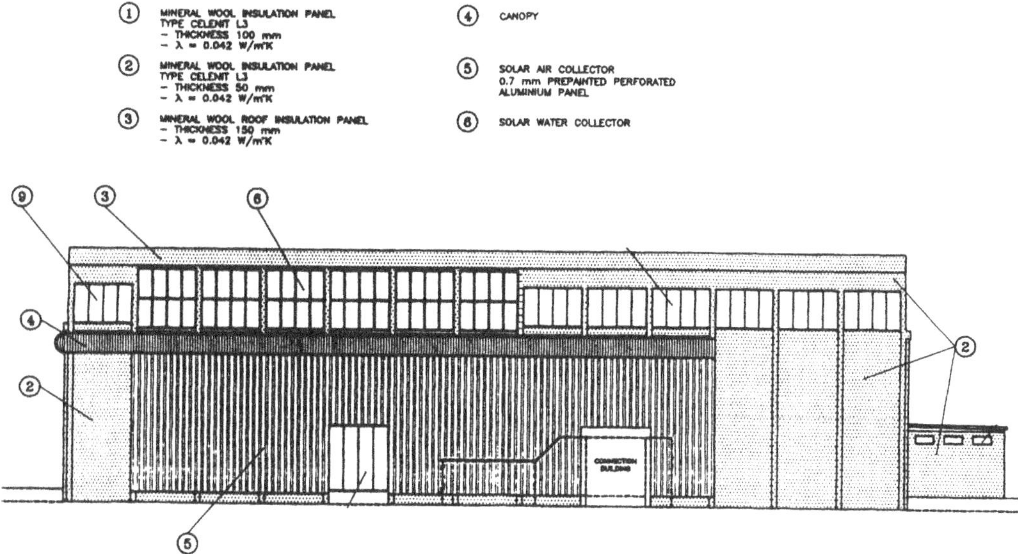


*Figure VI.6.2. The south facade with the solar panels*

facade. The collector was designed to be architecturally attractive with a dark grey colour and a curved canopy along the top, which also collects the hot air and directs it to the intake of the HVAC unit. Ventilation air is preheated by a corrugated aluminium sheet with small holes (1.6 mm diameter) uniformly distributed across the surface. The absorber panels are mounted out from the main wall to create an air cavity. Outside air is drawn through the holes, picking up the heat in the metal, and then rises to the top of the air cavity. The air is collected in the curved canopy plenum and then ducted to the west side of the building and into the air-handling system. The solar facade has a total surface area of approximately 370 $m^2$ and an additional 180 $m^2$ of a semi-circular canopy. The air flow through the collector varies from 11 $m^3/hm^2$ to 55 $m^3/hm^2$.

### Storage

There is no storage designed in the system. The solar heat is used directly during the day, as it is available. The building is normally unoccupied at night, so there is no need to bring in fresh air during this time.

### Distribution

The solar-heated air is collected at the top of the solar panels and ducted across the top of the south wall to the west wall and into the intake of the existing HVAC unit. No additional fan power is required since the existing fan unit is used. The hot air rising in the solar panels creates a natural stack effect and this air is then pulled in by the fan system. In the building, the air is then distributed along the ceiling and allowed to diffuse out. Tropical-type fans mix this solar-heated air with the air at the ceiling to destratify the air and provide good mixing. There are six motor-controlled diffusers to improve the air mixing inside the building. An electronic diffuser controller continuously checks the difference between the floor and ceiling temperatures to ensure that stratification is minimized.

### Controls

The HVAC unit is controlled by an energy management system. Outside air is continuously drawn through the solar-air collectors whenever heating is required. When heating is not necessary, a by-pass damper opens allowing outside air to enter directly, by-passing the solar collectors. At night, the indoor temperature and ventilation rate are reduced to save energy.

## PERFORMANCE

The expected heating-energy savings have been estimated by computer simulation to be 25% of the heat consumption before the solar retrofit. The energy savings for the cooling part are estimated at 72%. The total reduction of energy consumption is 2,114 MWh/a. With an energy cost of US $0.064/kWh (ECU 0.058/kWh), the payback time is estimated at 9.5 years. The solar-air heating portion of the project cost approximately US $256/m² (ECU 230/m²) of facade area and the payback time is estimated at between seven to nine years.

## REMARKS

The retrofitting of the Technology Hall at ISPRA is considered to be very successful. Monitoring of the energy savings has confirmed the computer-simulated performance. One of the unique features of the solar air heating system is that it not only saves energy, but improves the appearance of the older building. The architect, Piero Gatti, wanted the solar system to be fully integrated into the building, both functionally and architecturally.

## ACKNOWLEDGEMENTS

Manufacturer of the solar panels: SOLARWALL Italia srl, Via Enrico Fermi 11, 12038 Savigliano (CN), Italy; Tel. + 39 172 711106; Fax +39 172 712512
Solar panel: High performance SOLARWALL®
Building owner: JRC ISPRA
Chapter authors: J.C. Hollick, SOLARWALL International, Toronto, Canada and Ove Morck, CENERGIA, Denmark

The Technology Hall retrofit was one of eight buildings at ISPRA selected for the Eco Centre retrofitting project.

The successful winning team from the open call for tenders was a joint venture with Cogein, the contracting firm in Italy, together with METEC, the engineering company in Turin, and CENERGIA, the energy consultant from Denmark.

# VI.7 US Army Hangar

## Fort Carson, Colorado, USA

*Fort Carson, Colorado*

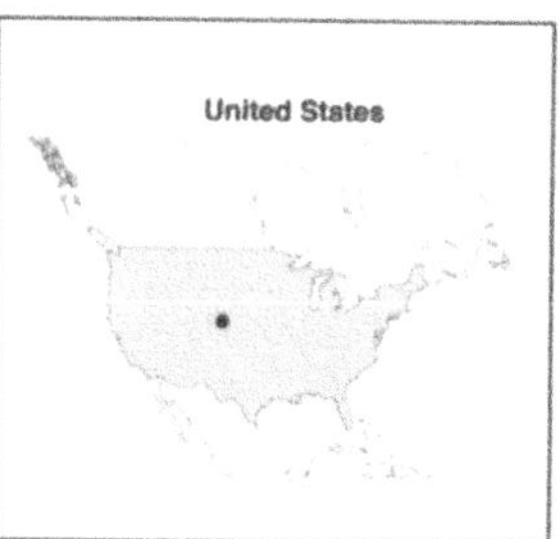


*System type 1*

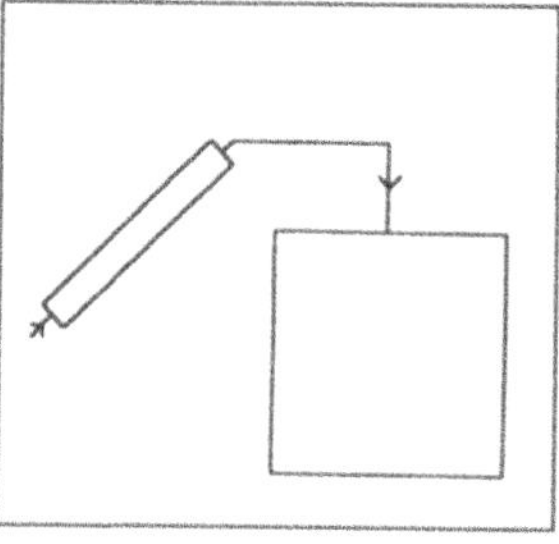

## PROJECT SUMMARY

The US Army's first solar-ventilated hangar is located at Fort Carson, Colorado. Fumes from the fuel tanks of up to 30 helicopters stored in the building are displaced with solar-warmed fresh air.

A conventional gas-heated ventilation system had been specified, but a value engineering analysis done for the Corps of Engineers showed that a solar-heated ventilation system would be comparable in cost to what was specified, so the design was changed. The fans were installed with the original building in 1992, but the solar cladding system was installed later, in 1995. The panels had to be supplied later as a retrofit project because of scheduling concerns at the time of construction. The solar-transpired collectors cover 725 m² of the south wall above the hangar doors and heat 107,000 m³/h of ventilation air. Cost savings have been calculated at US $14,000 (ECU 12,600) a year based on energy savings of 974,000 kWh a year.

### Summary statistics

| | |
|---|---|
| System type: | type 1 |
| Collector type: | unglazed transpired collector |
| Collector area: | 725 m² |
| Ventilation: | 107,000 m³/h (148 m³/m²h) |
| Solar system output: | 587,400 kWh/a (810 kWh/m²a) |
| Basis: | monitored |
| Destratification savings: | 386,200 kWh/a (533 kWh/m²a) |
| Heated floor area: | 4300 m² |
| Year solar system installed: | 1995 |

## SITE DESCRIPTION

Fort Carson is a large US Army military base located near Colorado Springs, CO, which has approximately 3,000 degree days to base 18°C. Heating is required for eight months of the year:

| | |
|---|---|
| latitude | 39°N |
| longitude | 105°W. |

## BUILDING PRESENTATION

The Avum hangar is a new building built to house approximately 30 army helicopters at the Fort Carson Army Base. A plan of the hangar and a cross-section are shown in Figure VI.7.1. After a value engineering report recommended switching to infra-red heaters and a solar heated ventilation system, it was decided to install the solar panels (as a retrofit) on the existing south wall above the hangar doors. Although the original proposal had included covering the entire south wall above the doors, the location of the six ventilation fans was better suited to the solar panels being in two sections above the two large doors (Figure VI.7.2).

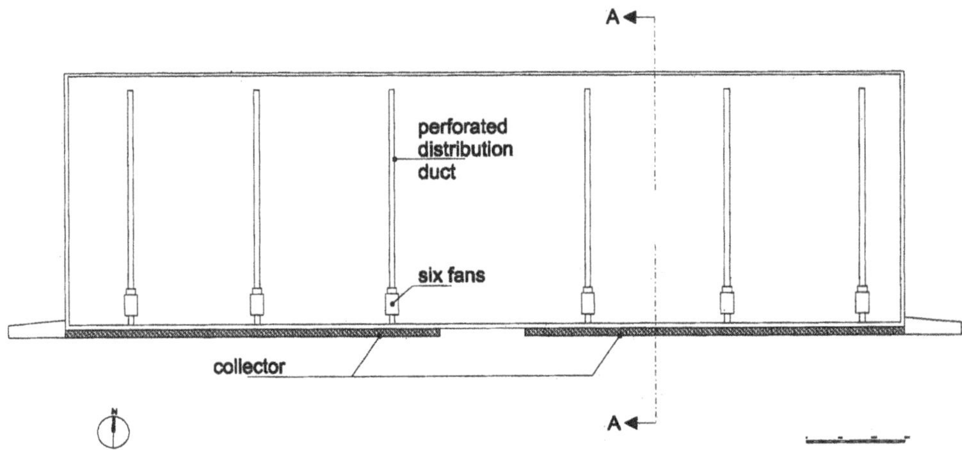


*(a) Floor plan*

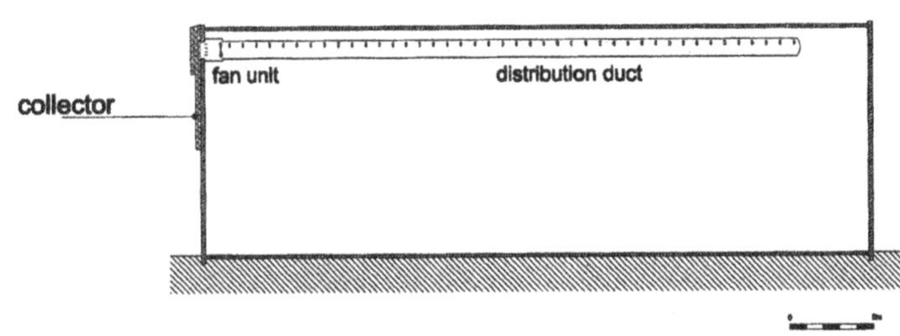


*(b) Section A-A*

*Figure VI.7.1. (a) Plan and (b) a section of the hangar*

## SOLAR SYSTEM

### The system

A study carried out by the National Renewable Energy Laboratory (NREL) in Colorado concluded that an unglazed transpired metal solar collector for heating ventilation air would be the most cost-effective solar system to heat the fresh air required in the building. The air could be heated by a combination of solar energy, heat recovery from the wall behind the solar panels and utilization of the stratified heat at ceiling level that is typically found in buildings with high ceilings.

### The collector

Solar cladding was mounted on a framework of steel clips and horizontal Z bars, which was installed over the existing metal wall, leaving an air space of 250 mm. An internal plenum was created between the top of the wall and the next horizontal Z bar. The air flow in the panel is through the perforations in the aluminium absorber, then vertically to the top interior plenum and then horizontally to the nearest fan. The army chose a dark bronze colour for the solar absorber to complement the shades of beige on the existing building. The solar absorber is unglazed and is perforated with 1.5 mm holes throughout its surface. Outside air is drawn

*Figure VI.7.2. South elevation showing the solar panels above the hangar doors*

through the openings, picking up the solar heat. The air flow rate through the solar collectors varies from 50 to 148 $m^3/hm^2$. A close-up of the collector is shown in Figure VI.7.3.

### Distribution

Six axial fans mounted below the ceiling behind the solar panels distribute the solar-heated air at ceiling level through six ducts that run from the south end to the north end of the building. Each fan is 810 mm in diameter and delivers approximately 17,900 $m^3/h$ of air, which can vary from 100% outside air to less than 30% depending on the temperature. The fresh air mixes with the stratified warm air at the ceiling and destratifies it, providing additional heating on days when there may be insufficient solar heat.

### Controls

The fans are on all the time during the heating season. Modulating dampers between the solar panels and each fan will self-adjust automatically to maintain a minimum temperature in the distribution ducts. If the air is too cold, dampers will modulate to mix ceiling air with outside air to maintain the minimum set point.

### Storage

There is no storage designed in the system. The solar heat is used during the day as it is available. At night the amount of outside air brought into the building is reduced and the system relies on the stratified heat to temper the fresh air.

## PERFORMANCE

The energy savings from the solar ventilation system are a combination of solar heat and heat recovery from the wall and destratification of the hot air under the ceiling. Buildings with high ceilings will have higher temperatures at the ceiling than at the floor since hot air rises. Outside air, at a temperature below room temperature, can be distributed at ceiling level. It passes through the warm stratified air. By the time it reaches floor level, it reaches room temperature.

The dark bronze colour absorbs approximately 91% of the solar radiation; black would absorb 95%. The ability to choose a colour other than black is important to ensure that the building will be attractive.

The US Army pays a very low price for its natural gas. The total savings are US $14,000 (ECU 12,600) a year and, with a project cost of approximately US $100,000 ($138/m²) [ECU 90,000 (ECU 124/m²)], the project had a payback of approximately seven years. Had this been built as part of the original building, the cost would have been comparable to what was spent on the existing wall system and the gas-heated air make-up fans and the payback time would have been minimal. The savings are summarized in Table VI.7.1.

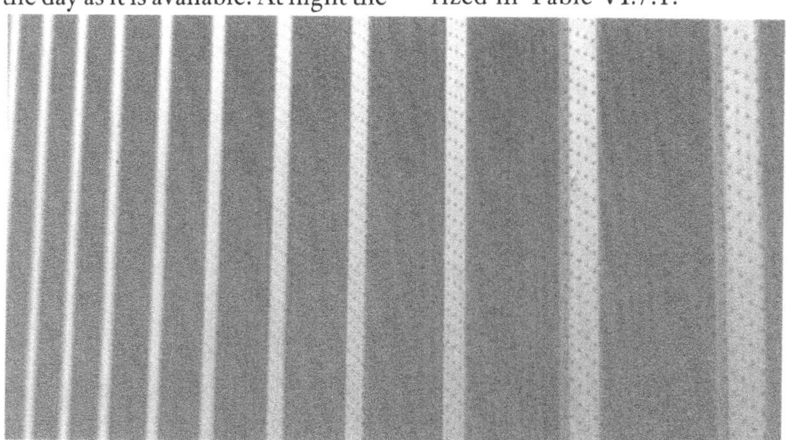

*Figure VI.7.3. A close-up view of the transpired solar collector*

*Table VI.7.1. Summary of the savings*

| | | |
|---|---|---|
| Solar gain and heat recovery from wall | 587,400 kWh | (810 kWh/m²a) |
| Destratification | 386,200 kWh | (533 kWh/m²a) |
| Total | 973,600 kWh | (1343 kWh/m²a) |

## REMARKS

Indoor air quality is an important issue in the USA and the solar ventilation system allows fumes from the fuel tanks to be displaced with fresh air.

The SOLARWALL system is being monitored by NREL. The US federal government has a mandate to consider renewable-energy technologies when the system has a payback of less than ten years. Two more army buildings have also received a SOLARWALL heating system and it is anticipated that many other military buildings will follow.

## ACKNOWLEDGEMENTS

Solar panel manufacturer: Conserval Systems Inc., 4242 Ridge Lea Rd. #1, Buffalo, NY 14426-1051, USA; Tel +1 716 835 4903; Fax +1 716 835 4904
Solar panel: High performance SOLARWALL®
Building owner: US Army, Fort Carson, CO, USA
Designer: Foltz Engineering, Estes Park, CO, USA
Chapter author: J.C. Hollick, Conserval Systems Inc., Buffalo, NY, USA

# VII Office buildings

# VII.1 Introduction

In low-energy office buildings more than 60% of the heat losses during the winter are caused by ventilation needs. This may lead to the impression that solar air systems are well suited for mechanically ventilated office buildings. What make the situation more complex are the high internal gains from computers, lighting and other electrical equipment. A conflict exists between solar-system gains and the useful internal gains. Therefore, a solar air system for an office building must be very carefully designed, considering the actual internal gains that can be expected. The first step is to reduce internal gains by increasing daylighting and specifying energy-saving equipment.

At night the internal gains are minimal, so that time-shifted solar gains by passive or active discharge may be used. The night setback of the room temperature has to be considered in this calculation.

Because of the already-mentioned high internal gains in office buildings, the summers are critical. A combination of solar air systems with a natural ventilation system, in which forced night flushing is used, should be considered. Naturally driven solar chimneys and storage walls having day charge and night discharge can save fan energy.

For high-rise office buildings a combination of facade and a solar air systems should be considered. Wind and street noise may make opening windows impossible. External shading is difficult because it is exposed to wind and freezing. For these reasons, double facades are very much a current topic. A double facade consists of an inner and outer glazing with an air gap of at least 30 cm. In this gap, an 'external' shading device can be placed, where it is secure from wind and the elements. In summer the gap can perform as a solar chimney; in winter the sun-tempered air in the gap can reduce the energy consumed for heating intake ventilation air.

**ACKNOWLEDGEMENTS**

Chapter author: M. Schuler, Transsolar

# VII.2 WAT Office Building

## Karlsruhe, Germany

*Karlsruhe*

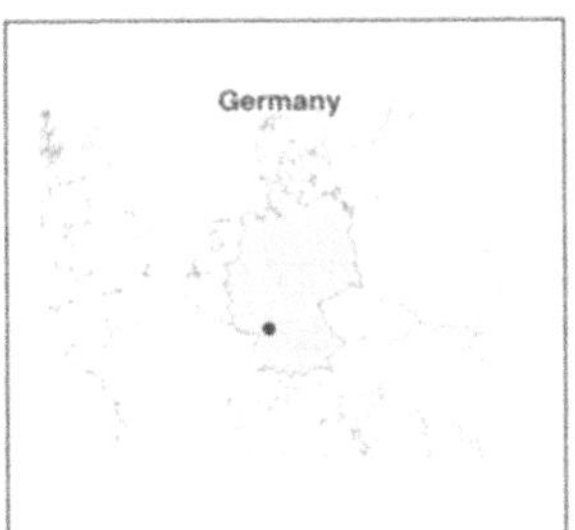


*System type 1*

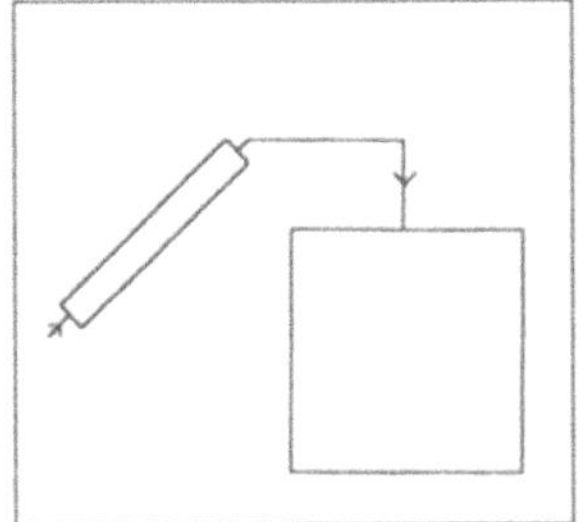

### PROJECT SUMMARY

In 1994, the WAT (Wasser- und Abfalltechnik GmbH) office building, with various energy-saving features, was erected in Karlsruhe, Germany. The office building, net area 1,500 $m_2$, incorporates:

- facade-integrated solar air collectors preheating the ventilation air;
- exhaust air extraction through a solar chimney
- heat recovery from ventilation air;
- concrete floors with integrated air ducts and no suspended ceilings.

During the design phase, TRNSYS was used to determine the performance of the building and the solar system.

#### Summary statistics

| | |
|---|---|
| System type: | type 1 |
| Collector type: | glazed, opaque, facade integrated |
| Total collector area: | 59 $m_{coll}^2$ |
| Storage type, volume, capacity: | building structure, 535 $m^3$, 310 kWh/K |
| Annual contribution from the solar air system: | 3.5 kWh/$m^2$(heated floor area) |
| Total heating load: | 37 kWh/$m^2$a; measured 40.5 kWh/$m^2$a; calculated |
| Floor area, volume: | 1900 $m^2$, 5600 $m^3$ |
| Year solar system built: | 1994 |

The solar gains of the air collector are low because the system was originally optimized for the night flushing/cooling of the building and then, in a second step, an air collector was added to the system. Furthermore, the south facade with the collectors is shaded by another building and the collector area is small compared to the net heated floor area (59 $m^2$/ 1500 $m^2$).

### SITE DESCRIPTION

The WAT office building is located in the centre of a new industrial district called Grossoberfeld, near Karlsruhe in Germany:

latitude 49°N<br>
longitude 8°E<br>
altitude: 112 m above sea level.

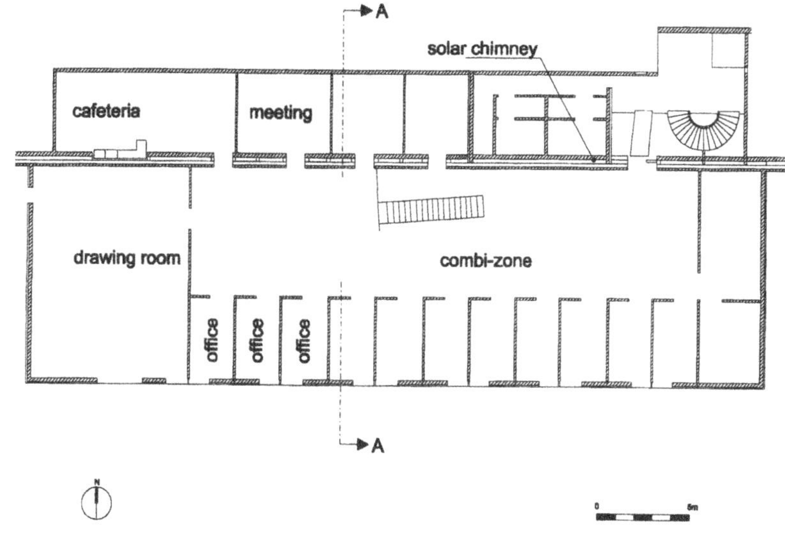


*(a) Floor plan*

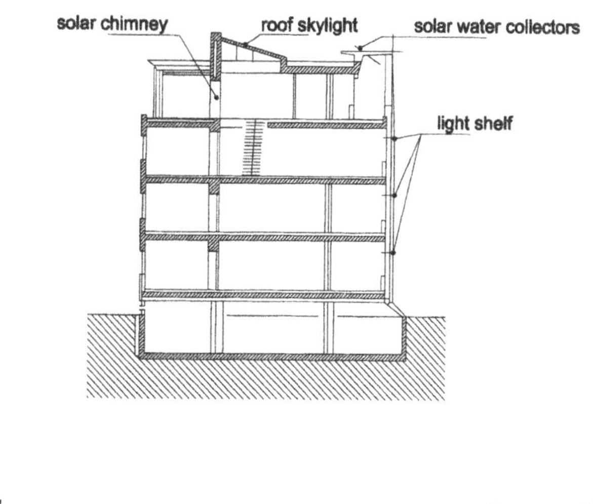


*(b) Section A-A*

*Figure VII.2.1. (a) Plan view of the main floor and (b) section A–A*

## BUILDING PRESENTATION

The building is oriented with a north–south long axis. It has a basement, three upper working levels and an attic.

The ventilation air of the southern part of the building is preheated by facade-integrated solar air collectors and buffered in the floor by air ducts. The inlet air for the northern part is preheated by a heat-recovery system using the exhaust air from the whole building. The building is heated by a condensing value boiler using natural gas. The usual heating period is short (November–February) owing to the high insulation level and the energy concept of the building. The hot-water system is supplemented by solar water collectors (9 m$^2$) on the roof.

A main feature of the building is a big black-coated wall separating the north and the south zones. The wall stores solar heat gains and amplifies the chimney effect to ventilate the building. The double shell also serves as an installation duct. The components of the control system for ventilation and heating are linked by a bus system. In this way, it is possible to optimize heating and ventilation strategies by using the data from different reference points inside and outside.

The building is cooled naturally in the summer period by night flushing. There is no mechanical cooling device installed. Figure VII.2.1 shows a plan of the main floor and a section of the building, Figure VII.2.2 a plan of an office on the south side of the building and Figure VII.2.3 a plan of the roof.

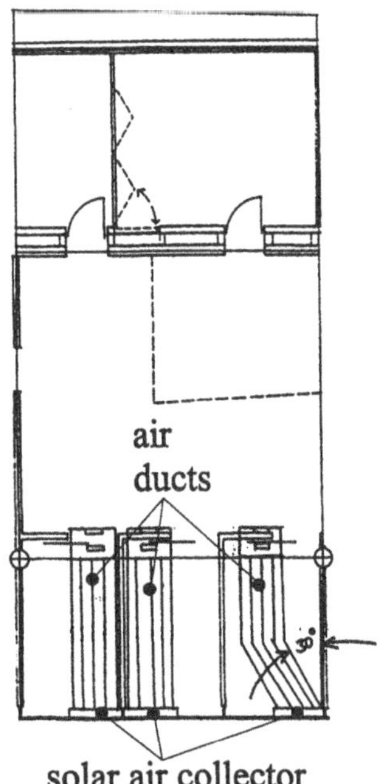


*Figure VII.2.2. A plan of an office on the south side of the building*

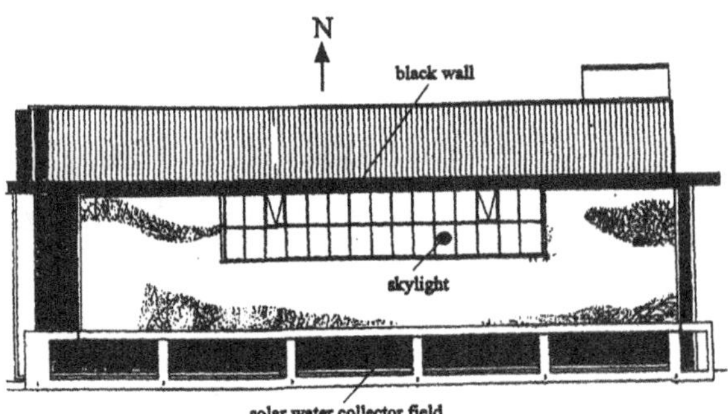


*Figure VII.2.3. A plan of the roof*

Light shelves are integrated into the facade to improve the daylight situation in the rooms and they also work as a fixed shading device for the south facade. Because of the presence of the light shelves, the internal gains and the energy consumption are reduced.

Building statistics are given in Table VII.2.1 and *U*-values in Table VII.2.2. The contribution of the auxiliary space heating has been measured as 37 kWh/m$^2$a.

## SOLAR SYSTEM

The solar air collector integrated in the south facade heats up outside air, which is sucked in by a small fan located in the floor. The user can vary the power of the fan; this user control increases acceptance of the system tremendously. Solar preheated air is fan-forced through air ducts in the concrete floor. The massive floor provides heat storage. A small auxiliary heater in the air outlet avoids too low inlet air temperatures in winter.

Figure VII.2.4 shows the preheating system of the south-oriented rooms. There are two outlet air openings, one directly to the room and one to the floor area. Figure VII.2.5(a) illustrates the functioning of the ventilation system during the heating period.

The ventilation air is tempered by the solar gains absorbed in the black-coated wall separating the south and north parts of the building.(see Figure VII.2.3). The ventilation air of the northern rooms is preheated by a heat-recovery system. The inlet air of the northern part is moved by fans.

*Table VII.2.1. Building statistics*

| | |
|---|---|
| Gross heated floor area | 1900 m$^2$ |
| Gross heated volume | 5600 m$^2$ |
| Net heated floor area | 1500 m$^2$ |
| Net heated volume | 4800 m$^2$ |

*Table VII.2.2. U-values*

| | *U*-value (W/m$^2$K) |
|---|---|
| Roof | 0.2 |
| Floor | 0.2 |
| Wall | 0.2 |
| Glazing | 1.3 |

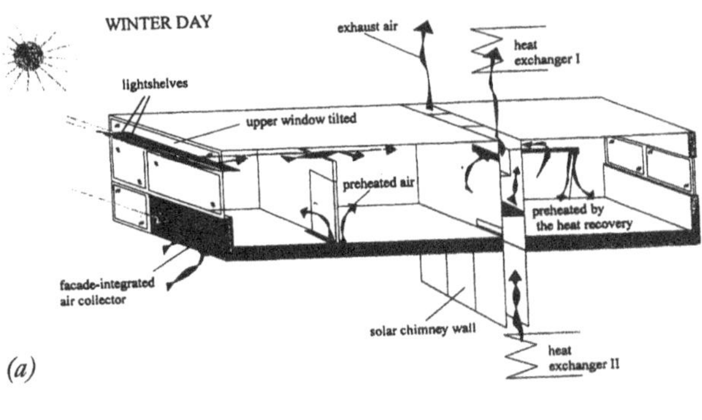


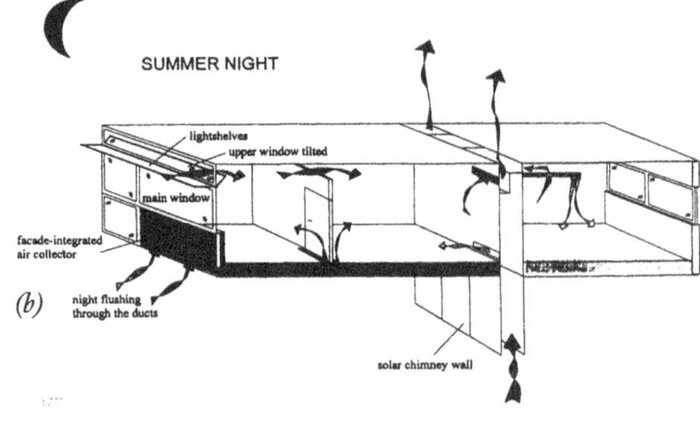


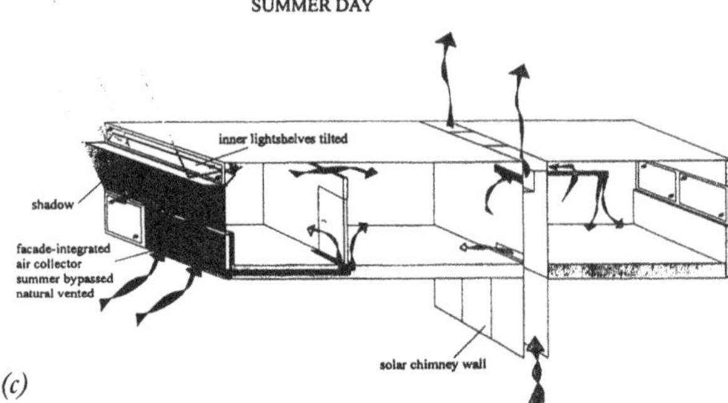


*Figure VII.2.5. The ventilation system: (a) on a winter day; (b) on a summer night; (c) on a summer day*

*Table VII.2.3. The performance of the system*

| | |
|---|---|
| Solar air collector | |
| number of collectors | 59 |
| net total area | 59 $m^2$ |
| air flow rate | 30–180 $m^3/m_{coll}^2h$ |
| solar contribution | 85 $kWh/m_{coll}^2a$, calculated |
| storage | |
| building structure | 310 $kWh/m_{coll}^2K$ |
| distribution | |
| total fan power | Max. 1420 W, average 650 W |
| collector fan power | Max. 24 $W/m_{coll}^2$, average 11 $W/m_{coll}^2$ |
| consumption | 1650 kWh/a, calculated |

A critical issue for office buildings today is usually not the energy consumption for heating but the risk of overheating in summer. The usual solution is mechanical cooling. However, in the WAT building, the ventilation system in combination with the massive building structure substitutes for a mechanical cooling system. The building is ventilated during summer nights, in order to decrease the temperatures of the mass, i.e. the concrete parts of the building (see Figure VII.2.5(b)). By using the air ducts in the floor construction, it is possible to directly cool the massive floors of the south-oriented rooms, benefiting from the lower night ambient air temperatures. The cool floor reduces the amplitude of the room temperature over the day.

The northern part of the building and the corridor are supplied with fresh air by air ducts in the black wall. The inlet opening for air is in the northern part of the building over a pool.

The largest part of the energy needed for the night flushing is delivered by the warm black wall, charged during the day. The ventilation is driven by the difference between the ambient and chimney temperature.

The windows are closed during the day. The inner light shelf is adjustable so that it provides solar protection (see Figure VII.2.5(c)).

## PERFORMANCE

The performance of the system is summarized in Table VII.2.3. Both timings and temperature are controlled. As a wireless system is used, any future changes in room design could be easily carried out.

## REMARKS

In the first few months after the building was occupied in January 1995, some occupants did not understand or accept the new low energy concept. Some wanted to raise the room temperature above 23°C because they wear summer clothes during the winter. Every morning, the cleaning team opened all windows in order to ventilate the rooms.

In some rooms without plants and with high ventilation rates the air becomes too dry. In 1995, the measured auxiliary heating demand was about 38 kWh/a. This is about 9% less than the calculated value.

It is very important that all the occupants be instructed about their new building. It is also necessary to provide support for the occupants during the first few months.

Wherever it is possible, standard components should be used to reduce costs.

## ACKNOWLEDGEMENTS

Architect: Guenter Leonhard, Friedhofstrasse 71, Pf 107, 70191 Stuttgart, Germany; Tel. +49 711 2569656

Energy concept: Transsolar

Project Engineer: Peter Voit, Matthias Schuler, Nobelstrasse, 15,70569 Stuttgart, Germany; Tel. +49 711 679760; Fax: +49 711 6797611

Chapter author: Helmut Meyer, Nobelstrasse, 1570569 Stuttgart, Germany

# Appendices

# Appendix A: The International Energy Agency/Solar Heating and Cooling Agreement

## THE INTERNATIONAL ENERGY AGENCY

The IEA, founded in 1974, is an autonomous body within the framework of the Organization for Economic Cooperation and Development (OECD). Twenty four member countries and the European Commission carry out a comprehensive programme of energy cooperation. Policy goals include the ability to respond promptly and flexibly to energy emergencies; environmentally sustainable provision and efficient use of energy; research, development and market deployment of new and improved energy technologies; and cooperation among energy market participants.

These goals are addressed within the framework of 40 Implementing Agreements headed by the Committee on Energy Research and Technology. CERT is supported by a small Secretariat staff in Paris. Four Working Parties (Conservation, Fossil Fuels, Renewable Energy and Fusion) monitor the Implementing Agreements, identify new areas for cooperation and advise the CERT on policy.

## THE SOLAR HEATING & COOLING PROGRAMME (SHC)

The SHC Programme was one of the first R&D Implementing Agreements of the IEA. Since 1977, its Participants have been conducting a variety of projects in active, passive and photovoltaic solar technologies, primarily for building applications. The nineteen members are:

| | | |
|---|---|---|
| Australia (AUS) | Finland (FIN) | Norway (N) |
| Austria (A) | France (F) | Spain (E) |
| Belgium (B) | Germany (D) | Sweden (S) |
| Canada (CDN) | Italy (I) | Switzerland (CH) |
| Denmark (DK) | Japan (J) | United Kingdom (UK) |
| European Comm. (EC) | Netherlands (NL) | United States (USA) |
| New Zealand (NZ) | | |

Twenty-six projects or 'Tasks' have been undertaken since the Programme began. These are led by Operating Agents and monitored by an Executive Committee consisting of a representative from each member country. Following is the list of Tasks and the countries supplying Operating Agents:

| | |
|---|---|
| Task 1*: | Investigation of the Performance of Solar Heating and Cooling Systems (DK) |
| Task 2*: | Coordination of Research and Development on Solar Heating and Cooling (J) |
| Task 3*: | Performance Testing of Solar Collectors (D / UK) |
| Task 4*: | Development of an Insolation Handbook and Instrument Package (USA) |
| Task 5*: | Use of Existing Meteorological Information for Solar Energy Application (S) |
| Task 6*: | Solar Systems Using Evacuated Collectors (USA) |
| Task 7*: | Central Solar Heating Plants with Seasonal Storage (S) |
| Task 8*: | Passive and Hybrid Solar Low Energy Buildings (USA) |
| Task 9*: | Solar Radiation and Pyranometry Studies (CDN / D) |
| Task 10*: | Solar Material Research and Testing (J) |
| Task 11*: | Passive and Hybrid Solar Commercial Buildings (CH) |
| Task 12*: | Building Energy Analysis and Design Tools for Solar Applications (USA) |
| Task 13*: | Advanced Solar Low Energy Buildings (N) |
| Task 14*: | Advanced Active Solar Systems (CDN) |
| Task 15: | Not initiated |
| Task 16*: | Photovoltaics in Buildings (D) |
| Task 17*: | Measuring and Modeling Spectral Radiation (D) |
| Task 18*: | Advanced Glazing Materials (UK) |
| Task 19: | Solar Air Systems (CH) |
| Task 20: | Solar Energy in Building Renovation (S) |
| Task 21: | Daylighting in Buildings (DK) |
| Task 22: | Building Energy Analysis Tools (USA) |
| Task 23: | Optimization of Solar Energy Use in Large Buildings (N) |
| Task 24: | Solar Procurement (S) |
| Task 25: | Solar Assisted Air Conditioning of Buildings (D) |
| Task 26: | Solar Combisystems (A) |

*Completed

# Appendix B: IEA SHC Task 19 and the Experts

## IEA SHC TASK 19

Solar air systems can deliver space heating and temper ventilation air in winter, as well as heating domestic hot water in summer. Air systems, unlike water systems, need no frost protection, nor are leaks damaging to the building structure or its contents. In contrast to passive systems, active air systems provide better heat distribution and improved comfort and make fuller use of solar gains. Solar air systems are a natural fit to mechanically ventilated buildings and mechanical ventilation is increasingly common, not only in commercial and institutional buildings, but also in very-low-energy residences.

The spread of this technology is slowed down, however, because designers lack experience in planning, analysing and constructing solar air systems. Furthermore, documentation on built prototypes that can be used to convince building clients is scarce. For this reason, the Solar Heating & Cooling Programme (SHC) of the International Energy Agency (IEA) initiated this four-year research Task in October 1993 to address these needs. The specific goals are to:

- Demonstrate the effectiveness of solar air systems via example buildings, in which the solar air system is an integral aspect of the architecture.
- Provide designers with guidance on selecting, dimensioning and detailing a solar air system.
- Support industry in developing components that simplify construction, cut costs and perform better.

The work resulted in the following:

- A book of example buildings with diverse solar air systems in different countries.
- An engineering handbook based on component testing, building monitoring and computer modelling.
- A simple analysis tool using TRNSYS with an easy user interface for air systems.
- A catalogue of manufactured components for laying out a solar air system.

As in all IEA SHC Tasks, the work of the national experts has been financed separately by agencies within each participating country. The experts meet every six months to review the work to date and coordinate the next work to be done. The Operating Agent presents a six-monthly progress report to the Executive Committee, which oversees and coordinates all solar research tasks.

**The participating experts:**

S. Robert Hastings
Solararchitektur
ETH-Hönggerberg
CH-8093 Zürich
Switzerland
hastings@orl.arch.ethz.ch

Hubert Fechner
Fach 8
Oest. Forsch. zentrum
Faradygasse 3
A-1031 Wien
Austria
fechner@email.arsenal.ac.at

Thomas Zelger & M. Bruck
Kanzlei Dr. Bruck
Prinz-Eugen-Strasse 66
A-1040 Wien
Austria
bruck@magnet.at

Sture Larsen
Architekt
Lindauerstrasse 33
A-6912 Hörbranz
Austria
solarsen@computerhaus.at

John Hollick
Solar Wall Internat. Ltd.
200 Wildcat Road
Downsview ONT M3J 2N5
Canada
conserval@globalserve.net

Peter Elste
Charles Filleux
Basler & Hofmann
Forchstrasse 395
CH-8029 Zürich
Switzerland
chfilleux@bhz.ch

Karel Fort
Ing. Büro
Weiherweg 19
CH-8604 Volketswil
Switzerland
k_fort@CompuServe.COM

Gerhard Zweifel
Abt. HLK
Zentralschw. Tech. Luz.
CH 6048 Horw
Switzerland
GZweifel@ztl.ch

Hans Erhorn
Heike Kluttig
Johann Reiss
Fraunhofer Institut für Bauphysik
Nobelstrasse 12
D-70569 Stuttgart
Germany
erh@ibp.fhg.de
hk@ibp.fhg.de
RE@ibp.fhg.de

Joachim Morhenne
Ing.büro Morhenne GbR
Schülkestrasse 10
D-42277 Wuppertal
Germany
IBMorhenne@t-online.de

Frank Heidt
Fachbereich Physik & Solar.
Universität GH Siegen
Adolf Reichweinsstrasse
D-57068 Siegen
Germany
heidt@physik.uni-siegen.de

Matthias Schuler
Alexander Knirsch
Helmut Meyer
Trans Solar GmbH
Nobelstrasse 15
D-70569 Stuttgart
Germany
schuler@transsolar.com
knirsch@transsolar.com meyer@transsolar.com

Siegfried Schröpf
Solar Env. Tech.
Grammer KG
Wernher v. Braunstrasse 6
D-92224 Amberg
Germany

Soren Østergaard Jensen
Solar Energy Centre Denmark
Danish Tech. Institute
Postbox 141
DK-2630 Taastrup
Denmark
soren.o.jensen@dti.dk

Ove Mørck
Cenergia ApS
Sct. Jacobs Vej 4
DK-2750 Ballerup
Denmark
cenergia@compuserve.com

Gianni Scudo
DPPPE
Politecnico di Milano
Via Bonardi 3
I-20133 Milano
Italy
scudo@cdu8g5.cdc.polimi.it

Giancarlo Rossi
I.U.A.V. Dept. Building Construction
Università di Venezia
S. Croce, 191
I-30135 Venezia
Italy
rossigc@brezza.iuav.unive.it

Harald N. Røstvik
Sunlab/ABB
Alexander Kiellandsgt 2
N-4009 Stavanger
Norway
sunlab@rl.telia.no

Torbjörn Jilar
Dept. of Biosys. & Tech.
Swedish Univ. of Agric.
PO Box 30
S-23053 Alnarp
Sweden
Torbjorn.Jilar@jbt.slu.se

Christer Nordström
Ch. Nordström Arkitektkontor AB
Asstigen 14
S-43645 Askim
Sweden
cna@cna.se

Kevin Lomas
Institute of Energy and Sustainable Development
De Montfort University
The Gateway
Leicester LE1 9BH
UK
klomas@dmu.ac.uk

George Löf
6 Parkway Drive
80110 Englewood CO
USA

Printed and bound by CPI Group (UK) Ltd, Croydon, CR0 4YY
22/10/2024
01777391-0001